PROGRESS IN IMMUNOLOGY RESEARCH

PROGRESS IN IMMUNOLOGY RESEARCH

BARBARA A. VESKLER
EDITOR

Nova Biomedical Books
New York

NOTICE TO THE READER

The Publisher has taken reasonable care in the preparation of this book, but makes no expressed or implied warranty of any kind and assumes no responsibility for any errors or omissions. No liability is assumed for incidental or consequential damages in connection with or arising out of information contained in this book. The Publisher shall not be liable for any special, consequential, or exemplary damages resulting, in whole or in part, from the readers' use of, or reliance upon, this material.

This publication is designed to provide accurate and authoritative information with regard to the subject matter covered herein. It is sold with the clear understanding that the Publisher is not engaged in rendering legal or any other professional services. If legal or any other expert assistance is required, the services of a competent person should be sought. FROM A DECLARATION OF PARTICIPANTS JOINTLY ADOPTED BY A COMMITTEE OF THE AMERICAN BAR ASSOCIATION AND A COMMITTEE OF PUBLISHERS.

Library of Congress Cataloging-in-Publication Data

Progress in immunology research / Barbara A. Veskler (editor).
        p. ; cm.
Includes index.
ISBN 1-59454-380-1 (hardcover)
1. Immunology.
[DNLM: 1. Immune System. 2. Anti-Infective Agents--immunology. 3. Immunity, Cellular. 4. Peptides--secretion. QW 504 P964 2005] I. Veskler, Barbara A.
QR181.P766   2005
616.07'9--dc22   2005005451

Published by Nova Science Publishers, Inc. ✤ New York

Contents

Preface

Immunology is the study of the body's protection from foreign macromolecules or invading organisms and the responses to them. These invaders include viruses, bacteria, protozoa or even larger parasites. In addition, immune responses are developed against our own proteins (and other molecules) in autoimmunity and against our own aberrant cells in tumor immunity. The first line of defense against foreign organisms are barrier tissues such as the skin that stop the entry of organism into our bodies. A second line of defense is the specific or adaptive immune system which may take days to respond to a primary invasion (that is infection by an organism that has not hitherto been seen). This new book brings together new research from around the globe dealing with this extremely important subject.

An effective immune response depends on the ability of specialized immune cells to recognize foreign molecules (antigens), to become activated, and to proliferate and differentiate into mature effector cells. Many types of effector molecules play a role in this process, including protein kinase C (PKC), leading eventually to the activation of various transcription factors and the deployment of T-cell effector functions. Chapter I will focus on T-cell signaling, and specially on the role of one of the PKC members, PKCθ, which is required for the activation of the transcription factors nuclear factor-kappaB (NF-κB) and activator protein-1 (AP-1), which in turn regulate the production of interleukin-2 (IL-2), a necessary step for the deployment of T-cell effector functions. The chapter will also focus on one of the important aspects of PKCθ activation, its phosphorylation.

As explained in Chapter II, understanding the development of the human immune system is of interest to basic scientists as well as to clinicians. It is particularly of interest to researchers trying to develop effective means of fetal therapy based on cellular transplantation or gene therapy. The anatomic development of the thymus at the end of the first trimester of gestation is a major milestone in T-cell development. However, analysis of younger tissues has found T cells to be present as early as 8 weeks' gestation. Although the size of the T-cell pool is small at the end of the first trimester, a diverse repertoire of α/β T-cell receptors develop in the ensuing weeks. A high percentage of γ/δ T-cells are also present in the developing fetus. Despite the protective environment of the womb and the naïveté of the fetal immune system, a surprisingly large frequency of fetal T-cells were found to express cell-surface markers associated with activation by antigen.

As a conglomerate of immunology and zoology comparative immunology has gained wide acceptance in biology. There is much interest in the immune system of invertebrates as representing early models or precursors of the innate system of vertebrates. Although a small discipline, developmental and comparative immunology has started to take advantage of modern analytical tools such as mass spectrometry especially in proteomics. Bioscience applications of the technique in organic and inorganic analysis based on annelida research are discussed in Chapter III introducing the variability of mass spectrometry.

Antimicrobial peptides are ancient host defense molecules widely represented across the animal and plant kingdoms. Animals possess numerous antimicrobial peptides, which are expressed in many tissues and cells, including neutrophils, macrophages and mucosal epithelial cells. A prominent family of mammalian antimicrobial peptides is the cathelicidins. These peptides have broad-spectrum antimicrobial activity and are an important innate immune defense. However, in addition to their well-recognized antimicrobial activity, cathelicidins function in initiating and amplifying host innate and adaptive immune responses. Chapter IV reviews recent data illustrating the nonmicrobicidal activities of cathelicidins.

Keratinocytes participate in immune response and inflammation by secreting cytokines and chemokines. Membrane-bound peptidases control local concentrations of signalling peptides and recently have been proposed as an additional mechanism of cell-to-cell interaction and signal transmission. In Chapter V, the authors examined expression of three membrane-bound peptidases: aminopeptidase N (APN; EC 3.4.11.2; CD13), neutral endopeptidase (NEP; EC 3.4.24.11; CD10) and dipeptidyl-peptidase IV (DPPIV; EC 3.4.14.5; CD26) on cultured keratinocytes obtained from normal human skin. The data suggest a role of APN in regulation of keratinocyte growth.

As presented in Chapter VI, the authors suggest a concept for local deposition of therapeutically active proteins using secretory lysosomes of hematopoietic cells as delivery vehicles. During hematopoietic differentiation secretory proteins are stored in granules such as secretory lysosomes for release upon activation e.g. in areas of inflammation and malignancy. The protein of interest is expressed in progenitor cells, granule targeted, and deposited into a tumor or site of inflammation, to which the mature cells are migrating. Results from *in vivo* investigations will determine the suitability of this local protein delivery principle in inflammation and tumors.

Cytokines are potent pleiotropic immunomodulatory molecules with extended roles in the direction of immune regulation pathways. Polymorphisms of cytokine genes have been identified and can significantly influence cytokine production levels, thus affecting the outcome of immune balance. Considering that cytokines are inter-dependable and function within complex networks, it is impossible to evaluate cytokines individually in order to assess their involvement with a certain disease condition. As reported in Chapter VII, an investigative approach recently employed in the authors laboratory utilizes the strategy of cross-tabulation of cytokine polymorphisms to determine the significance of one cytokine in the context of another. In this manner the development of genetic patterns of cytokine polymorphism combinations, otherwise referred to as immunogenetic profiles, were established. A better understanding of how cytokines may function in concert with one another is warranted and might elucidate their precise roles in disease pathogenesis.

Parasitic infections are prevalent in both tropical and sub-tropical areas. Most of the affected areas are in the developing countries of the world where control measures are lacking or are intermittently applied. Vaccinations against parasitic diseases have been the major talk over the past decades yet there has been very little advancement in developing effective vaccines for diseases such as malaria and schistosomiasis. Chapter VIII gives insight into the understanding of infection and exposure to schistosome infections necessary in the development and current search for vaccines to schistosome infections.

Trypanosoma cruzi, the causative agent of Chagas'disease is a parasitic protozoa that infects more than eighteen million people in South and Central America. Chagas'disease is characterized by a heart defect and megaviscera in a proportion of patients, and these clinical signs are associated with extensive destruction of parasympathetic, enteric and other neurons and degeneration of cardiac muscle. Chagas'disease is often associated with the presence of autoantibodies (autoAb) against host tissues, including specialized components of striated muscle, neurons and connective tissue, which make autoimmune reactions likely. Interestingly, a remarkable polyspecificity of the autoAb was observed. Indeed, the autoAb have been found to cross-react with animal erythrocytes and distinct structural basement membrane proteins such as laminin and collagen. The purpose of chapter IX is to summarize some of the current data on the pathophysiology of *T. cruzi* infection. Special attention is given to recent data mainly from our own laboratory illustrating the important role of an immunomodulatory factor released by the parasite, the Tc52 protein, in the induction and perpetuation of chronic disease.

Porin was purified to homogeneity from *Shigella dysenteriae* type 1. The protein formed hydrophilic diffusion pores by incorporation into artificial liposome vesicles and exhibited significant porin activity. The molecular weight of the native porin molecule was 130,000, consisting of 38,000 monomer. Murine anti-porin antibody raised against the purified porin reacted with whole cell preparation of *S. dysenteriae* type 1 suggesting that porin possessed surface component. Porin could also be visualized on the bacterial surface by immunoelectron microscopy. The anti-porin antibody of *S. dysenteriae* cross-reacted with porin preparations of *S. flexneri*, *S. boydii, and S. sonnei,* indicating that porins are antigenically related among *Shigella* species. As explained in chapter X, porins are of particular interest because they have been characterized as potent adjuvants and have great potential as a novel component of vaccines. Understanding the mechanism of adjuvanticity of porin of *S. dysenteriae* type 1 is a necessary step towards the development of a better adjuvant against shigellosis.

In: Progress in Immunology Research
Editor: Barbara A. Veskler, pp. 1-21

ISBN 1-59454-380-1
©2005 Nova Science Publishers, Inc.

Chapter I

PKCθ and T-Cell Signaling

S. Thébault[*]

Laboratoire Infections Rétrovirales et Signalisation Cellulaire, CNRS UMR 5121, Institut de Biologie, 4 boulevard Henri IV, 34060 Montpellier, France

Abstract

An effective immune response depends on the ability of specialized immune cells to recognize foreign molecules (antigens), to become activated, and to proliferate and differentiate into mature effector cells. Many types of effector molecules play a role in this process, including protein kinase C (PKC), leading eventually to the activation of various transcription factors and the deployment of T-cell effector functions. This review will focus on T-cell signaling, and specially on the role of one of the PKC members, PKCθ, which is required for the activation of the transcription factors nuclear factor-kappaB (NF-κB) and activator protein-1 (AP-1), which in turn regulate the production of interleukin-2 (IL-2), a necessary step for the deployment of T-cell effector functions. This review will also focus on one of the important aspects of PKCθ activation, its phosphorylation.

Abbreviations

TCR	= T-cell antigen receptor
APC	= antigen-presenting cell
MHC	= major histocompatibility complex
IS	= immunological synapse
SMAC	= supramolecular activation cluster
LFA-1	= leukocyte function-associated molecule-1

[*] Laboratoire Infections Rétrovirales et Signalisation Cellulaire, CNRS UMR 5121, Institut de Biologie, 4 boulevard Henri IV, 34060 Montpellier, France; Phone: (33) 4 67 60 86 60; Fax: (33) 4 67 60 44 20; E-mail: sabine.thebault@univ-montp1.fr

ICAM-1   = intercellular cell adhesion molecule-1
PTK   = protein tyrosine kinase
PLCγ   = phospholipase Cγ
DAG   = diacylglycerol
IP$_3$ = inositol (1, 4, 5)-triphosphate
PKC   = protein kinase C
NF-κB   = nuclear factor-kappaB
AP-1   = activator protein-1
NF-AT   = nuclear factor of activated T cells
IL-2   = interleukin-2
PI$_3$K   = phosphatidylinositol 3-kinase
PBMC   = peripheral blood mononuclear cell
MTOC   = microtubule-organizing center
CREB   = cAMP response element-binding protein
CREM   = cAMP response element modulator
bZIP   = basic leucine zipper
TRE   = TPA response element
JNK   = c-Jun N-terminal kinase
IKK   = I-κB kinase

T-Cell Signaling and Immunological Synapse

An effective immune response depends on the ability of specialized immune cells to recognize foreign molecules (antigens), to become activated, and to proliferate and differentiate into mature effector cells. This process is mediated by a cell-surface antigen recognition apparatus and a complex intracellular receptor-coupled signal-transducing machinery, both operating with a high fidelity to discriminate self from nonself antigens.

The T-cell membrane-bound antigen-specific receptor (TCR) is composed of two covalently bound polymorphic subunits, which provide antigen specificity, associated to at least four different types of invariant chains that are necessary for signal transduction. Antigen is presented to T lymphocytes by antigen-presenting cells (APCs) in association with cell-surface major histocompatibility complex (MHC) molecules (Fig. 1). A stable interaction between the TCR and the MHC-antigen complex results in a temporal and spatial reorganization of multiple cellular elements in the contact region, a specialized region referred to as the immunological synapse (IS) [1] or the supramolecular activation cluster (SMAC) [2].

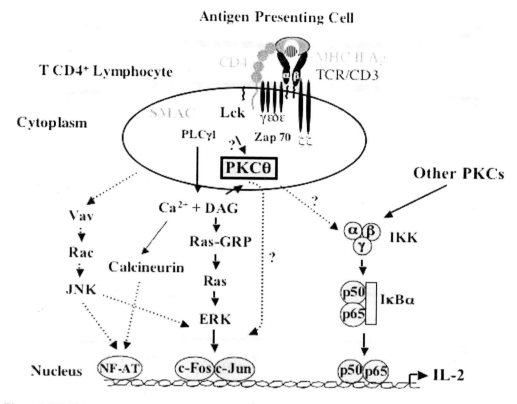

Figure 1. PKCθ in T-Cell signalling

The TCR / peptide-MHC interactions first accumulate in a ring surrounding a central cluster of leukocyte function-associated molecule-1 (LFA-1) / intercellular cell adhesion molecule-1 (ICAM-1) interactions, creating an immature T-cell synapse, which later inverts such that a ring of integrin, the peripheral supramolecular activation cluster (p-SMAC), surrounds a central cluster of TCR / peptide-MHC, the c-SMAC, at a mature IS [3]. This process of T-cell / APC conjugation was first observed in vitro using dynamic imaging techniques, which showed that, on contact with a cell surface bearing appropriate antigen receptor ligands, naïve T lymphocytes rapidly extend lamellipodia that spread along the surface membrane of the opposing cell [4, 5]. The modified shape of the T-cell is then maintained as long as the T-cell / APC conjugate exists. The central accumulation of TCR / peptide-MHC is thus clearly dependent on actin cytoskeletal processes and might be regulated in part by ezrin-radixin-moesin protein phosphorylation controlling the link between some membrane proteins and the cytoskeleton [6].

Many types of effector molecules play a role in the TCR-mediated signal. Reversible tyrosine phosphorylation and the related enzymes (protein tyrosine kinases and phosphatases) play a key role in regulating T-cell activation. In particular, protein tyrosine kinases (PTKs) of the Src and Syk families are activated. Activated PTKs phosphorylate multiple cellular proteins, including phospholipase Cγ (PLCγ) which thus becomes activated [7, 8]. PLCγ activation involves its translocation to the cell membrane where it cleaves phosphatidylinositol (4, 5)-biphosphate to produce the second messenger diacylglycerol (DAG) and inositol (1, 4, 5)-triphosphate (IP$_3$). These second messengers DAG and IP$_3$ stimulate an elevation of intracellular Ca^{2+} concentration and transduce the signal, among others, to protein kinase C (PKC), leading eventually to the activation of transcription factors and the deployment of T-cell effector functions.

Members of the PKC family play an important role in T-cell activation [9]. This has been well documented in studies using dominant negative and/or constitutively active mutants of PKC [10, 11] as well as studies using pharmalogical inducers or inhibitors of PKC [12].

PKC Family

PKC represents a family of at least twelve Ser/Thr kinase isoforms (Table 1 and Fig. 2) that play critical roles in the regulation of differentiation and proliferation in many cell types and in response to diverse stimuli [13]. For a long time, PKC has been handled almost as one single entity of redundant protein kinases. And even now, their non-redundant functions remain mostly unresolved [14]. Differences in structure and substrate requirement have permitted division of the isoforms into three groups [15, 16]. The conventional PKC isoforms (cPKC) α, ßI, ßII and γ are Ca^{2+}-dependent and require phosphatidylserine (PS) and DAG to become activated. The novel PKCs (nPKC) δ, ε, η and θ are Ca^{2+}-independent and also require PS and DAG for activation. Finally, the atypical PKCs (aPKC) ζ and ι/λ are Ca^{2+}-independent and do not require DAG for activation although PS regulates their activity (Table 2).

Table 1. The human PKC proteins

Gene	Protein amino acid number	Predominant tissue expression
Conventional subfamily		
α	671	Ubiquitous, high in T cells
ß	672	Ubiquitous, high in B cells
γ	696	Brain
Novel subfamily		
δ	675	Ubiquitous, high in T cells
ε	736	Ubiquitous, high in T cells
η	681	Ubiquitous, high in T cells
θ	705	T cells and skeletal muscle
Atypical subfamily		
ζ	591	Ubiquitous
ι/λ	586	Ubiquitous

Table 2. PKC proteins requirements for activation

Subfamily	Ca^{2+}-dependent	PS-dependent	DAG-dependent
Conventional subfamily	+	+	+
Novel subfamily	-	+	+
Atypical subfamily	-	+	-

Ca^{2+} : calcium
PS : phosphatidylserine
DAG : diacylglycerol

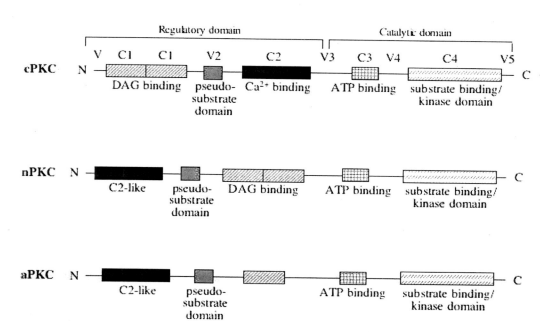

Figure 2. The Human PKC structures

The PKC signaling system appears to be highly versatile. Complex PKC regulatory mechanisms are thought to operate at the transcriptional, translational and post-translational levels. This flexibility is supported by a number of distinct and functionally non-redundant PKC isotypes, yet largely undefined upstream regulatory molecules, that are phospholipases, scaffolds and serine/threonine protein phosphatases, that mediate negative feedback by reversing protein phosphorylation on PKC itself as well as on its substrate proteins. The activity of PKC enzymes is effectively regulated by phosphorylation and binding of defined cofactors. PKC-mediated signaling responses appear to be strictly dependent on the given cellular differentiation status, that is due to modulation of the cellular subsets of distinct PKC isotypes.

Scaffolding proteins mediate the efficiency and specificity of PKC/effector reactions by coordinating their interaction along a cytoskeletal infrastructure in a spatial and temporal manner. Hence, subcellular location is believed to be a key element in regulating PKC isotype-specificity in vivo [17]. Consistently, little PKC substrate specificity has been observed in vitro, and recombinant PKC is able to phosphorylate various recombinant protein substrates such as receptors, other membrane proteins, contractile and cytoskeletal proteins and enzymes. In addition, enzyme activation is associated with its redistribution among different cellular compartments, commonly from the cytosolic to the particulate (membrane) fraction. Indeed, PKC is thought to reside in the cytosol in an inactive conformation and to translocate to the plasma membrane/cytoskeleton/rafts upon cell activation. However, PKCs, as well as other serine/threonine and tyrosine kinases, may regulate each other [18], thereby forming complex functional protein kinase networks.

For some time, the contribution of the different PKC isoforms to TCR-mediated activation was not clearly understood. Recent studies on PKCθ have uncovered its critical role in T-cell activation [19 , 20 , 21].

PKCθ Protein in T-Cell Activation

Role of PKCθ in T-Cell Activation

PKCθ is a member of the novel PKC subfamily [22]. The highest expression levels of this kinase have been detected in skeletal muscle and haematopoietic cells, but an increasing number of cell types, including endothelial, melanoma and mast cells, hepatocytes and nerve cell lines have also been shown to express PKCθ. This kinase has distinct functions according to the cell type. In non-haematopoietic cells, it is involved in mechanisms underlying actin polymerisation and cell motility [23 , 24]. Conversely, in mature T-cells, PKCθ plays a non-redundant role in cell activation and proliferation processes. Thus, due to its special role in T-cell receptor-mediated activation of T-lymphocytes, PKCθ is now the subject of concentrated investigation.

In particular, PKCθ has been found to act as an intermediate step between TCR proximal events and the activation of transcription factors such as nuclear factor-kappaB (NF-κB) and the activator protein-1 (AP-1) [25, 26]. Previous transient-overexpression studies in T-cell lines had suggested that PKCθ plays a key role in activating the JNK pathway, which functions to phosphorylate and activate members of the c-Jun family within the AP-1 complex [27, 28]. Interestingly, genetic ablation of PKCθ does not impair JNK activation, suggesting that PKCθ regulates AP-1 activity independently of the JNK pathway, possibly at the level of c-Fos and c-Jun transcription [21]. PKCθ has also been reported to synergize with calcineurin to activate nuclear factor of activated T-cells (NF-AT) and interleukin-2 (IL-2) transcription [28]. In addition, although the TCR stimulation alone can lead to PKCθ activation, it is clear that costimulatory CD28 signals augment PKCθ activation [29].

The relevance of PKCθ to T-cell activation was confirmed in PKCθ-deficient mice [21, 30]. Peripheral T cells from mice deficient in PKCθ expression are unable to proliferate when stimulated with antibodies against CD3 and CD28, even though there is no apparent defect in proximal signal transduction or ERK/JNK/p38-MAP-kinase activation [21]. The lack of proliferation of PKCθ$^{-/-}$ T cells results in part from a deficiency in IL-2 production as well as from a failure to upregulate CD25, the high affinity subunit of the IL-2 receptor. There is a controversy concerning NF-AT activity, which has been described as normal [21] or impaired

[30]. However, there is a clear selective defect involving the transcription factors AP-1 and NF-κB, which may account for the defective production of IL-2 [21, 30].

PKCθ Translocation to the Immunological Synapse and Lipid Rafts

Activation of PKCθ, as is the case for other PKC isoforms, requires translocation to cellular membranes. Among the T-cell-expressed PKC family members, PKCθ provides the example of a non-redundant recruitment of PKC isotypes in cells, because PKCθ selectively co-localizes with the TCR at the center of the SMAC during antigen stimulation [31], and more specifically of c-SMAC [2]. This localization occurs at a high stoichiometry and lasts for at least 2-4 h, suggesting that it could play an important role in propagating and extending activation signals that are required for T-cell commitment to IL-2 production. The c-SMAC clustering of PKCθ correlates with its catalytic activation and only occurs upon productive activation of T-cells. Co-clustering of talin and tubulin, and formation and reorientation of the microtubule-organizing center (MTOC) are also observed under these conditions. Furthermore, it has been reported that, during T-cell activation, PKCθ is recruited to particular membrane microdomains known as lipid rafts, and that this recruitment is required for T-cell activation [29]. These lipid rafts are detergent-insoluble membrane microdomains that concentrate critical signaling mediators. Similar to PKCθ itself, these lipid rafts also cluster at the IS in an antigen-stimulated T-cell but, unlike PKCθ, the rafts are not restricted to the c-SMAC. Receptor-induced clustering of lipid raft may serve as an important driving force that promotes the initial translocation of PKCθ to the IS, where additional mechanisms may function to selectively recruit it into a specific region of the IS, the c-SMAC.

The precise mechanism by which PKCθ is recruited to the IS during antigen stimulation is not totally understood. However, Vav and Rac selectively promote the membrane and cytoskeleton translocation of PKCθ, and mediate its enzymatic activation by CD3/CD28 costimulation (that is widely used as a TCR activation mimic) in a process that depends on actin cytoskeleton reorganization [32, 33]. PI3K (phosphatidylinositol 3-kinase) also plays a role in regulating the membrane and lipid raft translocation of PKCθ, acting upstream of Vav. Effectively, PI3K-generated lipid products activate Vav and recruit it to the membrane by binding to its pleckstrin-homology domain [34]. However, using three different approaches (a selective PLC inhibitor, a PLCγ-deficient T-cell line, and a dominant negative PLCγ mutant), the CD3/CD28-induced membrane recruitment and the carboxy-terminal phosphorylation of PKCθ were shown to be partially PLCγ-independent [35]. In conclusion, unlike the conventional PLC-dependent DAG-mediated translocation of other PKC isoforms, translocation of PKCθ is partly PLCγ-independent, but PI3K- and Vav-dependent, underscoring the importance of the cytoskeleton in T-cell activation. In addition, in T-cells, the Src family protein tyrosine kinase p56[lck] has also been shown to be critical in TCR-induced PKCθ translocation [36].

PKCθ Activation Regulation

PKCθ is composed of several functional modules that are conserved among the PKC isoforms and are implicated in regulating its recruitment and activation (Fig. 2). The C1 domain binds the membrane-embedded second messenger DAG. The plasma-membrane association of PKCθ may also be stabilized by the amino-terminal C2-like domain, which lacks the canonical Ca^{2+}-binding residues found in the conventional subfamily members but which may retain the ability to interact weakly with anionic phospholipids. PKCθ may be subject to at least four modes of regulation (Fig. 3). As shown in Fig 3A, PKCθ may be regulated intramolecularly due to the ability of its pseudo-substrate domain to sequester the kinase domain [37]. As depicted in Fig. 3B, following TCR-mediated PLCγ activation, DAG is generated, leading to membrane association of PKCθ via its C1 domain. The C2 domain may also contribute to strengthen the membrane association by interacting with membrane phospholipids (PL). In addition, as shown in Fig. 3C, PKCθ is subject to several regulatory phosphorylations (see the next section), in particular by Lck on Tyr-90 implicated in raft localization, by PDK1 on Thr-538 involved in its catalytic function, and on carboxy-terminal residues. To finish, as shown on Fig. 3D, other regulatory proteins such as Vav, Cbl, Fyn or Akt, can interact with PKCθ, regulating its subcellular localization, the accessibility of its kinase domain and its ability to interact with some of its subtrates [32].

Despite all the knowledge on PKCθ activation mechanisms, gaps remain on the molecular events that regulate its kinase activity and its ability to interact with upstream or downstream signaling mediators. Phosphorylation would play a key role.

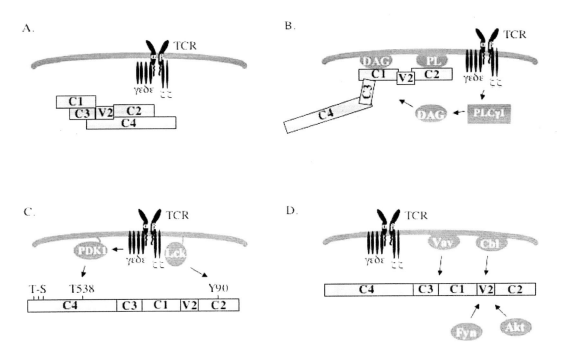

Figure 3. Regulation of PKCθ activation

PKCθ Regulation by Phosphorylation

Phosphorylation of PKC isoforms on Ser/Thr/Tyr residues is known to play a pivotal role in the regulation of their activity [38]. Sites of regulatory phosphorylation of PKCs have only been extensively studied in the conventional subfamily of PKCs (which depend on calcium and diacylglycerol for activation). There are three well-studied sites in the carboxy-terminal-halves of PKCα and PKCβ that play fundamental roles in regulating protein function, including catalytic activity, localization and degradation [39]. The possibility of systematic differences in the roles of regulatory phosphorylation between the novel subfamily (which includes PKCθ) and the conventional subfamily has been raised by findings with PKCδ, the only novel PKC isoform whose phosphorylation has been relatively well studied. Comparison of PKCθ sequence with other PKC isozymes identified a number of conserved phosphorylation sites implicated in the regulation of its activity (Fig. 4).

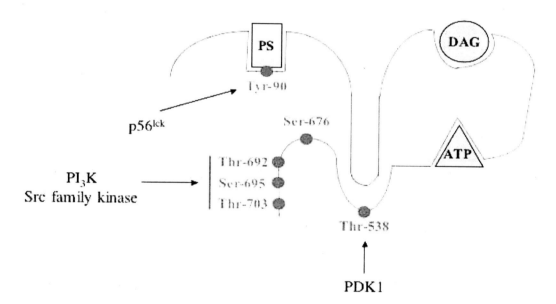

Figure 4. PKCθ phosphorylation sites

As mentionned earlier, p56[lck] is critical in TCR-induced PKCθ translocation. T-cell activation is followed by tyrosine phosphorylation of PKCθ, predominantly on Tyr-90 in the regulatory domain [40]. This phosphorylation is mediated by the kinase p56[lck] that interacts directly with the PKCθ regulatory domain. p56[lck] association with PKCθ can be observed in resting cells, increases following T-cell activation, involving the SH2 and SH3 domains of p56[lck]. A Tyr-90-mutation (PKCθ Y90F) leads to a markedly decrease in the T-cells proliferation rate and in the NF-AT activity induction in the presence of ionomycin [40]. These results suggest that the physical association of p56[lck] with PKCθ and the p56[lck]-induced tyrosine phosphorylation of PKCθ represent physiological events that regulate PKCθ during TCR-induced T-cell activation. Another member of the Src family kinases, p59[fyn], is also able to interact with and to phosphorylate PKCθ on a tyrosine in vitro [41].

PDK1, a kinase recruited to the signalsome by PI_3K, has been shown to phosphorylate a critical Thr residue in the activation loop of PKCδ and PKCζ, which leads to significant enhancement of basal kinase activity [39]. It has also been shown that PDK1 can physically interact with PKCθ in fibroblasts and that PDK1 constitutively phosphorylates Thr-538 in the activation loop [42 , 43]. Mutation of the activation loop threonine residue to alanine (PKCθ T538A) results in more than a 100-fold decrease in activity. At the contrary, mutation to glutamic acid (PKCθ T538E) preserves much of the kinase activity. This phosphorylation would be required for the subsequent phosphorylation events in the carboxy-terminal part of PKCθ, and would also be essential for the catalytic function and for the ability of PKCθ to activate NF-κB when overexpressed. The phosphorylation state of Thr-538 also plays a crucial role in the control of PKCθ intracellular localization [44].

Following Thr-538 phosphorylation, PKCθ is also phosphorylated in its carboxy-terminal "tail". Conventional PKC isoforms are known to undergo auto-phosphorylation (reviewed in [13, 39]). In the case of PKCθ, it has been reported that endogenous protein from PBMCs (peripheral blood mononuclear cells) or Jurkat cells was recognized by a phospho-Thr specific antibody upon activation with anti-TCR antibodies or PDBu [45]. The PKCθ phosphorylation event in this report was identified as an auto-phosphorylation reaction, but the particular residue recognized by the antibody was not defined. However, we described a phosphorylation event in mouse primary T-cells that was not prevented by pre-treatment with Gö6983, a potent PKC inhibitor that blocks TCR-mediated NF-κB activation, and therefore was not an auto-phosphorylation reaction [43]. It has also been reported that PKCθ transfected into HEK-293T cells is constitutively phosphorylated at the activation loop Thr-538, the turn motif Ser-676 and the hydrophobic motif Ser-695 [42]. The constitutive phosphorylations of Thr-538 and Ser-676 are in agreement with our results on mouse PKCθ, whereas the phosphorylation of Ser-695 that we characterized in mouse primary T-cells was T-cell activation dependent [43]. Finally, it has also been reported that endogenous PKCθ from Jurkat cells becomes phosphorylated at Ser-695 upon activation with αCD3+αCD28 [35]. This phosphorylation event is accompanied by migration of PKCθ from the cytoplasm to the membrane/cytoskeleton. This is in accordance with our data demonstrating that, following T-cell activation, there is a Ser/Thr phosphorylation event on murine PKCθ that takes place simultaneously with its recruitment to cellular membranes. By analyzing Ser and Thr point mutants, we found that phosphorylation of Ser-695 appears to be required to trigger further phosphorylation events at adjacent Thr-692 and Thr-703 residues. These phosphorylations require Src family kinase and PI_3K activities [43]. Interestingly, the T-cell activation-induced PKCθ phosphorylation event can, in some cases, lead to a change in the electrophoretic mobility of the protein, whether it is murine PKCθ as shown by stimulation of murine $CD4^+CD8^+$ thymocytes or spleen T cells [43, 46] or human PKCθ as shown upon αCD3+αCD28 stimulation in Jurkat cells [47]. An emerging theme in studies of phosphorylations of the "tail" of PKCs is their role in regulating PKCs localization. It would be interesting to initiate the same studies for PKCθ. It has also been postulated that phosphorylations at the "tail" of PKCθ (Ser-695 and possibly Thr-692 and Thr-703) would structure the active site so that it binds ATP and substrate with higher affinity [42]. Finally, the carboxy-terminal phosphorylation of PKCθ could also play a role in its interaction with downstream signaling effectors.

PKCθ-Induced IL-2 Production
Following T-Cell Activation

Productive T-cell activation leads to the synthesis and secretion of multiple cytokines, including IL-2, which functions as a major growth-promoting factor for lymphocytes. The *IL-2* gene promoter possesses consensus binding sites for several known transcription factors, including AP-1, NF-κB, NF-AT, Oct and CREB (cAMP response element-binding protein) / CREM (cAMP response element modulator) [48, 49]. A cooperative interaction between these factors is necessary to efficiently induce the *IL-2* gene. The critical role of PKCθ in induction of IL-2 production and the concomitant T-cell proliferation was confirmed by the findings that mature T-cells of PKCθ$^{-/-}$ mice produced very little IL-2 and underwent minimal proliferation in response to TCR/CD28 costimulation [21].

AP-1 in IL-2 Production

The transcription factor AP-1 is composed primarily of members of the Jun (c-Jun, JunB and JunD) and Fos (c-Fos, FosB, Fra-1 and Fra-2) families of proteins. These proteins are characterized by a highly charged, basic DNA-binding domain immediately adjacent to an amphipathic dimerization domain referred to as the "leucine zipper" and constituted of a heptad repeat of leucine residues [50]. They are part of the so-called bZIP family (basic leucine zipper) of transcription factors. Jun proteins can form both homo-and heterodimers among themselves. Jun proteins also dimerize with Fos proteins and other bZIP factors such as members of the ATF/CREB family and the Maf transcription factors. Dimerization is required for efficient interaction with specific DNA sequences, TREs (TPA response elements), found in the promoter of various target genes. These target genes are important for regulating many biological processes including proliferation, differentiation, apoptosis and transformation [51, 52].

The *IL-2* promoter contains a TRE sequence and therefore is a target for AP-1. In addition, AP-1 is known to be activated by phorbol esters (which use PKC as a cellular receptor) [51]. PKCθ, specifically, is able to activate AP-1. But a dominant negative (kinase-inactive) PKCθ mutant blocks phorbol ester-stimulated AP-1 activity [11]. Stable overexpression of PKCθ in T-cells followed by stimulation with an anti-CD3 antibody plus PMA promotes a strong IL-2 production [11]. Similarly, an overexpression of constitutively active PKCθ and calcineurin leads to a synergistic activation of the *IL-2* gene, whereas dominant negative PKCθ selectively inhibits *IL-2* promoter activation [28, 53]. Consistent with this finding, PKCθ, but not other PKC isotypes, activates in a T-cell specific manner the c-Jun N-terminal kinase (JNK), which contributes to AP-1 activation by phosphorylating c-Jun on Ser-63 and Ser-73 [27, 28]. Furthermore, PKCθ activates MKK4, the intermediate MAP kinase that phosphorylates and activates JNK [27]. However, the physiological relevance of PKCθ-mediated JNK activation is unclear since peripheral T cells from PKCθ$^{-/-}$ mice display intact TCR/CD28-induced JNK activation in the face of defective AP-1 activation [21]. These findings suggest not only that PKCθ plays an important and non-redundant role in AP-1 activation, but also that its function in JNK activation can be compensated by another PKC isoform or even by a PKC-independent pathway.

NF-κB in IL-2 Production

NF-κB/Rel transcription factors play a central role in the regulation of genes involved in a variety of cellular processes [54]. They are crucial mediators of immune and stress responses, exert pro- and antiapoptotic effects and are important in cell proliferation and differentiation. Dysregulation of the NF-κB system is thought to be involved in acute and chronic inflammatory processes as well as cancer [55]. The regulatory NF-κB complex consists of two subunits, which can form homo- or heterodimers. The prototypic and best understood NF-κB complex is composed of p50 and p65 (RelA). Other subunits have been identified, namely p52, c-Rel and relB, as well as the precursor proteins of p50 and p52, respectively p105 and p100 (Fig. 5). All NF-κB proteins share a conserved region of 300 amino acids known as the Rel homology domain (RHD) that contains the nuclear localization signal as well as the dimerization and DNA binding functions. In the cytosol of unstimulated cells, the dimeric NF-κB complex is present in an inactive state bound to inhibitory I-κB proteins that contain ankyrin repeat domains. Several inhibitors have been identified, namely I-κBα, I-κBβ, and I-κBε [56]. In addition, the aforementioned precursor proteins p105 (I-κBγ) and p100 (I-κBδ) may also act as inhibitors [57]. Association of p50/p65 with I-κB occludes the nuclear localization sequence of p50 and p65, leading to cytoplasmic sequestration [58, 59]. Stimulation of cells induces the phosphorylation of the I-κB proteins followed by poly-ubiquitination and subsequent proteolysis of these molecules (Fig. 6). Phosphorylation of the inhibitory proteins, the initial regulated step of I-κB degradation by the proteasome and thus NF-κB activation, is mediated by cytosolic high molecular weight complexes, collectively named the I-κB kinase IKK complex [60]. Three subunits of this complex have been identified: two kinase-active proteins, IKKα (or IKK1) and IKKβ (or IKK2), which are the I-κB phosphorylating kinases and are able to form homo- or heterodimers, as well as IKKγ (or NF-κB essential modulator NEMO, IKK associated protein 1 IKKAP1) which is a regulatory adaptor protein.

PKCθ stimulates and is required for TCR/CD28-induced activation of the CD28 response element (CD28-RE) in the *IL-2* gene promoter [20], a site that binds a combination of NF-κB and AP-1 [61]. Effectively, TCR/CD28 co-stimulation induces phosphorylation and activation of the IKKs, leading to IκB degradation and NF-κB activation [62]. In addition, as demonstrated using transient-overexpression approaches, PKCθ plays an important and selective role in activating the transcription factor NF-κB in T-cells [63]. Overexpression of a constitutively active form of PKCθ activates both NF-κB and CD28-RE reporter genes in T-cells, whereas a kinase-deficient PKCθ or a PKCθ antisense vector inhibits TCR/CD28 co-stimulation and NF-κB activation in T-cells [20]. In agreement with these studies, primary mouse T-cells lacking PKCθ display a severe defect in receptor-induced NF-κB activation [21].

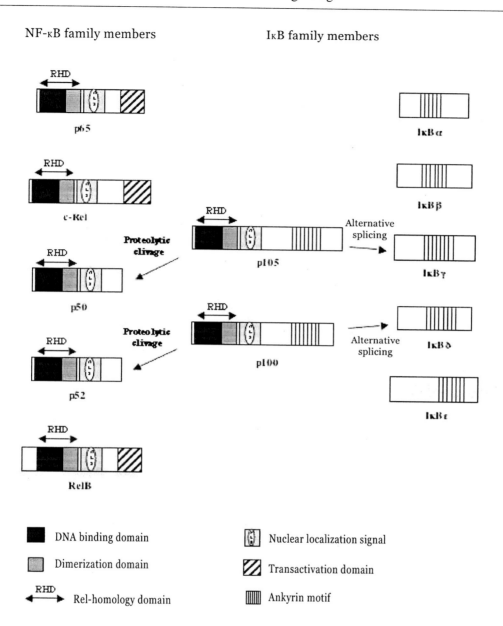

Figure 5. Schematic representation of the NF-κB and IκB family members

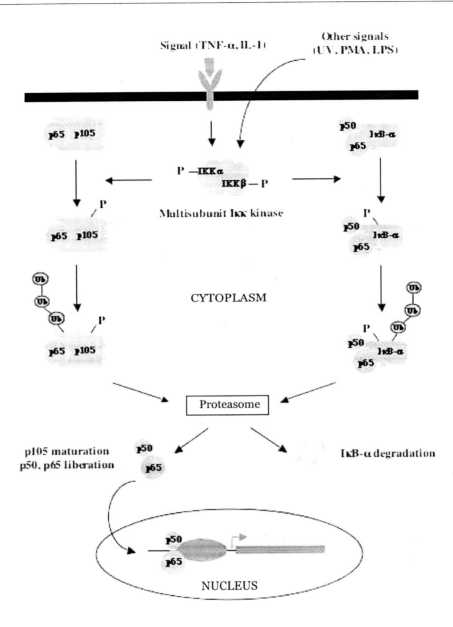

Figure 6. Regulation of NF-κB activation

NF-AT in IL-2 Production

Transcription factors of the NF-AT family, which are expressed in most immune-system cells, play a pivotal role in the transcription of cytokine genes and other genes critical for the immune response [64]. At least ten isoforms exist, some of them derived by alternative splicing (Fig. 7). Three different steps of activation have been defined: dephosphorylation, nuclear translocation and increase in DNA affinity [65, 66]. The activity of NF-AT proteins is tightly regulated by the calcium/calmodulin-dependent phosphatase calcineurin. Calcineurin controls the translocation of NF-AT proteins from the cytoplasm to the nucleus of activated cells by interacting with an amino-terminal regulatory domain conserved in the NF-AT family (the NF-AT homology domain). The DNA-binding domains of NF-AT proteins resemble those of Rel-family proteins, and Rel and NF-AT proteins show some overlap in their ability to bind to certain regulatory elements in cytokine genes. NF-AT is also notable for its ability to bind cooperatively with transcription factors of the AP-1 family to composite NF-AT:AP-1 sites, found in the regulatory regions of many genes that are inducibly transcribed by immune-system cells.

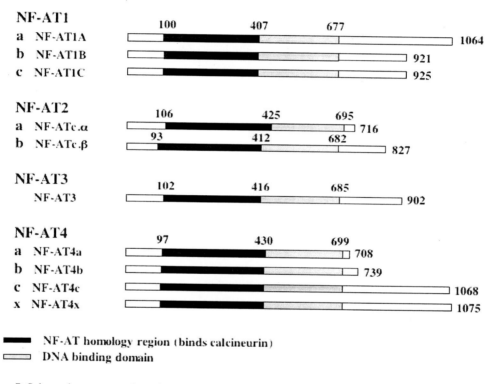

Figure 7. Schematic representation of NF-AT proteins

PKCs in general can cooperate with calcineurin to induce a downstream event important for T-cell activation as demonstrated by the synergistic effects of PMA and Ca^{2+} ionophore on both JNK activation and NF-AT activity [67]. The specificity of PKCθ to modulate the transcriptional activity of NF-AT has been clearly demonstrated by using a constitutively active PKCθ (PKCθ A148E) that preferentially increases the activity of NF-AT in synergism with calcineurin [28] or in the presence of ionomycin [68]. The role of PKCθ as a selective upstream regulator of NF-AT was finally substantiated in Nef-expressing cells, in which Nef-induced NF-AT-dependent gene transcription was found to require PKCθ [69].

Oct in IL-2 Production

The transcription factors of the POU family have come to play essential roles in the development of highly specialized tissues, such as complex neuronal systems, but also in more general cellular housekeeping. Members of the POU family recognize defined DNA sequences, and a well-studied subset have specificity for a motif known as the octamer element which is found in the promoter region of a variety of genes. The structurally bipartite POU domain has intrinsic conformational flexibility and this feature appears to confer functional diversity to this class of transcription factors. The POU domain for which there are more structural data is from Oct-1, which binds an eight base-pair target (ATGCAAAT) and variants of this octamer site. There are at least six Oct isoforms described to have very diversified functions. Oct factors act in synergy with other transcription factors such as AP-1.

The octamer proteins regulate the *IL-2* promoter by binding to two octamer-binding sites within it [70]. Mutation of either of these sites results in a partial loss of *IL-2* promoter activity, but mutation of both sites abolishes promoter function [71]. Both Oct-1 and Oct-2 proteins bind to the proximal Oct site and cooperate functionally with Fos- and Jun-family proteins [72, 73]. Although Oct-2 is clearly superior to Oct-1 as a transactivator at this site, Oct-1 and Oct-2 differ in their expression patterns: whereas Oct-1 is constitutively expressed, Oct-2 is newly synthesized in cells activated with antigen [74]. However, despite its striking activity in transient transfection assays, Oct-2 is not essential for *IL-2* gene transcription since mice deficient in Oct-2 show no gross abnormalities in peripheral T-cell function [75].

CREB/CREM in IL-2 Production and T-Cell Anergy

The ATF/CREB family of transcription factors represents a group of the bZIP proteins. These ATF/CREB proteins are able to bind to the CRE site TGACGTCA [76]. They can be classified into subgroups on the basis of their amino acid similarity: the CREB, CREB-2, ATF-2, ATF-3, ATF-6 and ATF-7 subgroups (Table 3).

Contrary to what has been described above, PKCθ appears to also regulate *IL-2* gene transcription in a different, potentially negative, manner. This regulation involves a site in the *IL-2* promoter located about 180 bp from the transcriptional start site. This site was found to bind CREB/CREM, ATF-2/c-Jun and Jun-Jun/Oct complexes [49]. However, induction of T-cell anergy leads to an increase in binding of only the CREB/CREM complex. Moreover, only mutation of the CREB/CREM-binding site reduces the susceptibility of T-cells to anergy induction [49]. These findings suggest that the -180 region of the *IL-2* promoter is the target of a CREB/CREM transcriptional inhibitor that contributes to the repression of IL-2 production in T-cell anergy [77]. Binding to this site is specifically regulated by PKCθ since T-cell activation induces a PKCθ-mediated phosphorylation of CREB, and TCR/CD28 stimulation induces binding of phospho-CREB to the -180 site in the *IL-2* promoter [78]. These findings raise the possibility that PKCθ-mediated CREB phosphorylation negatively regulates IL-2 transcription, thereby driving the responding T-cells into an anergic state. It is possible that this represents a feedback regulatory mechanism to terminate IL-2 production by activated T-cells. Alternatively, the increased CREB phosphorylation observed in anergic

T-cells may be mediated by another Ser/Thr kinase that antagonizes the activating function of PKCθ, therefore inducing anergy.

Table 3 . ATF/CREB family

Subgroup	Members
CREB	CREB
	CREM
	ATF-1 (TREB-36, TCRATF-1)
CREB-2	CREB-2 (ATF-4, TAXREB-67)
	MATF4
	MTR67
	ApCREB-2
	C/ATF
ATF-2	ATF-2 (CRE-BP1, TCRATF-2, mXBP)
	ATFa
	CRE-Bpa
ATF-3	ATF-3 (LRF-1)
	JDP-2
ATF-6	ATF-6
	CREB-RP
ATF-7	ATF-7

CREB : cAMP response element binding protein
CREM : cAMP response element modulator
ATF : activating transcription factor

Conclusion

Recent studies on PKCθ greatly improved our understanding of the selective function of this particular PKC isoform in T-cell activation and established its role as the second critical TCR signal that cooperates with calcineurin to activate the *IL-2* gene. The relatively selective expression and essential function of PKCθ in T-cell signaling and survival suggest that pharmacological or genetic strategies designed to selectively block the function of PKCθ in cells may be therapeutically useful in several potential scenarios. However, it would be important to use regimens that temporarily block PKCθ function at a critical time without inducing overt general immunosuppression. Beyond this prospective use of PKCθ as a drug target, several fundamental questions remain to be answered on PKCθ function and regulation.

References

[1]   Grakoui, A., Bromley, S. K., Sumen, C., Davis, M. M., Shaw, A. S., Allen, P. M. and Dustin, M. L. (1999) *Science* 285, 221-227

[2]   Monks, C. R., Freiberg, B. A., Kupfer, H., Sciaky, N. and Kupfer, A. (1998) *Nature* 395, 82-86

[3]   Davis, D. M. and Dustin, M. L. (2004) *Trends Immunol* 25, 323-327

[4]   Negulescu, P. A., Krasieva, T. B., Khan, A., Kerschbaum, H. H. and Cahalan, M. D. (1996) *Immunity* 4, 421-430

[5]   Delon, J., Bercovici, N., Liblau, R. and Trautmann, A. (1998) *Eur J Immunol* 28, 716-729

[6]   Faure, S., Salazar-Fontana, L. I., Semichon, M., Tybulewicz, V. L., Bismuth, G., Trautmann, A., Germain, R. N. and Delon, J. (2004) *Nat Immunol* 5, 272-279

[7]   Cantrell, D. (1996) *Annu Rev Immunol* 14, 259-274

[8]   Wange, R. L. and Samelson, L. E. (1996) *Immunity* 5, 197-205

[9]   Altman, A., Mustelin, T. and Coggeshall, K. M. (1990) *Crit Rev Immunol* 10, 347-391

[10]   Genot, E. M., Parker, P. J. and Cantrell, D. A. (1995) *J Biol Chem* 270, 9833-9839

[11]   Baier-Bitterlich, G., Uberall, F., Bauer, B., Fresser, F., Wachter, H., Grunicke, H., Utermann, G., Altman, A. and Baier, G. (1996) *Mol Cell Biol* 16, 1842-1850

[12]   Thorp, K. M., Verschueren, H., De Baetselier, P., Southern, C. and Matthews, N. (1996) *Immunology* 87, 434-438

[13]   Newton, A. C. (1997) *Curr Opin Cell Biol* 9, 161-167

[14]   Baier, G. (2003) *Immunol Rev* 192, 64-79

[15]   Liu, W. S. and Heckman, C. A. (1998) *Cell Signal* 10, 529-542

[16]   Mellor, H. and Parker, P. J. (1998) *Biochem J* 332 (Pt 2), 281-292

[17]   Dempsey, E. C., Newton, A. C., Mochly-Rosen, D., Fields, A. P., Reyland, M. E., Insel, P. A. and Messing, R. O. (2000) *Am J Physiol Lung Cell Mol Physiol* 279, L429-L438

[18]   Kampfer, S., Hellbert, K., Villunger, A., Doppler, W., Baier, G., Grunicke, H. H. and Uberall, F. (1998) *Embo J* 17, 4046-4055

[19]   Altman, A., Isakov, N. and Baier, G. (2000) *Immunol Today* 21, 567-573

[20]   Lin, X., O'Mahony, A., Mu, Y., Geleziunas, R. and Greene, W. C. (2000) *Mol Cell Biol* 20, 2933-2940

[21]   Sun, Z., Arendt, C. W., Ellmeier, W., Schaeffer, E. M., Sunshine, M. J., Gandhi, L., Annes, J., Petrzilka, D., Kupfer, A., Schwartzberg, P. L. and Littman, D. R. (2000) *Nature* 404, 402-407

[22]   Baier, G., Telford, D., Giampa, L., Coggeshall, K. M., Baier-Bitterlich, G., Isakov, N. and Altman, A. (1993) *J Biol Chem* 268, 4997-5004

[23]   Tang, S., Gao, Y. and Ware, J. A. (1999) *J Cell Biol* 147, 1073-1084

[24]   Stapleton, G., Malliri, A. and Ozanne, B. W. (2002) *J Cell Sci* 115, 2713-2724

[25]   Arendt, C. W., Albrecht, B., Soos, T. J. and Littman, D. R. (2002) *Curr Opin Immunol* 14, 323-330

[26]   Isakov, N. and Altman, A. (2002) *Annu. Rev. Immunol.* 20, 761-794

[27]   Avraham, A., Jung, S., Samuels, Y., Seger, R. and Ben-Neriah, Y. (1998) *Eur J Immunol* 28, 2320-2330

[28]   Werlen, G., Jacinto, E., Xia, Y. and Karin, M. (1998) *Embo J* 17, 3101-3111

[29]   Bi, K., Tanaka, Y., Coudronniere, N., Sugie, K., Hong, S., van Stipdonk, M. J. and Altman, A. (2001) *Nat Immunol* 2, 556-563

[30] Pfeifhofer, C., Kofler, K., Gruber, T., Tabrizi, N. G., Lutz, C., Maly, K., Leitges, M. and Baier, G. (2003) *J Exp Med 197*, 1525-1535

[31] Monks, C. R., Kupfer, H., Tamir, I., Barlow, A. and Kupfer, A. (1997) *Nature 385*, 83-86

[32] Villalba, M., Coudronniere, N., Deckert, M., Teixeiro, E., Mas, P. and Altman, A. (2000) *Immunity 12,* 151-160

[33] Dienz, O., Hehner, S. P., Droge, W. and Schmitz, M. L. (2000) *J Biol Chem 275*, 24547-24551

[34] Han, J., Luby-Phelps, K., Das, B., Shu, X., Xia, Y., Mosteller, R. D., Krishna, U. M., Falck, J. R., White, M. A. and Broek, D. (1998) *Science 279*, 558-560

[35] Villalba, M., Bi, K., Hu, J., Altman, Y., Bushway, P., Reits, E., Neefjes, J., Baier, G., Abraham, R. T. and Altman, A. (2002) *J Cell Biol 157*, 253-263

[36] Huang, J., Lo, P. F., Zal, T., Gascoigne, N. R., Smith, B. A., Levin, S. D. and Grey, H. M. (2002) *Proc Natl Acad Sci U S A 99*, 9369-9373

[37] Newton, A. C. (1995) *J Biol Chem 270*, 28495-28498

[38] Newton, A. C. (2003) *Biochem J 370*, 361-371

[39] Parekh, D. B., Ziegler, W. and Parker, P. J. (2000) Embo J 19, 496-503

[40] Liu, Y., Witte, S., Liu, Y. C., Doyle, M., Elly, C. and Altman, A. (2000) *J Biol Chem 275,* 3603-3609

[41] Ron, D., Napolitano, E. W., Voronova, A., Vasquez, N. J., Roberts, D. N., Calio, B. L., Caothien, R. H., Pettiford, S. M., Wellik, S., Mandac, J. B. and Kauvar, L. M. (1999) *J Biol Chem 274,* 19003-19010

[42] Liu, Y., Graham, C., Li, A., Fisher, R. J. and Shaw, S. (2002) *Biochem J 361*, 255-65.

[43] Thébault, S. and Ochoa-Garay, J. (2004) *Molecular Immunology 40*, 931-942

[44] Sparatore, B., Passalacqua, M., Pedrazzi, M., Ledda, S., Patrone, M., Gaggero, D., Pontremoli, S. and Melloni, E. (2003) *FEBS Lett* 554, 35-40

[45] Bauer, B., Krumbock, N., Fresser, F., Hochholdinger, F., Spitaler, M., Simm, A., Uberall, F., Schraven, B. and Baier, G. (2001) *J Biol Chem 276*, 31627-31634

[46] Asada, A., Zhao, Y., Komano, H., Kuwata, T., Mukai, M., Fujita, K., Tozawa, Y., Iseki, R., Tian, H., Sato, K., Motegi, Y., Suzuki, R., Yokoyama, M. and Iwata, M. (2000) *Immunology 101,* 309-315

[47] Cao, Y., Janssen, E. M., Duncan, A. W., Altman, A., Billadeau, D. D. and Abraham, R. T. (2002) *Embo J 21*, 4809-4819

[48] Jain, J., Loh, C. and Rao, A. (1995) *Curr Opin Immunol 7,* 333-342

[49] Powell, J. D., Lerner, C. G., Ewoldt, G. R. and Schwartz, R. H. (1999) *J Immunol 163*, 6631-6639

[50] Landschulz, W. H., Johnson, P. F. and McKnight, S. L. (1988) *Science 240*, 1759-1764

[51] Karin, M., Liu, Z. and Zandi, E. (1997) *Curr Opin Cell Biol 9,* 240-246

[52] Shaulian, E. and Karin, M. (2001) *Oncogene 20*, 2390-2400

[53] Ghaffari-Tabrizi, N., Bauer, B., Villunger, A., Baier-Bitterlich, G., Altman, A., Utermann, G., Uberall, F. and Baier, G. (1999) *Eur J Immunol 29*, 132-142

[54] Baeuerle, P. A. and Baltimore, D. (1996) *Cell 87*, 13-20

[55] Baldwin, A. S. (2001) *J Clin Invest 107*, 241-246

[56] May, M. J. and Ghosh, S. (1998) *Immunol Today 19*, 80-88

[57] Thanos, D. and Maniatis, T. (1995) *Cell 80*, 529-532

[58] Blank, V., Kourilsky, P. and Israel, A. (1991) *Embo J 10*, 4159-4167

[59] Haskill, S., Beg, A. A., Tompkins, S. M., Morris, J. S., Yurochko, A. D., Sampson-Johannes, A., Mondal, K., Ralph, P. and Baldwin, A. S., Jr. (1991) *Cell 65*, 1281-1289

[60] Israel, A. (2000) *Trends Cell Biol 10*, 129-133

[61] McGuire, K. L. and Iacobelli, M. (1997) *J Immunol 159*, 1319-1327

[62] Harhaj, E. W. and Sun, S. C. (1998) *J Biol Chem 273*, 25185-25190

[63] Coudronniere, N., Villalba, M., Englund, N. and Altman, A. (2000) *Proc Natl Acad Sci U S A 97*, 3394-3399

[64] Rao, A. (1994) *Immunol Today 15*, 274-281

[65] Wesselborg, S., Fruman, D. A., Sagoo, J. K., Bierer, B. E. and Burakoff, S. J. (1996) *J Biol Chem 271,* 1274-1277

[66] Loh, C., Shaw, K. T., Carew, J., Viola, J. P., Luo, C., Perrino, B. A. and Rao, A. (1996) *J Biol Chem 271*, 10884-10891

[67] Crabtree, G. R. and Clipstone, N. A. (1994) *Annu Rev Biochem 63*, 1045-1083

[68] Bauer, B., Krumbock, N., Ghaffari-Tabrizi, N., Kampfer, S., Villunger, A., Wilda, M., Hameister, H., Utermann, G., Leitges, M., Uberall, F. and Baier, G. (2000) *Eur J Immunol 30*, 3645-3654

[69] Manninen, A., Huotari, P., Hiipakka, M., Renkema, G. H. and Saksela, K. (2001) *J Virol 75*, 3034-3037

[70] Kamps, M. P., Corcoran, L., LeBowitz, J. H. and Baltimore, D. (1990) *Mol Cell Biol 10*, 5464-5472

[71] Zhang, L. and Nabel, G. J. (1994) *Cytokine 6*, 221-228

[72] Ullman, K. S., Northrop, J. P., Admon, A. and Crabtree, G. R. (1993) *Genes Dev 7,* 188-196

[73] Pfeuffer, I., Klein-Hessling, S., Heinfling, A., Chuvpilo, S., Escher, C., Brabletz, T., Hentsch, B., Schwarzenbach, H., Matthias, P. and Serfling, E. (1994) *J Immunol 153,* 5572-5585

[74] Kang, S. M., Tsang, W., Doll, S., Scherle, P., Ko, H. S., Tran, A. C., Lenardo, M. J. and Staudt, L. M. (1992) *Mol Cell Biol 12,* 3149-3154

[75] Corcoran, L. M. and Karvelas, M. (1994) *Immunity 1,* 635-645

[76] Deutsch, P. J., Hoeffler, J. P., Jameson, J. L., Lin, J. C. and Habener, J. F. (1988) *J Biol Chem 263,* 18466-18472

[77] Powell, J. D., Ragheb, J. A., Kitagawa-Sakakida, S. and Schwartz, R. H. (1998) *Immunol Rev 165,* 287-300

[78] Solomou, E. E., Juang, Y. T. and Tsokos, G. C. (2001) *J Immunol 166*, 5665-5674

In: Progress in Immunology Research
Editor: Barbara A. Veskler, pp. 21-43

ISBN 1-59454-380-1
©2005 Nova Science Publishers, Inc.

Chapter II

Novel T-Cell Populations in the Human Fetus

Marcus O. Muench[*] *and Alicia Bárcena*

Department of Laboratory Medicine, University of California, 513 Parnassus Ave., Room
HSW-901B, San Francisco, CA 94143-0793, USA.

Abstract

Understanding the development of the human immune system is of interest to basic
scientists as well as to clinicians. It is particularly of interest to researchers trying to
develop effective means of fetal therapy based on cellular transplantation or gene
therapy. The anatomic development of the thymus at the end of the first trimester of
gestation is a major milestone in T-cell development. However, analysis of younger
tissues has found T cells to be present as early as 8 weeks' gestation. Although the size
of the T-cell pool is small at the end of the first trimester, a diverse repertoire of α/β T-
cell receptors develop in the ensuing weeks. A high percentage of γ/δ T-cells are also
present in the developing fetus. Despite the protective environment of the womb and the
naïveté of the fetal immune system, a surprisingly large frequency of fetal T-cells were
found to express cell-surface markers associated with activation by antigen. A notable
number of fetal T-cells were observed to have the memory T-cell phenotype of
$CD45RO^+CD45RA^-$ in contrast to the paucity of these T cells observed in the umbilical
cord blood at birth. Activation markers CD25, CD122, CD69, CD80 and CD95 were also
present on fetal T-cells but were present to a much lower extent on neonatal T-cells.
CD95 (fas) expression defined a number of unique subpopulations of fetal T-cells. CD95
was expressed at a higher level on $CD45RO^-CD45RA^+$ T-cells in the fetus than in the
adult, were $CD95^+$ T-cells are primarily of the memory phenotype. A unique $CD95^-$
population was also observed among the pool of $CD45RO^+CD45RA^-$ fetal T-cells. These
findings indicate that in the initial weeks following thymic maturation, a high frequency
of T-cells with an activated phenotype exist in the fetal circulation. An explanation for

---

[*] University of California at San Francisco; 513 Parnassus Ave., Room HSW-901B; San Francisco, CA
94143-0793. E-mail: muench@itsa.ucsf.edu, telephone: (415) 476-8420, fax: (415) 476-2956.

these findings remains to be discerned. Possibilities include activation by exposure to external antigens (via the maternal circulation), maternal antigens or response to self-antigens. Alternatively, expression of activation markers may not be the result of antigen activation, but rather the result of systemic signals promoting the rapid development of the fetal immune system. Another possible explanation is that these novel T-cells represent immature migrants from the thymus.

Introduction

Development of the cellular immune system begins in the human embryo and continues through fetal development. Lymphocytes represent about 80% of fetal peripheral blood leukocytes during midgestation. The percentage of lymphocytes declines during gestation, but the concentration of lymphocytes in umbilical cord blood are comparable to adult values at the time of birth when the neonate is exposed for the first time to the full onslaught of human pathogens [1]. The reason for the early appearance of T cells, the subject of this report, is not clear. At least three possible reasons for the early development of T cells can be envisioned. A straightforward explanation is that the thymus undergoes organogenesis in the embryo and growth in the fetus, as do all other organs, so that at birth sufficient numbers of T cells are present to provide cellular immunity in the neonate. The suggestion in this scenario is that the early emergence of T cells in the embryo serves no function other than marking the beginning of the buildup of the T-cell pool. Another possibility is that the presence of T cells during more than 6 months of gestation does serve to protect the fetus from the rare pathogens that can infect a fetus via the maternal circulation. A third possibility is that the fetal T-cells help protect the fetus from maternal engraftment. Exchange of cells between a mother and her fetus is known to occur. Since a mother and her fetus share only about half of their major histocompatibility complex (MHC) antigens, the possibility for invading maternal lymphocytes to cause a graft-versus-host like reaction in an fetus can be envisioned. None of these three possibilities is mutually exclusive; the fetus' cellular immune system may go unchallenged in many pregnancies. However, in some cases the presence of fetal T-cells may provide critical protection from pathogens or maternal engraftment.

The majority of what is known regarding T cells in the human fetus has come from studies of neonatal umbilical cord blood. The T-cell repertoire, based on examination of T-cell receptor (TCR) β-chain diversity, appears fully formed at birth. However, the repertoire appears to be composed of primarily naïve T-cells since none of the skewing of the subfamilies of β-chain variable regions, associated with oligoclonal expansion of activated T cells, was observed among neonatal T-cells [2]. Additionally, expression of CD45 isoforms by neonatal T-cells supports the contention that these cells are naïve [3]. Memory T-cells that have undergone previous antigen activation express CD45RO but lack CD45RA expression. These memory T-cells are common in adult blood but are rare in neonatal blood. Most T cells in umbilical cord blood are CD45RO⁻CD45RA⁺, the phenotype of naïve T-cells [4-8]. Other differences between neonatal and adult T-cells have also been reported. Signaling by activation of CD3 and CD28 molecules is attenuated in neonatal T-cells as accessed by the failure of neonatal T-cells to increase CD25, CD154 and CD178 (Fas ligand) expression [9]. A reduced proliferative response by neonatal T-cells has also been observed after allostimulation [4, 10-12]. Umbilical cord blood T-cells also have greatly reduced perforin expression, suggesting a reduced cytolytic capacity of neonatal T-cells [13]. These findings suggest that despite T-cell development beginning in the first trimester of gestation, T-cell functions appear attenuated in the neonate in addition to the naïveté of the fetal immune system owing to a lack of foreign antigen exposure.

Our laboratory's interest in the development of cellular immunity derives from an interest in fetal medicine. We are interested in the potential of maternal chimerism to cause disease in the fetus and neonate and what the role of the fetal immune system is in protecting from the invasion of maternal cells. We are also actively developing methods of in utero transplantation [14] and fetal gene therapy [15]. Both pursuits are similar to the passage of maternal cells into a fetus in that they represent the introduction of foreign antigens to the fetus. Knowing the functional capacity of fetal T-cells is critical to understanding the barriers that these cells pose to the development of novel fetal therapies. This report represents our initial examination of the emergence of T cells in the human embryo, development of TCR diversity and expression of cell-surface molecules important in T-cell growth and function. Our findings point to an early emergence of T-cell diversity and a surprising frequency of T cells with the phenotypic characteristics of activated or memory T-cells.

Thymic Development and the Emergence of T Cells in the Embryo

The cellular immune system begins its development early in gestation around the end of the first trimester. The fetal thymus is not anatomically completely mature until the 15th week of gestation [16]. At this time, single-positive $CD4^+CD8^-CD3^+$ and $CD4^-CD8^+CD3^+$ T-cells begin to accumulate rapidly in the periphery [17]. The maturation of the thymus and increased production of T cells represents a milestone in the field of fetal transplantation. Most recent attempts at transplantation of hematopoietic stem cells into fetuses afflicted with a hereditary disease have occurred before 15 weeks' gestation and the emergence of T cells in the fetus [14]. The rational for this is that the fetus is immunodeficient prior to 15 weeks' gestation and will accept the transplantation of foreign cells.

We have analyzed late embryonic and midgestation fetal tissues obtained from elective abortions for the presence of $CD3^+$ T-cells [18-20]. In Fig. 1, we show the results of flow cytometric analyses of 46 liver specimens of gestational ages between 8 and 23 weeks. Before 15 weeks' gestation the liver, the primary hematopoietic organ in the fetus at this age, contained less than 1×10^6 $CD3^+$ cells. After this time, the number of peripheral T-cells that were present in the liver increased markedly owing to the maturation of the thymus. However, intrathymic CD3+ T-cells have been detected as early as the 8th week of gestation, long before the complete maturation of the thymus [17, 21]. That the thymus can produce mature T-cells even before its anatomical maturation is supported by our findings of T cells in late first trimester livers (Fig 2.). Additionally, we have observed $CD3^+$ cells in umbilical cord blood as early as 8 weeks' gestation [20].

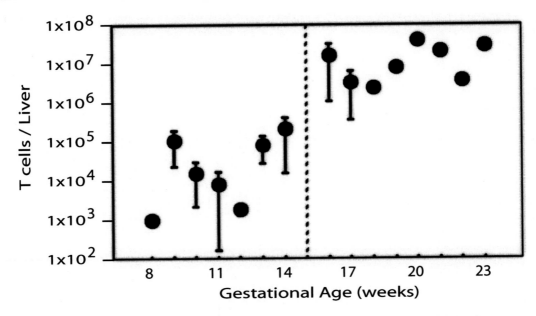

Figure 1. Presence of T cells in the human liver between 8 and 23 weeks' gestation. T-cell numbers were determined by flow cytometric analysis of individual liver specimens. Liver specimens that appeared nearly intact were selected for analysis, but the measurements may represent a modest underestimate owing to the delicate nature of the tissue and the disruption that often occurs because of the abortion process. A vertical line at 15 weeks' gestation marks the time of anatomic maturation of the thymus, after which peripheral T-cell numbers rise sharply. Note that the numbers of T cells are indicated on a logarithmic scale. The scattergram of hepatic CD3+ T-cells was compiled from data reported in three publications [18-20].

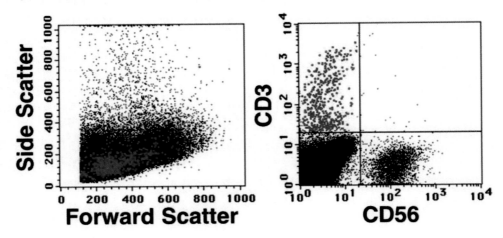

Figure 2. Presence of lymphocytes in the liver at the end of the first trimester of gestation. Both CD3+ T-cells and CD56+CD3- NK-cells are present among light-density liver cells at 13 weeks' gestation. To enrich the T-cell population for display, an electronic gate was set around the CD3+ population as indicated by the red dots. The right dot-plot displays the light scatter characteristics of these CD3+ events. An electronic gate defining low forward and side light-scatter cells (a lymphocyte gate) was drawn around the CD3+ events and was used to display the events in the left plot.

The total numbers of T cells present during the first trimester of gestation appear very few and are unlikely to present sufficient diversity to provide for a vigorous response to foreign antigens. There may simply be too few T-cell clones to respond to a viral antigen, but responses to allogeneic MHC generally involve a greater portion of the T-cell repertoire and may be possible in the late first trimester of gestation. Therefore, the possibility exists that T cells present in the fetus before the end of the first trimester may help protect the fetus from invasion by maternal cells. The functional capacity of fetal T-cells has only just begun to be studied. Some initial observations come from Renda et al., who demonstrated that T-cell clones obtained from 16 weeks' gestation livers were capable of responding to allogeneic peripheral blood mononuclear cells (PBMCs) as well as to phytohemagglutinin, immobilized anti-CD3 mAb and anti-TCR mAb [22]. Additionally, Lindton et al. have studied the responses, in mixed lymphocyte cultures, of T-cells to alloantigen and found evidence of significant T-cell responses from liver tissues as young as 9 weeks' gestation [23, 24].

TCR Gene Expression and Development of TCR Diversity in the Fetus

To gain a better understanding of the diversity of the T-cell repertoire in fetal development, we analyzed the expression of TCR genes on fetal splenocytes ranging in age from 19 to 23 weeks' gestation (n = 5). The α/β TCR was expressed on 80.2±4.3% of fetal T-cells whereas the γ/δ chains were expressed on 22.2±4.5% of T cells [25]. The frequency of γ/δ TCR expression of fetal T-cells was significantly greater than the 3.0±0.6% expression observed on neonatal T-cells (n=4) and the 8.3±0.8% expression measured on adult peripheral blood T-cells (n=6). A representative example of the flow cytometric analysis of TCR expression is shown in Fig. 3. γ/δ T lymphocytes are considered innate T cells that represent, in conjunction to natural killer (NK) cells, a first line of defense against infection [26]. γ/δ T cells perform effector functions that protect tissues from infection and malignancy [27, 28] as well as maintain tissue integrity [29]. In addition, they play important regulatory roles in autoimmunity and inflammation [30, 31]. As a component to the innate immune system, γ/δ T cells often recognize antigens of self-origin that are induced or expression altered upon damage, stress or transformation. Their early presence during human fetal development has been reported by several groups [32, 33] and their higher frequency during fetal life might be due to the necessary establishment of the innate immune system during its development.

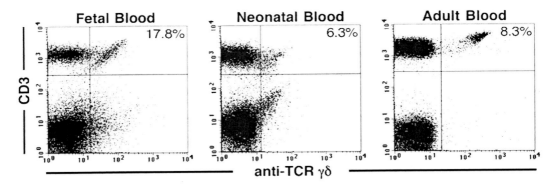

Figure 3. Expression of γδ TCR by T cells at different stages of ontogeny. T cells were obtained from a 20 weeks' gestation spleen, term umbilical-cord blood and from the peripheral blood of a 52-year adult. Dot plots display events gated using a lymphocyte gate as defined in Fig. 2. Percentages indicate the percent of all CD3$^+$ events that express γδ TCR as defined by the quadrants shown.

The TCR Vβ chain repertoire expressed by fetal T-cells was analyzed by flow cytometry to determine the diversity of TCR expression in the emerging immune system (Fig. 4). Splenic CD3$^+$ cells ranging in age from 16 to 24 weeks' gestation were analyzed for Vβ chain expression [25]. A diverse repertoire was observed. Indeed, the expression of the various Vβ chains on fetal T-cells fell within the ranges of expression observed on adult specimens reported by the manufacturer of the test reagents. These data demonstrate that a diverse repertoire of αβ TCRs is present shortly after anatomical maturation of the thymus. A similar conclusion was reached by Renda et al. [22]. These investigators studied TCR VDJ β-chain gene transcripts in liver and blood cells ranging from 7 to 20.5 weeks' gestation. They found variability in the Vβ8 gene sequence among fetal samples as young as 9 weeks' gestation that was comparable to that detected in adults. Together, these findings indicate that a diverse repertoire of T cells arises very early in development, suggesting that the fetal T-cells have the capacity to respond to a variety of foreign antigens.

CD45 Isoform Expression on T Cells at Various Stages of Ontogeny

The fetus is protected from most environmental pathogens and antigens, in general, by its confinement in the womb. For this reason, fetal T-cells are assumed to be naïve, an assumption borne out by studies showing neonatal T-cells to be comprised of primarily CD45RA$^+$ T-cells [4-8]. Interestingly, we observed a large number of CD45RO$^+$CD45RA$^-$ cells among midgestation T-cells (Fig. 5). This subset of memory T-cells represented a mean of 16.2±6.1% and 15.5±2.4% of T cells in the fetal blood and spleen, respectively (Fig. 6). This was significantly higher than the mean 1.7±0.4% CD45RO$^+$CD45RA$^-$ T-cells in neonatal blood obtained from full term newborns. Thus, our findings of a high frequency of CD45RO expression by midgestation T-cells indicated a surprising change in CD45 isoform expression occurs on T cells in the approximately 20 weeks that separate midgestation and full-term birth.

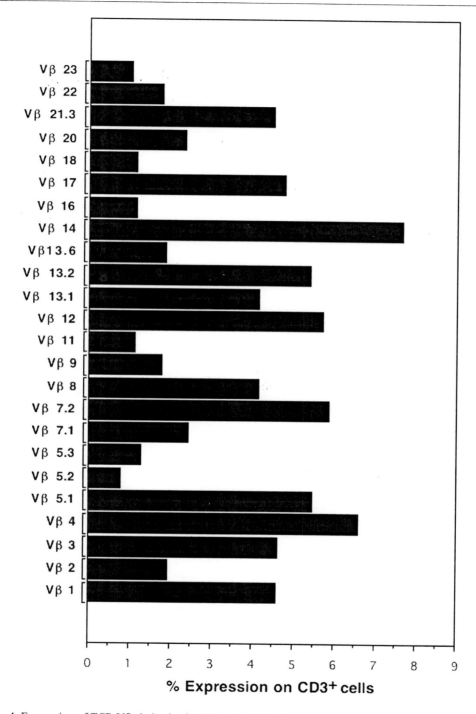

Figure 4. Expression of TCR Vβ chains by fetal T-cells. TCR Vβ chain expression was analyzed by 4-color flow cytometry. T cells were identified by their expression of CD3 as well as by their light scatter profile as indicated in Fig. 2. Results show Vβ chain expression analyzed on T cells isolated from a 16 weeks' gestation spleen.

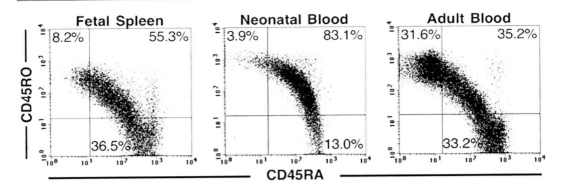

Figure 5. Expression patterns of CD45 isoforms by CD3$^+$ T-cells at different stages of ontogeny. Dot plots show the results of analyses performed on light-density cells isolated from a 20 weeks' gestation spleen, term umbilical cord blood and a 38 year-old male. The percentages of CD3$^+$ events are indicated for three of the quadrant regions.

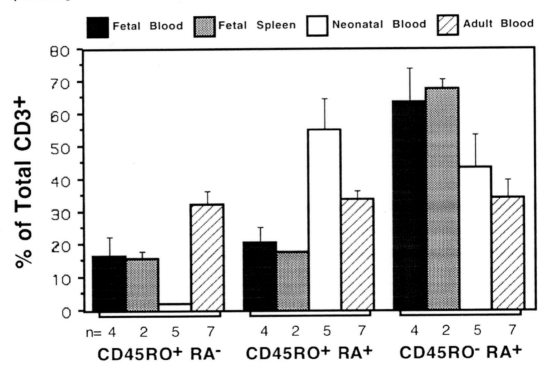

Figure 6. Expression of CD45 isoforms by T cells at different stages of ontogeny. The mean + SE percentages of CD45RO$^+$CD45RA$^-$, CD45RO$^+$CD45RA$^+$ and CD45RO$^-$CD45RA$^+$ T-cells in fetal tissues, neonatal umbilical-cord blood collected at term and adult blood are shown. The numbers (n) of tissue samples analyzed are indicated at the bottom of the plot.

In adult blood the frequency of CD45RO$^+$CD45RA$^-$ T-cells (32.3±4.0%) was still higher than that observed in the fetus (Figs. 5 and 6). Both fetal blood and spleen also had significantly reduced numbers of CD45RO$^+$CD45RA$^+$ T-cells compared to adults. Consequently, naïve CD45RO$^-$CD45RA$^+$ T-cells were more prevalent in the fetus than in the adult. In contrast, the majority of T cells in neonatal blood were CD45RO$^+$CD45RA$^+$ and there was a similar frequency of CD45RO⁻CD45RA⁺ T-cells in neonatal and adult blood. Thus, the expression of CD45 isotypes is more polarized in the fetus than at birth. Both the naïve CD45RO⁻CD45RA⁺ and memory CD45RO⁺CD45RA⁻ T-cell subsets are enriched in the fetus compared to the time of birth. Our findings are supported by another report that also documented a higher frequency of CD45RO⁺CD45RA⁻ T-cells in the fetus than the in the neonate [34].

The significance of the CD45RO⁺CD45RA⁻ T-cell subset in the developing immune system is presently unclear. The expression of CD45RO by fetal T-cells suggests that they may have been previously activated. Although it has been suggested that the fetus can be exposed to external antigens [35], the prevalence of the CD45RO⁺ subpopulation suggests a more abundant source of antigen. One possibility is that these cells are responding to maternal antigens that have crossed into the fetal circulation. Another possibility is that CD45RO⁺CD45RA⁻ T-cells are responding to autologous antigen. Indeed, we speculate that a wave of activation of auto-reactive T-cells early in the development of the cellular immune system could be an important step in the establishment of suppressor T-cell populations and peripheral tolerance. Another possible origin of the fetal CD45RO⁺CD45RA⁻ T-cells is that these are recent thymic emigrants that did not loose CD45RO and gain CD45RA expression before exiting the thymus. Yet another possibility is that CD45RO expression on fetal T-cells is not the result of antigen activation, but rather is the result of non-specific growth stimuli unique to the fetal environment that mimic antigen-specific activation. To begin to discern between these possibilities, we analyzed the expression of TCR Vβ chain expression and various activation antigens on subpopulations of fetal T-cells to better characterize these cells.

Comparison of TCR Vβ Chain Expression on Splenic and Thymic CD45RO⁺ T-Cells in the Fetus

We examined TCR Vβ chain expression on fetal CD45RO⁺ T-cells since differences or similarities in Vβ chain expression between splenic and thymic CD45RO⁺ T-cells could indicate if the peripheral T-cells are recent emigrants from the thymus. Most developing T-cells in the thymus express CD45RO, which is expressed in an inverse relationship to CD45RA. Before exiting the thymus, T cells lose CD45RO expression and become CD45RA⁺ [36-38]. Hence, the possibility exists that the CD45RA⁻ T-cells in the spleen are thymocytes that have not gained CD45RA expression before exiting the thymus. To accomplish the required four-color flow cytometric analyses, the CD45RO⁺CD3⁺ T-cells were defined by their lack of CD45RA staining rather than their expression of CD45RO. This is because many CD45RO⁺ cells can also express CD45RA, whereas all CD45RA⁻ cells are CD45RO⁺. Accordingly, Vβ chain expression was compared on splenic and thymic CD45RA⁻CD3⁺ cells [25].

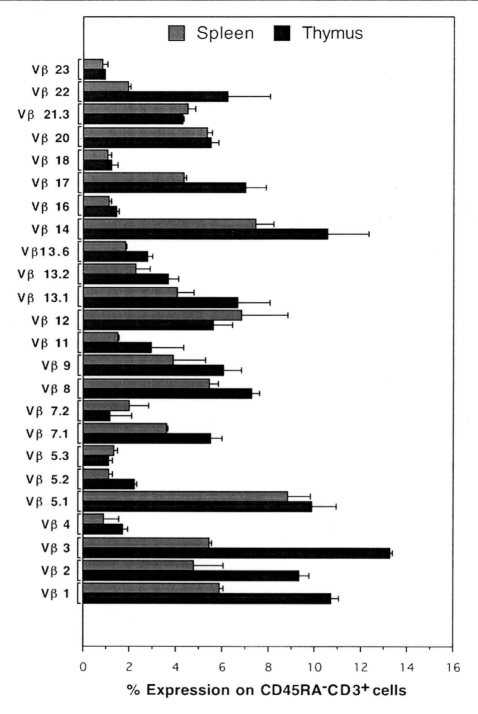

Figure 7. Expression of TCR Vβ chains by splenic and thymic CD45RA- T-cells. TCR Vβ chain expression was analyzed as indicated in Fig. 4. Splenocytes and thymocytes from the same fetal specimens, of 19 and 22 weeks' gestation, were analyzed for Vβ chain expression on T-cells lacking expression of CD45RA. Results are presented as the mean±SE.

TCR Vβ chain expression was diverse in both the thymic and splenic populations of CD45RA⁻CD3⁺ cells (Fig. 7). Hence, the CD45RO⁺ T-cells are polyclonal and do not seem to represent the expansion of only a small number of clones that have undergone activation and expansion. These observations argue against the possibility that the activated T-cells are responding to external antigen or a viral pathogen, since a broad polyclonal response is less likely under such circumstances. Some dissimilarity between the splenic and thymic populations was apparent. Mainly, a significantly lower representation of Vβ 1, Vβ 3, Vβ 5.2 and Vβ 13.6 was observed on the splenic T-cells. These differences indicate that the splenic CD45RA⁻ T-cells are not an exact match to the corresponding thymic population and suggest that the splenic cells are not premature emigrants from the thymus.

The expression of Vβ chains was further compared between splenic CD45RA⁻ and CD45RA⁺ T-cells to determine if any differences exist between these populations of T cells. Some differences in Vβ chain expression were observed, which may indicate selective expansion of T-cell clones among the CD45RA⁻ subset. The two subsets of splenic T-cells displayed the following significant differences: The CD45RA⁻ subset had a higher representation of Vβ 5.1 and Vβ 11, whereas the CD45RA⁻ subset had a lower representation of Vβ 3, Vβ 14, Vβ 16 and Vβ 21.3 (Fig. 8). Based on these data, we can tentatively conclude that the CD45RO⁺ T-cells in the peripheral circulation of the fetus have been selected. If signals in the fetal environment exist that promote non-specific activation of T-cells, perhaps because of rapidly expanding the pool of peripheral T-cells, then we would expect the Vβ-chain repertoire to be similar between the activated (CD45RO⁺) and the resting (CD45RO⁻) T-cells. The possibility remains that a specific subset of T cells, such as CD4⁺ or CD8⁺ T-cells, is non-specifically activated in the periphery and that this subset of T-cells has a pattern of Vβ chain expression that differs from the remainder of T cells expressing TCR αβ. Another interesting observation is the similar pattern in Vβ chain expression among the three fetal specimens analyzed despite the different genetic backgrounds of the specimens.

Unique Pattern of CD95 Expression on Fetal T-Cells

CD95, also known as fas antigen, is a key regulator of immune system homeostasis. CD95 expression is increased with T-cell activation and is highest on CD45RO⁺ T-cells [39]. Triggering of programmed cell death by CD95 activation, through interaction with CD178 (fas ligand), leads to clearance of activated T-cells and thereby limits the immune response [40, 41]. Neonatal T-cells are known to express less CD95 than their adult counterparts consistent with the paucity of CD45RO⁺ T-cells in neonatal blood [39, 42].

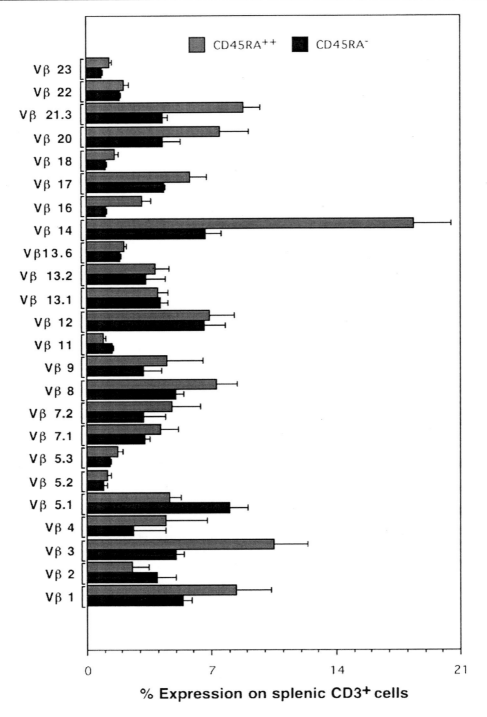

Figure 8. Expression of TCR Vβ chains by splenic CD45RA⁻ and CD45RA⁺⁺ T-cells. T cells from three spleens, of 16, 19 and 22 weeks' gestation, were analyzed for Vβ chain and CD45RA expression. T-cells were subdivided based on the expression of CD45RA. Vβ chain expression is shown for cells not expressing CD45RA (CD45RO⁺ cells) and those expressing high levels of CD45RA (mostly CD45RO⁻ cells). Results are presented as the mean±SE.

We analyzed the expression of CD95 on T-cells to gain further insight into the activation status of these cells. Our examination of midgestation fetal tissues indicated that the frequency of T cells that express CD95 in these tissues is comparable to that of adult T-cells with $65.6\pm10.0\%$ of fetal splenic T-cells expressing CD95 and $63.8\pm4.2\%$ of adult peripheral blood T-cells expressing CD95. We also confirmed a low expression of CD95 on neonatal T-cells (Fig. 9). The levels of CD95 expression on fetal T-cells, isolated from the blood or spleen, were less than half those on adult peripheral blood T-cells. Fetal T-cells consisted of a major population of cells that expressed low levels of CD95 and a small subpopulation of T cells that expressed higher levels of CD95 (Fig. 9). A small subpopulation that expressed high levels of CD95 was also observed among neonatal T-cells. In contrast, adult T-cells tended to be polarized into two subsets consisting of either CD95$^+$ or CD95$^-$ cells.

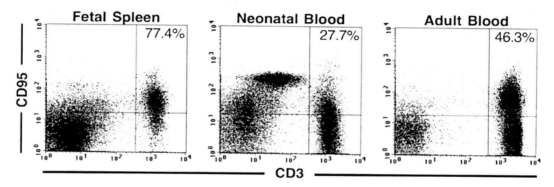

Figure 9. Expression patterns of CD95 on CD3$^+$ T-cells at different stages of ontogeny. Dot plots show the results of analyses performed on light-density cells isolated from a 20 weeks' gestation spleen, term umbilical cord blood and a 38 year-old male. The percentages of CD3$^+$ events are indicated for three of the quadrant regions.

We further analyzed CD95 expression on fetal T-cells to determine if CD95 was co-expressed with CD45RO as has been reported for adult T-cells [39]. CD95 expression was found on both CD45RA$^+$ and CD45RO$^+$ fetal T-cells (Fig. 10). Differences between fetal and adult T-cells were also apparent. The median frequency of fetal CD45RA$^+$ T-cells that expressed CD95 was 62.0% in the blood and 85.4% in the spleen. The median frequencies of CD95 expression on adult and neonatal CD45RA$^+$ T-cells were significantly lower at 45.1% and 37.1%, respectively. CD95 expression was significantly reduced on CD45RO$^+$ T-cells from fetal blood (median 59.7%), fetal spleen (median 72.8%) and neonatal blood (median 54.9%) compared to the adult (median 89.1%). Additionally, examination of CD95 expression on fetal T-cells revealed a subpopulation of CD45RO$^+$CD45RA$^-$ T-cells that was CD95$^-$. As mentioned, these cells were best defined by their lack of CD45RA staining rather than their expression of CD45RO. The CD95$^-$CD45RA$^-$ T-cell population was not present to any appreciable degree in either term neonatal blood or adult peripheral blood. We are not aware of any previous description of a CD95$^-$CD45RO$^+$CD45RA$^-$ T-cell population, and the role of this subset in the developing immune system is presently unclear.

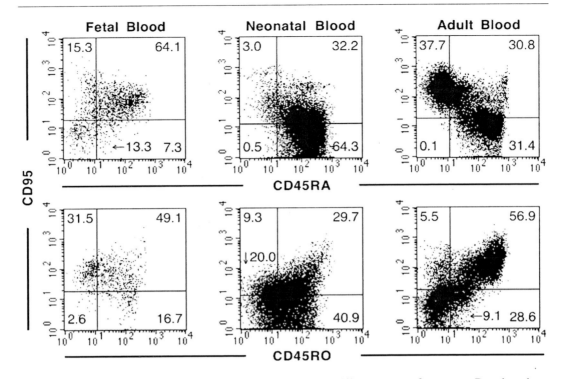

Figure 10. Expression of CD95 and CD45 isoforms by T-cells at different stages of ontogeny. Dot plots show the results of analyses performed on light-density cells isolated from a 16 weeks' gestation blood, term umbilical cord blood and a 37 year-old. Numbers represent the percentage of events in the corresponding quadrants.

The expression of CD95 on some CD45RO⁺ fetal T-cells is further evidence that these cells may have been previously activated. However, the existence of CD95⁻CD45RO⁺ T-cells is equally intriguing. Indeed, the lack or decreased expression of CD95 on CD45RO⁺ T-cells would mean that these cells can not be cleared by a CD95-mediated apoptotic mechanism, the usual means by which an immune response is tempered [34]. These findings raise the interesting possibility that the activated fetal T-cells are not meant to be destroyed and that the preservation of these cells serves an important purpose in the development of the immune system. Alternatively, the loss of CD95 expression may still be associated with the removal of these cells. There are reports that signaling through CD95 may support T-cell growth rather than apoptosis in some circumstances and, thus, the lack of CD95 expression by fetal CD45RO⁺ T-cells could limit the further expansion of this cell population [43, 44].

Expression of Cytokine Receptors and Activation Markers by Fetal and Adult T-Cells

To find support for our hypothesis that a sizable portion of fetal T-cells has undergone activation, we analyzed the expression of various cytokine receptors and other cell-surface markers associated with T-cell growth and activation (Table 1). T-cell activation results in upregulation of the α and β subunits of the interleukin (IL)-2 receptor, CD25 and CD122 respectively. Both components of the IL-2 receptor were found on fetal T-cells, with a notably higher number of CD122$^+$ T-cells in the fetus compared to the adult. CD25 was expressed on a similar portion of fetal T-cells as on adult T-cells. Very little CD25 and CD122 expression was observed on neonatal T-cells. The expression of CD25 and CD122 by fetal T-cells suggests these cells may be activated, although it is possible that some of these cells represent CD4$^+$ suppressor/regulatory T-cells [45]. This subset of regulatory T-cells has been described in neonatal and adult blood and is characterized in part by CD25 and CD122 expression [46]. CD4$^+$CD25$^+$ T-cells can suppress the response of single positive T-cells by secretion of inhibitory cytokines such as IL-4, IL-10 and transforming growth factor as well as by cell-contact dependent mechanisms [45].

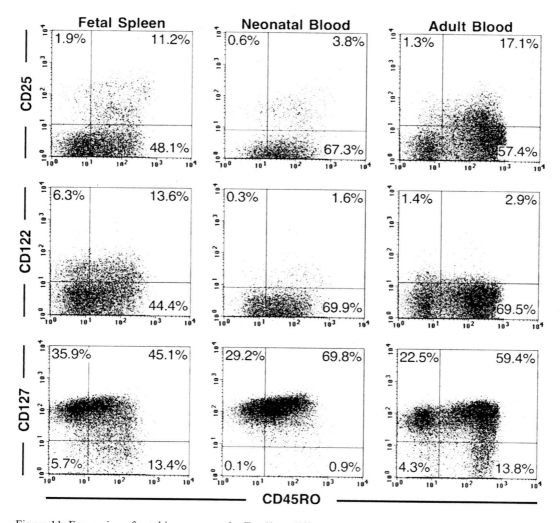

Figure 11. Expression of cytokine receptors by T-cells at different stages of ontogeny. Dot plots show the results of analyses performed on light-density cells isolated from a 23 weeks' gestation splenocytes, term umbilical cord blood and a 61-year female. Numbers represent the percentage of events in the corresponding quadrants.

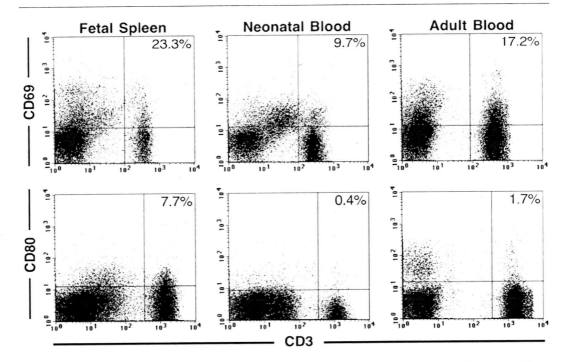

Figure 12. Expression of CD69 and CD80 activation markers by T-cells at different stages of ontogeny. Dot plots show the results of analyses performed on light-density cells isolated from a 23 weeks' gestation splenocytes, term umbilical cord blood and a 61-year female. Numbers represent the percentage of events in the corresponding quadrants.

Table 1. Expression of cell-surface markers on fetal, neonatal and adult CD3$^+$ T-cells[a]

	Total T-cells			CD45RO$^+$ T-cells			CD45RO$^-$ T-cells		
Marker	Fetal	Neonatal	Adult	Fetal	Neonatal	Adult	Fetal	Neonatal	Adult
CD25	11.0[b]	3.4[c]	14.8	17.7	5.4[c]	22.2	3.8	2.6	3.4
CD122	14.3[b, c]	0.4	1.9	16.2[b, c]	0.7	2.2	11.5[b, c]	0.3[c]	2.0
CD127	77.8[b]	97.5[c]	84.3	67.5[b, c]	82.4	84.8	84.3[b]	100[c]	80.6
CD132	99.6	95.3[c]	100	92.5	96.3	98.4	100[b]	96.5	100
CD56	9.9	4.3	9.7	8.2	5.7	11.5	9.6	3.7	6.2
CD69	21.0[c]	12.5	10.7	ND[d]	ND	ND	ND	ND	ND
CD80	3.7	0	0.3	5.2	0	0.6	0.6	0	0

[a]Light-density cells isolated from fetal spleens and mononuclear cells isolated from neonatal umbilical cord blood and adult blood were analyzed for the expression of CD3, CD45RO and the indicated marker. T cells were defined by their expression of CD3 and their characteristic light-scatter profile. Values represent the median level of expression observed on 5 fetal, 4 neonatal and 6 adult samples.

[b]$P \leq 0.05$ versus neonatal T-cells.

[c]$P \leq 0.05$ versus adult T-cells.

[d]ND = Not Determined

IL-7 plays a critical role in the maintenance of the naïve T-cell pool through interaction with its receptor, CD127/CD132 [47]. After T-cell activation, expression of the α-chain subunit of the IL-7 receptor, CD127, is lost. Thus, low expression of CD127 serves as another marker of T-cell activation [48]. CD127 was widely expressed on T-cells from all tissue sources with neonatal T-cells expressing the most CD127 (Table 1). However, we observed CD127⁻ T-cells in both the fetus as well as in the adult, and there was a higher portion of CD45RO⁺ T-cells that lacked CD127 expression in the fetus than in the adult. In contrast, the common γ-chain subunit of the IL-2, IL-4, IL-7, IL-9 and IL-15 receptors, CD132, was expressed on nearly all T-cells at all stages of ontogeny.

The two activation markers, CD69 and CD80, are known to be expressed on T cells that have become activated [49-51]. CD69 is an activation antigen expressed early in T-cell activation [49]. CD80 is a stimulatory molecule for T cells expressed by various leukocytes, which can be expressed on T cells during the later phase of activation [50, 51]. The frequency of fetal T-cells expressing CD69 was nearly twice that of neonatal and adult T-cells. CD80 was also expressed at a higher frequency on fetal T-cells than on neonatal and adult T-cells. CD80 expression was particularly prevalent on fetal CD45RO⁺ T cells, consistent with at least some of these cell representing activated T-cells.

Conclusions

T cells arise at the end of embryonic life in the human fetus, about 8 weeks before the thymus has anatomically matured. A diverse repertoire of T cells is present soon after thymic maturation. Both TCR αβ and TCR γδ T-cells are present in the fetus. Initial measurements indicate that these T-cells are capable of responding to antigen and, thus, are believed to be functional. Surprisingly, many fetal T-cells appear to have undergone activation based on the expression of a panel of antigens. In particular, the expression of CD45RO by a large portion of peripheral T-cells was unexpected in the nearly pathogen-free environment of the womb.

The patterns of cytokine-receptor and activation-antigen expression on fetal T-cells support the hypothesis that a sizable portion of fetal T-cells has undergone activation. These findings are in contrast to observations we made on neonatal T-cells, which display a predominantly naïve/resting phenotype. The nature of the activation signal responsible for the activation phenotype that we have documented is not known. Nor are the sequence of events understood that transform the activated state of the peripheral T-cell pool at midgestation into the resting phenotype observed at birth. We speculate that the activated status of fetal T-cells is the result of interaction with maternal or auto-antigens and may play an important role in either protecting the fetus from maternal engraftment or the establishment of peripheral tolerance, respectively. The presence of the novel CD95⁻CD45RO⁺CD45RA⁻ T-cell subset in the fetus perhaps favors the latter hypothesis, since this population appears unique to the fetus and may play a role in tolerance induction.

Further research into the development and functional status of the human fetal immune system is likely to provide insights into the steps required in the development of the immune system and the establishment of peripheral tolerance towards autologous antigens. These insights may lead to better therapies for immune compromised patients and transplant recipients in which the clinical goal is to rebuild the immune system. Efforts at developing fetal therapies, such as gene therapy or cellular therapy, would also benefit greatly from a clearer understanding of the functional capacity of the immune system at various stages of development. Fetal life may provide a unique opportunity for the introduction of foreign antigens or cells to repair birth defects, but uncertainty surrounding the functional capacity of the fetal immune system has hampered progress in developing effective fetal therapies.

Materials and Methods

Isolation of Human T-Cells from Adult, Neonatal and Fetal Tissues

Male and female human tissues were obtained and studied under the approval of the Committee on Human Research at our institute. Tissues were harvested at our institution or were obtained from Advanced Bioscience Resources Inc. (Alameda, CA). Consent was obtained from the adults donating their peripheral blood. Neonatal umbilical cord blood and fetal hematopoietic tissues were obtained with consent of the women prior to delivery or elective abortion, respectively.

Adult peripheral blood was obtained by venipuncture from healthy volunteers ranging in age from 24 to 61 years. Approximately 7 ml of blood was drawn into a vacutainer tube containing ethylene diamine tetraacetic acid (EDTA). The blood was diluted to a total volume of 50 ml in washing buffer consisting of phosphate buffered saline (PBS) containing 0.3% fraction-V ethanol-extracted bovine serum albumin (BSA) (Roche Applied Science, Indianapolis, IN) and 50 µg/ml gentamicin sulfate (Life Technologies, Grand Island, NY). The cells were pelleted by centrifugation and erythrocytes were depleted by a 1 minute chemical lyses using ACK lyses buffer, pH 7.2-7.4, consisting of 0.15 M NH_4Cl, 1.0 mM $KHCO_3$ and 0.1 mM Na_2 EDTA (Sigma Chemical Company, St. Louis, MO). The cells were pelleted by a 7-minute centrifugation and suspended in washing buffer. If lysis of the erythrocytes was incomplete, the lysis procedure was repeated. Peripheral blood mononuclear cells (PBMC) were isolated by centrifugation at 600xg for 25 minutes on a layer of 1.077 g/ml LymphoPrep (Life Technologies). The light-density cells were harvested and washed twice before being suspended in blocking buffer consisting of PBS with 5% normal mouse serum (Gemini Bio-Products, Inc., Woodland, CA) and 0.01% NaN_3.

PBMC were isolated from neonatal umbilical cord blood (33 weeks' gestation to full term) by immunomagnetic bead depletion of CD235a$^+$ erythrocytes, performed as previously described for fetal liver cells [52], and density separation using 1.077 g/ml Nycoprep (Life Technologies). The isolation of light-density neonatal blood cells by centrifugation was performed as described for the adult PBMC.

Fetal tissues were obtained shortly following the abortion and were transported to the laboratory in washing buffer held in sterile containers on ice. The tissues ranged in gestational age from 7 to 24 weeks, as determined by the foot length of the fetus. Experiments were performed on cells obtained from individual specimens; tissues were not pooled for analysis.

Fetal blood T-cells were harvested from umbilical cords, placental vessels and/or hearts obtained from elective abortions. Blood was harvested from the vessels of umbilical cords after first washing the cords with washing buffer and then cutting the cord into 2cm pieces with scissors. Blood was washed from the vessels with washing buffer injected using a 28 gage insulin syringe (Becton Dickinson & Co., Franklin Lakes, NJ). Alternatively, fetal umbilical cord blood was squeezed out through the fresh cut end using forceps. The blood samples were filtered using 70μ nylon-mesh cell-strainers (BD Biosciences, San Jose, CA) as needed to remove large debris. An alternative technique to isolate fetal blood was by direct venopuncture of surface vessels near the placenta-umbilical cord junction. Placental blood was drawn into a syringe containing heparin. A third source of fetal blood was form fetal hearts, which were collected with the pericardial sack. The surface was cleaned with washing buffer to remove potential contaminating maternal blood. The pericardial sack was then removed and, in a clean petri dish, the heart was cut into about 10 pieces. Blood was rinsed from the heart tissue using washing buffer. Fetal PBMC were prepared using LymphoPrep as described above. In some small samples from younger specimens, erythrocytes were depleted by chemical lysis or immunomagnetic bead depletion of $CD235a^+$ cells as an alternative to density separation [52, 53].

Fetal splenocytes and liver cells were isolated by dissociating the spleen through a wire-mesh cell strainer (Sigma Chemical Company) in washing buffer. Light-density $CD235a^-$ liver cells were prepared by immunomagnetic-bead depletion and centrifugation over a layer of 1.077 g/ml Nycoprep as previously described in detail [54]. Light-density splenocytes were isolated by centrifugation using LymphoPrep as described above. Alternatively, splenic erythrocytes were depleted by chemical lysis or immunomagnetic bead depletion of $CD235a^+$ cells [52].

Monoclonal Antibodies (mAbs)

The following fluorescein isothiocyanate (FITC) and phycoerythrin (PE) labeled mAbs were purchased from BD Biosciences (www.bdbiosciences.com): CD3-FITC (SK7), CD8-FITC (SK1), CD45RO-PE (UCHL1), CD95-PE (DX2), CD122-PE (TU27), mouse IgG_1-FITC, mouse IgG_{2a}-FITC and mouse IgM-FITC. CD56-FITC and CD56-PE (C5.9) were purchased from Exalpha Corporation (Boston, MA). CD4-tricolor (TC) (S3.5), CD25-PE (CD25-3G10), CD45-PE (HI30), mouse IgG_1-FITC, mouse IgG_1-PE, mouse IgG_{2a}-PE, mouse IgG_{2b}-FITC and mouse IgM-FITC were purchased from Caltag (Burlingame, CA). The following conjugated mAb were purchased from Beckman-Coulter (Miami, FL): CD3-phycoerythrin-cyanine 5 (PC5) (UCHT-1), CD45-PC5 (J33), CD45RA-FITC (ALB11), CD45RO-FITC (UCHL1), CD56-PC5 (N901), CD69-PC5 (TP1.55.3), CD80-PE (MAB104), CD127-PE (R34.34) and mouse IgG_1-PC5. A kit containing a panel of FITC-, PE- and a mixture of FITC- and PE-conjugated mAb recognizing different TCR Vβ chains was purchased from Beckman-Coulter and was used according to the manufacturers recommendations. PE-labeled mAbs recognizing TCR α/β (BMA031) and TCR γ/δ (5A6.E9) were purchased from Endogen (Woburn, MA) and a FITC-conjugated mAb recognizing TCR α/β was obtained from T Cell Diagnostics, Inc. (Cambridge, MA).

Flow Cytometric Analysis of Cell Surface Markers

Approximately 2×10^5 cells suspended in up to 200µl blocking buffer were incubated in 96-well Costar V-bottom plates (Corning Inc., Corning, NY) with saturating amounts of mAbs for at least 30 minutes on ice. Cells were washed twice with 250µl washing buffer with 0.01% NaN_3 added (Sigma Chemical Co.). The cells were suspended in the same buffer with the addition of 2µg/ml propidium iodide (PI), purchased from Sigma Chemical Co. PI was used to stain the nucleic acids of dead cells, so that they could be excluded from the analysis. PI was omitted in 3-color analyses using PC5-labeled mAbs. Flow cytometric analysis was performed using either a FACScan or a FACSCalibur flow cytometer (BD Biosciences). Analyses of results were performed using CellQuest software (BD Biosciences).

Data Presentation and Statistical Analysis

Results are presented as the median or mean ± 1 standard error of measurements made on multiple tissue samples. The significance of differences observed between fetal, neonatal and adult cells was determined using an unpaired Student's t-test. Results are considered significant when $P \leq 0.05$. Data were analyzed using StatView 4.5 software run on an Apple Computer.

References

[1]   Zhao Y, Dai ZP, Lv P, Gao XM. Phenotypic and functional analysis of human T lymphocytes in early second- and third-trimester fetuses. *Clin. Exp. Immunol.* Aug 2002;129(2):302-308.

[2]   Garderet L, Dulphy N, Douay C, et al. The umbilical cord blood alphabeta T-cell repertoire: characteristics of a polyclonal and naive but completely formed repertoire. *Blood.* Jan 1 1998;91(1):340-346.

[3]   Clement LT. Isoforms of the CD45 common leukocyte antigen family: markers for human T-cell differentiation. *J. Clin. Immunol.* Jan 1992;12(1):1-10.

[4]   Harris DT, Schumacher MJ, Locascio J, et al. Phenotypic and functional immaturity of human umbilical cord blood T lymphocytes. *Proc. Natl. Acad. Sci. USA.* Nov 1 1992;89(21):10006-10010.

[5]   Lai R, Visser L, Poppema S. Postnatal changes of CD45 expression in peripheral blood T and B cells. *Br. J. Haematol.* Jun 1994;87(2):251-257.

[6]   Roncarolo MG, Bigler M, Ciuti E, Martino S, Tovo PA. Immune responses by cord blood cells. *Blood Cells.* 1994;20(2-3):573-585.

[7]   Bofill M, Akbar AN, Salmon M, Robinson M, Burford G, Janossy G. Immature CD45RA(low)RO(low) T cells in the human cord blood. I. Antecedents of CD45RA+ unprimed T cells. *J. Immunol.* Jun 15 1994;152(12):5613-5623.

[8]   Keever CA, Abu-Hajir M, Graf W, et al. Characterization of the alloreactivity and anti-leukemia reactivity of cord blood mononuclear cells. *Bone Marrow Transplant.* Mar 1995;15(3):407-419.

[9] Sato K, Nagayama H, Takahashi TA. Aberrant CD3- and CD28-mediated signaling events in cord blood T cells are associated with dysfunctional regulation of Fas ligand-mediated cytotoxicity. *J. Immunol.* 1999;162(8):4464-4471.

[10] Harris DT, LoCascio J, Besencon FJ. Analysis of the alloreactive capacity of human umbilical cord blood: implications for graft-versus-host disease. *Bone Marrow Transplant.* 1994;14(4):545-553.

[11] Risdon G, Gaddy J, Horie M, Broxmeyer HE. Alloantigen priming induces a state of unresponsiveness in human umbilical cord blood T cells. *Proc. Natl. Acad. Sci. USA.* Mar 14 1995;92(6):2413-2417.

[12] Porcu P, Gaddy J, Broxmeyer HE. Alloantigen-induced unresponsiveness in cord blood T lymphocytes is associated with defective activation of Ras. *Proc. Natl. Acad. Sci. USA.* Apr 14 1998;95(8):4538-4543.

[13] Berthou C, Legros-Maida S, Soulie A, et al. Cord blood T lymphocytes lack constitutive perforin expression in contrast to adult peripheral blood T lymphocytes. *Blood.* Mar 15 1995;85(6):1540-1546.

[14] Muench MO, Bárcena A. Stem cell transplantation in the fetus. *Cancer Control.* 2004;11(2):105-118.

[15] Yang EY, Flake AW, Adzick NS. Prospects for fetal gene therapy. *Semin. Perinatol.* 1999;23(6):524-534.

[16] Haynes BF. Phenotypic characterization and ontogeny of the human thymic microenviroment. *Clin. Res.* Dec 1984;32(5):500-507.

[17] Haynes BF, Martin ME, Kay HH, Kurtzberg J. Early events in human T cell ontogeny: Phenotypic characterization and immunohistological localization of T-cell precursors in early human fetal tissues. *J. Exp. Med.* Sep 1 1988;168(3):1061-1080.

[18] Mychaliska GB, Muench MO, Rice HE, Leavitt AD, Cruz J, Harrison MR. The biology and ethics of banking fetal liver hematopoietic stem cells for *in utero* transplantation. *J. Pediatr. Surg.* 1998;33:394-399.

[19] Golfier F, Bárcena A, Cruz J, Harrison MR, Muench MO. Mid-trimester fetal livers are a rich source of CD34$^{+/++}$ cells for transplantation. *Bone Marrow Transplant.* 1999;24:451-461.

[20] Muench MO, Rae J, Bárcena A, et al. Transplantation of a fetus with paternal Thy-1+CD34+ cells for chronic granulomatous disease. *Bone Marrow Transplant.* 2001;27(4):355-364.

[21] Furley AJ, Mizutani S, Weilbaecher K, et al. Developmentally regulated rearrangement and expression of genes encoding the T cell receptor-T3 complex. *Cell.* Jul 4 1986;46(1):75-87.

[22] Renda MC, Fecarotta E, Dieli F, et al. Evidence of alloreactive T lymphocytes in fetal liver: implications for fetal hematopoietic stem cell transplantation. *Bone Marrow Transplant.* Jan 2000;25(2):135-141.

[23] Lindton B, Markling L, Ringden O, Kjaeldgaard A, Gustafson O, Westgren M. Mixed lymphocyte culture of human fetal liver cells. *Fetal Diagn. Ther.* 2000;15(2):71-78.

[24] Lindton B, Markling L, Ringden O, Westgren M. In vitro studies of the role of CD3+ and CD56+ cells in fetal liver cell alloreactivity. *Transplantation.* Jul 15 2003;76(1):204-209.

[25] Muench MO, Pott Bartsch EM, Chen J-C, Lopoo JB, Barcena A. Ontogenic changes in CD95 expression on human leukocytes: prevalence of T-cells expressing activation markers and identification of CD95⁻CD45RO⁺ T-cells in the fetus. *Dev. Comp. Immunol.* Dec 2003;27(10):899-914.

[26] Pardoll DM. Immunology. Stress, NK receptors, and immune surveillance. *Science.* Oct 19 2001;294(5542):534-536.

[27] Girardi M, Oppenheim DE, Steele CR, et al. Regulation of cutaneous malignancy by gammadelta T cells. *Science.* Oct 19 2001;294(5542):605-609.

[28] Carding SR, Egan PJ. Gammadelta T cells: functional plasticity and heterogeneity. *Nat. Rev. Immunol.* May 2002;2(5):336-345.

[29] Jameson J, Ugarte K, Chen N, et al. A role for skin gammadelta T cells in wound repair. *Science.* Apr 26 2002;296(5568):747-749.

[30] Jahng AW, Maricic I, Pedersen B, et al. Activation of natural killer T cells potentiates or prevents experimental autoimmune encephalomyelitis. *J. Exp. Med.* Dec 17 2001;194(12):1789-1799.

[31] Girardi M, Lewis J, Glusac E, et al. Resident skin-specific gammadelta T cells provide local, nonredundant regulation of cutaneous inflammation. *J. Exp. Med.* Apr 1 2002;195(7):855-867.

[32] Haynes BF, Denning SM, Singer KH, Kurtzberg J. Ontogeny of T-cell precursors: a model for the initial stages of human T-cell development. *Immunol. Today.* Mar 1989;10(3):87-91.

[33] Carding SR, McNamara JG, Pan M, Bottomly K. Characterization of gamma/delta T cell clones isolated from human fetal liver and thymus. *Eur. J. Immunol.* Jun 1990;20(6):1327-1335.

[34] Byrne JA, Stankovic AK, Cooper MD. A novel subpopulation of primed T cells in the human fetus. *J. Immunol.* Mar 15 1994;152(6):3098-3106.

[35] Devereux G, Seaton A, Barker RN. In utero priming of allergen-specific helper T cells. *Clin. Exp. Allergy.* Nov 2001;31(11):1686-1695.

[36] Pilarski LM, Gillitzer R, Zola H, Shortman K, Scollay R. Definition of the thymic generative lineage by selective expression of high molecular weight isoforms of CD45 (T200). *Eur. J. Immunol.* Apr 1989;19(4):589-597.

[37] Egerton M, Pruski E, Pilarski LM. Cell generation within human thymic subsets defined by selective expression of CD45 (T200) isoforms. *Hum. Immunol.* Apr 1990;27(4):333-347.

[38] Fujii Y, Okumura M, Inada K, Nakahara K, Matsuda H. CD45 isoform expression during T cell development in the thymus. *Eur. J. Immunol.* Jul 1992;22(7):1843-1850.

[39] Miyawaki T, Uehara T, Nibu R, et al. Differential expression of apoptosis-related Fas antigen on lymphocyte subpopulations in human peripheral blood. *J. Immunol.* Dec 1 1992;149(11):3753-3758.

[40] Siegel RM, Chan FK, Chun HJ, Lenardo MJ. The multifaceted role of Fas signaling in immune cell homeostasis and autoimmunity. *Nat. Immunol.* Dec 2000;1(6):469-474.

[41] Krammer PH. CD95's deadly mission in the immune system. *Nature.* Oct 12 2000;407(6805):789-795.

[42] McCloskey TW, Oyaizu N, Bakshi S, Kowalski R, Kohn N, Pahwa S. CD95 expression and apoptosis during pediatric HIV infection: early upregulation of CD95 expression. *Clin Immunol Immunopathol.* 1998;87(1):33-41.

[43] Alderson MR, Armitage RJ, Maraskovsky E, et al. Fas transduces activation signals in normal human T lymphocytes. *J. Exp. Med.* Dec 1 1993;178(6):2231-2235.

[44] Kabra NH, Kang C, Hsing LC, Zhang J, Winoto A. T cell-specific FADD-deficient mice: FADD is required for early T cell development. *Proc. Natl. Acad. Sci. USA.* May 22 2001;98(11):6307-6312.

[45] Shevach EM. CD4+ CD25+ suppressor T cells: more questions than answers. *Nat. Rev. Immunol.* Jun 2002;2(6):389-400.

[46] Wing K, Ekmark A, Karlsson H, Rudin A, Suri-Payer E. Characterization of human CD25+ CD4+ T cells in thymus, cord and adult blood. *Immunology.* Jun 2002;106(2):190-199.

[47] Webb LM, Foxwell BM, Feldmann M. Putative role for interleukin-7 in the maintenance of the recirculating naive CD4+ T-cell pool. *Immunology.* Nov 1999;98(3):400-405.

[48] Webb LM, Foxwell BM, Feldmann M. Interleukin-7 activates human naive CD4+ cells and primes for interleukin-4 production. *Eur. J. Immunol.* Mar 1997;27(3):633-640.

[49] Testi R, Phillips JH, Lanier LL. T cell activation via Leu-23 (CD69). *J. Immunol.* Aug 15 1989;143(4):1123-1128.

[50] Azuma M, Yssel H, Phillips JH, Spits H, Lanier LL. Functional expression of B7/BB1 on activated T lymphocytes. *J. Exp. Med.* Mar 1 1993;177(3):845-850.

[51] Wyss-Coray T, Mauri-Hellweg D, Baumann K, Bettens F, Grunow R, Pichler WJ. The B7 adhesion molecule is expressed on activated human T cells: functional involvement in T-T cell interactions. *Eur. J. Immunol.* Sep 1993;23(9):2175-2180.

[52] Bárcena A, Muench MO, Song KS, Ohkubo T, Harrison MR. Role of CD95/Fas and its ligand in the regulation of the growth of human $CD34^{++}CD38^{-}$ fetal liver cells. *Exp. Hematol.* Sep 1999;27(9):1428-1439.

[53] Bárcena A, Muench MO, Roncarolo MG, Spits H. Tracing the expression of CD7 and other antigens during T- and myeloid-cell differentiation in the human fetal liver and thymus. *Leuk. Lymph.* Mar 1995;17(1-2):1-11.

[54] Muench MO, Suskind DL, Bárcena A. Isolation, growth and identification of colony-forming cells with erythroid, myeloid, dendritic cell and NK-cell potential from human fetal liver. *Biol. Proced. Online.* Jun 11 2002;4:10-23.

In: Progress in Immunology Research
Editor: Barbara A. Veskler, pp. 45-67

ISBN 1-59454-380-1
©2005 Nova Science Publishers, Inc.

Chapter III

Mass Spectrometry in Developmental and Comparative Immunology

Simone Koenig[*]

Integrated Functional Genomics, Interdisciplinary Center for Clinical Research, Medical Faculty, University of Muenster, Germany

Abstract

As a conglomerate of immunology and zoology comparative immunology has gained wide acceptance in biology. There is much interest in the immune system of invertebrates as representing early models or precursors of the innate system of vertebrates. Although a small discipline, developmental and comparative immunology has started to take advantage of modern analytical tools such as mass spectrometry especially in proteomics. Bioscience applications of the technique in organic and inorganic analysis based on annelida research are discussed in this review introducing the variability of mass spectrometry. Thereby, protein analyses form a focus within the paper due to the high demand. However, protein analysis is hampered by the availability of sequence databases i.e. for annelida, which is, although in progress, still very limited. In addition, gas chromatography in combination with mass spectrometry for small molecule (lipids, toxins) detection is discussed. For metal and surface analysis (heavy metal contamination in tissue) ion microprobes and inductively-coupled plasma are refered to. Moreover, isotope ratio measurement for analyte quantification or radiocarbon dating are included due to their important in many bioscientific areas.

Key words: mass spectrometry, developmental immunology, annelida, earthworms, proteomics, SIMS, ICP-MS, AMS, GC-MS

[*] PD Dr. Simone Koenig; Integrierte Funktionelle Genomik; Roentgenstr. 21; 48149 Muenster, Germany; ph: xx-(0)251-8357164; fax: xx-(0)251-8357255; e-mail: koenigs@uni-muenster.de

Introduction

Mass spectrometry (MS) has gained wide acceptance in the life sciences in recent years due to the introduction of methods applicable to the very sensitive biological analytes. Appropriately, those were honoured with the Nobel Price for Chemistry to J. Fenn and K. Tanaka in 2002. MS has moved out of the specialists laboratory and become a research tool in many areas. However, as the impact of MS increases, so is the need to educate its users, and this article has been written to provide an overview about some of the various experimental possibilities. Those do not only include the - these days - well known proteomic analyses, but also small molecule detection, isotope ratio measurement, and inorganic MS.

This work was triggered by questions arising during collaborations with zoologists and researchers working in the field of developmental and comparative immunology and applications were chosen from this area. Nevertheless, the guide through MS methodology is of general usefulness. It is structured with respect to the analyte allowing the scientists to determine the best approach to study their sample.

As a conglomerate of immunology and zoology comparative immunology investigates the less complex immune defense mechanisms of invertebrates such as annelids as model systems for the more sophisticated immunity of vertebrates. Discrimination of self and nonself is one of the features of all animal species but the ways of elimination of nonself are different. Defense strategies of invertebrates, which lack antibodies and lymphocytes, are based on innate mechanisms and are described as natural non-specific, non-anticipatory, and non-clonal. This is in contrast to the macrophage T and B systems characterizing vertebrate adaptive immunity whose properties can be categorized as induced, specific, anticipatory, and clonal.[1-4]

Analytical requirements for investigations in this field involve mostly protein and peptide MS and this will be the focus of this review. Further experimental options are described which have also been successfully employed in annelida research. For an overview on MS in vertebrate immunology see ref.[5].

MS Basics

MS has evolved from a method of atom physics to a tool for the biosciences. It was discovered by physicists at ~1900 and it was continuously developed since then. While in the beginning only small ionic species such as H^+, O_2^+, or $COCl_2^+$ could be studied, instrumental advances soon allowed the measurement of hydrocarbons ~1940 and more complex, but volatile compounds ~1960. Further improvements eventually lead to applications on biological relevant molecules such as peptides ~1975 and proteins ~1990.[6, 7]

MS requires the ionisation of the analyte, because ions can be influenced (accelerated, focussed, separated) by electric and magnetic fields. Therefore, a typical mass spectrometer consists of a source for ion generation, an analyser region at high vacuum for ion separation, and an ion detector (Fig. 1). Mass spectrometers greatly differ in their design and geometry and they utilize various physical principles like the measurement of the flight time or scanning of quadrupoles which determine their experimental capabilities. There are high-end instruments built for few specialized experiments (i.e. accelerator spectrometers for isotope measurements or Fourier-transform ion cyclotron resonance instruments for maximum resolution) as well as flexible benchtop spectrometers for many applications. However, even those instruments become more and more dedicated to single routine procedures. Therefore, knowledge for measurements of certain analyte types is not necessarily available in every MS laboratory.

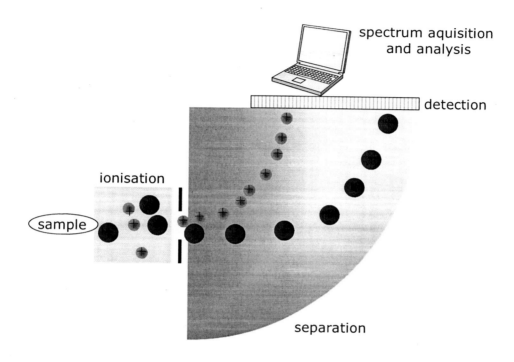

Figure 1. Schematic of a mass spectrometer. The sample is introduced and ionised. Ions are separated in high vacuum in electric and magnetic fields and detected.

Mass spectrometers measure molecular masses (also called molecular weight MW) of ions as mass/charge (m/z) ratios. MW is a unique property of the analyte which changes with modifications, reactions or fragmentation. The mass of a molecule is the sum of the atomic masses of all the atoms composing it. The symbol **u** (mass unit or unified atomic mass unit) represents by convention $1/12^{th}$ the mass of the most abundant naturally occurring stable isotop of carbon. By convention, in biochemistry it is often replaced by the unit dalton (Da)[8].

Instruments differ by their mass range, resolution, and accuracy. With respect to their main task, spectrometers may allow measurement up to only 600 m/z for small molecule detection, up to 2000 m/z as in early quadrupole devices, or up to hundreds of thousands m/z in time-of-flight tubes. Most modern mass spectrometers are capable of separating isotopes of peptides (Fig. 2a). This is necessary to search databases effectively. Examples for common atomic masses of importance in the life sciences are given in table 1. They are updated regularly by IUPAC[9]. Most elements found in biological compounds (^{12}C, ^{1}H, ^{14}N, ^{16}O) have low-abundant isotopes (^{13}C, ^{2}H, ^{15}N, ^{17}O, ^{18}O). It is important to realize that for masses larger than 2 kDa the ^{12}C-isotope is not the most abundant peak in an isotopic distribution of a biomolecule anymore. Isotopes of other elements alter the isotopic distribution significantly (i.e. Br, Cl). The monoisotopic mass characterizes the ion population of the lowest isotopes. The full width of the peak at half maximum (FWHM) describes the instrument resolution. Is the instrument only able to measure envelopes for isotope clusters which is usually the case for intact proteins, masses are determined as average as shown for the measurement of insulin B chain at different resolutions in Fig. 2b.

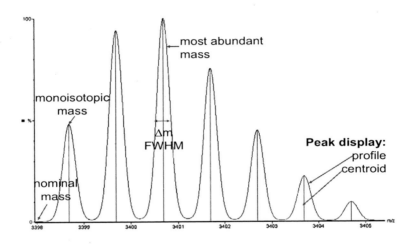

Figure 2. Spectrum of the isotope cluster of the protonated insulin B-chain ion (composition $C_{157}H_{233}N_{40}O_{41}S_2$]H^+ at m/z 3397.8, monoisotopic mass, and 3400.9, average mass). a) Peak annotations for monoisotopic resolution.

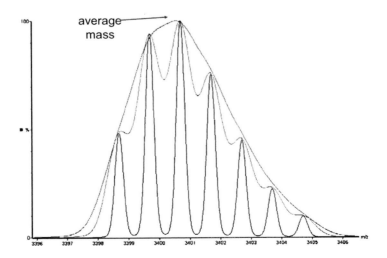

Figure 2b) Decreasing instrumental resolution and average mass.

Table 1. Atomic masses and abundances for biologically relevant isotopes[80]

Isotope	Atomic Weight	Natural Abundance %
^{12}C	12	98.93
^{13}C	13.003354826	1.07
^{1}H	1.007825035	99.9885
^{2}H	2.014101779	0.0115
^{14}N	14.003074002	99.632
^{15}N	15.00010897	0.368
^{16}O	15.99491463	99.757
^{17}O	16.9991312	0.038
^{18}O	17.9991603	0.205
^{31}P	30.9737620	100
^{32}S	31.97207070	94.93
^{33}S	32.97145854	0.76
^{34}S	33.96786665	4.29
^{36}S	35.96708062	0.02

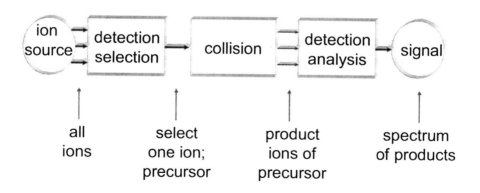

Figure 3. Schematic of tandem mass spectrometers. A collision cell allows fragmentation of analyte molecules *via* collisions with noble gas molecules at slightly inhanced pressure and/or increase of kinetic energy of the ions. The set-up combines two mass spectrometers in-line and is, therefore, called MS/MS.

Mass spectrometers can be operated in many ways to solve a specific problem. On one side, the MW of the analyte molecules can be determined. In that case, the spectrum is an MS image of the original sample. In principle, every compound present in the sample could be expected to generate a peak. This is not possible, however, because factors such as ion suppression, adduct formation, instrumental settings and others restrict MS output. In that respect sample preparation and purification are very important. Although buffers are tolerable to some extent, in general sample preparation must be adapted to the needs of MS. Therefore, on- and off-line coupling to chromatographic or electrophoretic methods is frequently employed. An important feature of some types of mass spectrometers is their capability to select certain ions and fragment them by collision in the gas phase (Fig. 3). In this way, peptides can be sequenced. Their product ions generate characteristic spectra, which can be searched against databases. Further options available in triple-quadrupole spectrometers are the neutral loss scan and the parent ion scan which are of importance for monitoring specific ions as is i.e. helpful in phosphorylation analysis.

The majority of experiments in the life sciences is now based on soft ionisation. The techniques electrospray (ESI) and matrix-assisted laser desorption (MALDI) gently ionise *via* the addition or removal of protons rather than the more destructive bombardment with high-energy particles (Fig. 4). In ESI, liquid samples are exposed to high electric fields while in MALDI the solid sample in an organic matrix (Fig. 5) is ionised using a laser (mostly ultraviolet UV, occasionally infrared). The matrix is a low molecular weight compound, which can absorb some of the lasers energy. In this way, it prevents decomposition of the sample and it also facilitates vaporization and ionization of the analyte molecules. ESI is characterized by the formation of multiply charged species detected in charge state envelopes for one analyte while charging is typically low in MALDI.[7]

The earlier technique fast atom bombardement (FAB) can occasionally still be found and it is applicable to sensitive analytes as was demonstrated for FMRF-amides of *Neireis virens*[10]. In FAB, analyte embedded in liquid matrix (i.e. glycerol) are ionised by bombardement with Ar or Xe atoms, or Cs^+ ions. Other analytes such as lipids or organic acids might require special ionisation sources such as atmospheric pressure chemical ionization or electron impact (EI). They will be described in the respective paragraph. For further reading on technical aspects of MS a number of books are available such as refs.[11, 12].

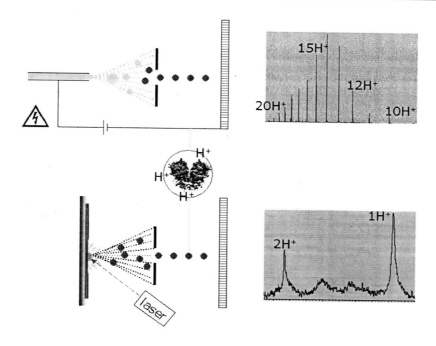

Figure 4. Principles of ionisation in ESI and MALDI. Top: Liquid sample is sprayed and ionised in a high electric field. Bottom: Sample which was co-crystallized with matrix is ionised using laser light. ESI generates charge states envelopes while in MALDI mostly one and two charges are observed. Spectra: relative ion abundance *versus* m/z.

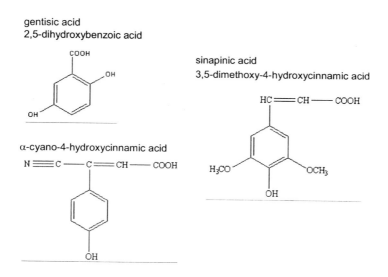

Figure 5. Common MALDI matrices. α-cyano-4-hydroxycinnamic acid is mainly used for peptide analysis. Gentisic acid shows advantages for some application such as phosphopeptides. Sinapinic acid is often used for protein analysis.

Functional Proteomic Analyses – Proteins and Peptides

Due to the capability of identifying proteins which were separated by classical biochemical tools such as gel electrophoresis or chromatography, MS has propelled protein analysis forward (Fig. 6)[13]. The term proteomics involves the investigation of large sets of proteins found for a certain organism at certain conditions while most functional studies are limited to few selected proteins and their modifications. The methodology, however, is the same apart from the fact that a large number of samples might require a high degree of automation.[14].

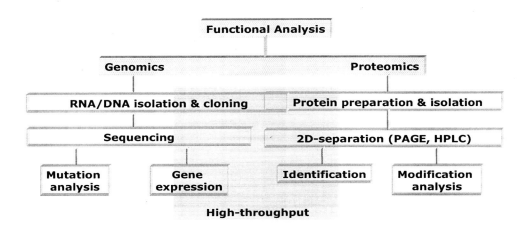

Figure 6. Functional genomic and proteomic analyses. Integrated functional genomics is based on separation techniques followed by detection methods for RNA/DNA and proteins. Some processes are amenable to automation.

Databases

Great synergism was achieved with the advances in genome sequencing and the maintenance and extension of huge public sequence databases on one hand and the development of the World-Wide-Web with appropriate web browsers and search engines on the other hand. Special software was necessary to effectively mine the databases and, therefore, the field of bioinformatics grew alongside. Basic tasks such as similarity analysis or pattern and profile prediction, but also advanced calculations like the modeling of the tertiary structure of proteins can now be performed with publicly available software.

The basis of protein analysis are the nucleotide and protein sequences compiled in worldwide accessible databases. The best known are maintained at the National Center for Biotechnology Information (NCBI; USA; http://www.ncbi.nlm.nih.gov/) and at the Swiss Institute for Bioinformatics (SIB; http://www.expasy.ch/). The NCBI Entrez Nucleotides database is a collection of sequences from several sources. The number of bases grows at an exponential rate and as of April 2004, there are over 38,989,342,565. The protein entries in the Entrez search and retrieval system have also been compiled from a variety of sources, including SwissProt and translations from annotated coding regions in GenBank. Although complete genomes of various organisms are available *via* Entrez, for annelida only 13525 nucleotide and 1572 protein entries can be found as of July 2004. The SIB Expasy Server allows access to the curated SwissProt database which strives to provide a high level of annotations and integration with other databases and a minimal level of redundancy. Here, only 123 entries are related to annelida, although release 44.0 of July 5, 2004 contains 153871 sequences. TrEMBL, which is a computer-annotated supplement of SWISS-PROT that contains all the translations of EMBL nucleotide sequence entries not yet integrated in SWISS-PROT suggests 922 entries (release 27.0; 7/5/2004). Since protein identification is achieved by direct comparison of measured and expected sequence data, the availability of annelida database entries is very important as was evident for earthworm hemolysins[15, 16]. There it had only been possible due to recent GenBank submissions to finally identify hemolytic proteins of coelomic fluid and coelomocytes which had been extensively characterized with classical immunological and biochemical methods before.

Initiatives to amend the lack of annelida sequencing projects include WormBase (http://www.wormbase.org/) which provides accessible information concerning the genetics, genomics and biology of *C. elegans* and some related nematodes. There are also recent efforts to expand LumbriBASE (http://convoluta.cap.ed.ac.uk/Lumbribase/lumbribase/lumbribase.html) to contain ~15,000 *Lumbricus rubellus* EST sequences. LumbriBase is a relational sequence database which can be queried by both sequence similarity and annotation. However, accessibility of this database with MS data still requires further bioinformatic input. For common proteins with sufficient homology across species such as carbonic anhydrase sequence and similarity analyses can still be possible as was shown for trophosome carbonic anhydrase of the symbiotic hydrothermal vent tubeworm *Riftia pachyptila* [17].

Protein Identification

Identification of known proteins is based on their enzymatic digestion, because peptides are easier to handle in extraction, separation, and detection procedures. Moreover, a set of peptide masses is characteristic for a given protein and can be searched against sequence databases. The protease used predominantly is trypsin, but in cases other cleavage methods can be more appropriate for specific analysis problems (Table 2).

Table 2. Proteolytic Agents

	Cleavage Site
Chemical	
BNPS skatol	Tryp...X
Cyanogen bromide	Met...X
Enzymatic	
Trypsin	Lys/Arg...X; X≠Pro
Chymotrypsin	Phe/Tyr/Trp...X and other hydrophobic residues
Endoproteinases	
Lys-C	Lys...X; X≠Pro
Glu-C	Glu...X
Asp-N	X...Asp
Endopeptidase	
Asp-C	Asp...X
Carboxypeptidase B	C-terminal residue especially Arg/Lys

It has been shown that proteins can be digested enzymatically in the gel after PAGE (polyacrylamide gel electrophoresis) and that the resulting peptide extract is sufficient for protein identification (Fig. 7)[18, 19]. Proteins can be digested after electroblotting with a similar procedure[20, 21] and there have also been efforts to combine in-gel and in-membrane digest and eliminate spot excision[22]. In general, Coomassie or colloidal Coomassie stained spots can be successfully analysed. For silver staining difficulties may arise, because it is very sensitive (1-10 ng limit) and the intensity of the staining does not necessarily reflect the amount of protein on the gel spot. Depending on the available instrumentation 50 % of all spots and more can usually be identified.

Peak lists generated with peptide mapping or MS/MS-fragmentation are the experimental data used to screen the databases for the protein in question. For that purpose, special search engines have been developed, which are publicly available over the internet (http://matrixscience.com, http://prospector.ucsf.edu/, http://prowl.rockefeller.edu/). The search programs differ in their algorithm and constraints and they are under constant development. With this approach only known proteins can be found. However, MS/MS spectra of peptides from unknown proteins serve to find short stretches of sequence (tags), which help to access the protein *via* cloning (Fig. 7). Structures of bioactive peptides are directly accessible to structure elucidation using MS/MS in combination with complimentary techniques such as Edman sequencing as was shown for a tachykinin isoform from midgut of desert locust *Schistocerca gregaria*[23] or enkephalin-related and prodynorphin-derived peptides from leech Theromyzon tessulatum[24, 25].

Success of the analysis means for the biologist the unambigious identification of the respective protein. This can, indeed, often be achieved solely by peptide mapping. However, the method is based on the assumption that the peaks in the spectrum represent tryptic peptides or known modifications. Therefore, contaminants, multiple proteins, or unexpected modifications might lead to false-positive results. In addition, difficulties in accessing database information due to properties of search algorithms or database entries may hinder the analysis. Although peptide mapping is a very fast procedure and amenable to automation (see example Fig. 9), it always needs verification either on the biochemical level or using MS sequencing (Fig. 10). Collision-induced dissociation predominantely causes fragmentation at the amide bonds of the polyamide backbone. Ion series b and y have been defined to describe cleavage from the C- or N-termini, respectively (Fig. 8, Table 3). Sequence information gained for only one peptide can be sufficient to confirm an assignment or allow identification in the first place.

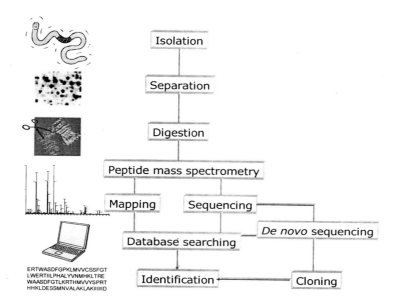

Figure 7. Routine protein identification is based on the digestion of gel electrophoretically separated proteins and the analysis of the peptides with MS. Peptide masses and fragment data are used to search database information. MS data can provide sequence tags for unknown proteins helping to access them *via* cloning.

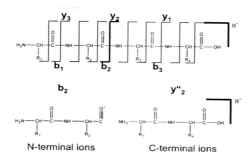

Figure 8. Peptide fragmention is enhanced at the backbone. Therefore, typical b and y ion series can be recognized indicating amino acid residue losses (Table 3).

Table 3. Amino acid residues and their mass

Amino acid	code	monoisotopic	average mass
Glycine	G / Gly	57.021	57.052
Alanine	A / Ala	71.037	71.079
Serine	S / Ser	87.032	87.078
Proline	P / Pro	97.053	97.117
Valine	V / Val	99.068	99.133
Threonine	T / Thr	101.048	101.105
Cysteine	C / Cys	103.009	103.145
Isoleucine	I / Ile	113.084	113.160
Leucine	L / Leu	113.084	113.160
Asparagine	N / Asn	114.043	114.104
Aspartic acid	D / Asp	115.027	115.089
Glutamine	Q / Gln	128.059	128.131
Lysine	K / Lys	128.095	128.174
Glutamic acid	E / Glu	129.043	129.116
Methionine	M / Met	131.040	131.199
Histidine	H / His	137.059	137.142
Phenylalanine	F / Phe	147.068	147.177
Arginine	R / Arg	156.101	156.188
Tyrosine	Y / Tyr	163.063	163.176
Tryptophane	W / Trp	186.079	186.213

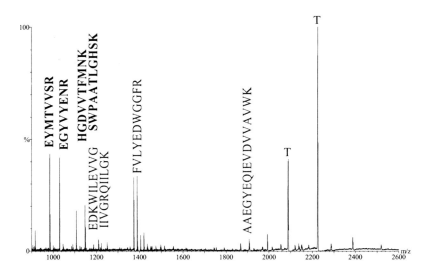

Figure 9. Peptide map of CL_{41} identified as lysenin from *E. fetida*. Sequences indicate peaks which were MS-fragmented in a second experiment. T labels trypsin autolysis peaks.

For the identification of proteins the detection of a sufficient number of peptides is necessary (Fig. 9). It is, in general, not important to find a terminal peptide. However, some research projects require exact knowledge of the termini. The basic mass mapping approach then often fails. For the determination of the N-terminus Edman sequencing can be used unless the terminus is blocked. In that case, MS experiments can be designed which specifically look for expected sequences[26]. For the C-terminus carboxypeptidase ladder sequencing has been shown[27].

Peptides are often separated with high-performance liquid chromatography (HPLC) using reversed phase (RP) material (silica particles covered with chemically-bonded hydrocarbon chains C_4 to C_{18}). In HPLC-MS/MS experiments, both the UV-absorption and the MS signal (total ion current) are used for peptide detection. In order to overcome some limitations of 2D-PAGE with respect to very acidic or basic proteins, excessively large or small and membrane proteins, the combination of HPLC-MS/MS is used and more recently multidimensional chromatographic methods have been created (2D-LC). Complex protein mixtures are digested and then fed into an HPLC-system coupled to a mass spectrometer for automated separation and data-dependent MS/MS[28, 29].

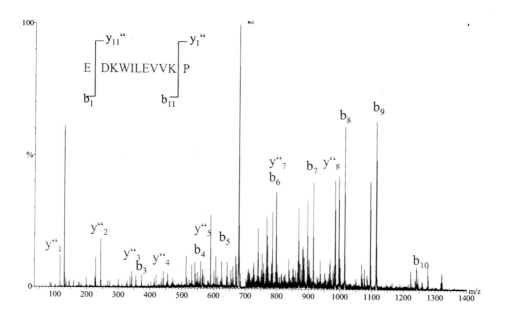

Figure 10. MS/MS-spectrum for a CL_{39} peptide assigned to the C-terminal end of fetidin from *E. fetida*.

Analysis of Modifications

The biological function of a protein is not only determined by its amino acid sequence but also by posttranslational chemical modifications of which more than 200 different types have been described to date. Major analysis task among those are the analysis of phosphorylation, glycosylation, and disulfide bridges. In general, the analysis is based on the standard proteomics approach explained above. Then the experiments have to be adapted to the specific modification. For disulfide bridges it might suffice to obtain fragmentation spectra of the linked peptides in order to assign the site, possibly in comparison with the reduced and alkylated protein.

Phosphorylation

Protein phosphorylation is a common and important modification of proteins, because nearly all aspects of cell life are influenced by reversible protein phosphorylation. Defining the sites of phosphorylation in a protein and the extent of phosphorylation at each specific site is an analytical challenge due to a number of reasons. Many phosphoproteins, especially those involved in signalling, are present in cells only in very few copies. Individual sites are often only partially phosphorylated. In addition, in MS, phosphorylated peptides usually show lower response in positive ion mode due to the electronegativity of phosphate groups. Solutions to improve MS detection involve the specific measurement of characteristic marker ions for phosphopeptides and the specific enrichment of phosphopeptides using immobilized metal ion affinity chromatography (IMAC). Once the phosphopeptide is isolated, MS/MS analysis can determine the modified sites.[30]

Glycosylation

Glycosylated proteins are very abundant among the secreted proteins and membrane-bound proteins[31]. The biological role of the attached sugar chains (glycans) involves conformational stability, protection against degradation as well as essential molecular and cellular recognition. Glycans are branched structures consisting of different carbohydrate residues. They can either be attached to asparagines residues in the consensus sequence Asn-Xxx-Ser/Thr/Cys (N-linked) or to serine or threonine (O-linked). Each site might vary in the glycan structures attached creating microheterogeneity. In addition, different sites may only be partially glycosylated. Depending on the information required (carbohydrate portion or protein part) different analysis strategies are available. They involve combinations of the release of N-glycans using peptide-N^4-(N-acetyl-β-glucosaminyl) asparagines amidase F (PNGase F), the cleavage of O-glycans on reductive β-elimination, and the separation of the peptide and sugar fractions using RP-HPLC[32]. An unusual example is shown in Fig. 11 for a *E. fetida* hemolysin designated H_1[15]. A high number of hexose losses were observed by chance during MS/MS sequencing experiments triggering further studies on the stimulation-dependent glyosylation of hemolysins in that organism.

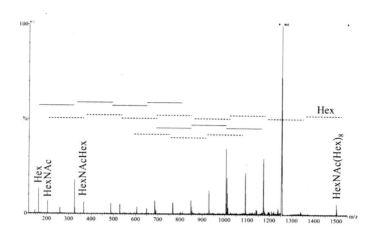

Figure 11. MS/MS spectrum of m/z 1253.43 ion of *E. fetida* hemolysin H_1 digest. A number of hexose losses are observed indicating high-mannose glycosylation. All peaks except the parent were magnified by 5.[15]

Access to Protein Tertiary Structure

In analogy to the detection of disulfide bridges, chemically cross-linked proteins can be analysed. Using a variety of linkers, amino acid distances are defined in folded proteins. The potential of this simple approach for the analysis of the protein tertiary structure has been recognized early on, but only the synergism of advances in bioinformatics, MS and structure databases has led to a strong renewed interest in the field[33]. The main advantage of MS is the low sample consumption in comparison to the traditional structure determinantion techniques (x-ray crystallography, nuclear magnetic resonance spectroscopy NMR) without the need for crystallization or intense labelling. Structure generation will be driven by minimization of three types of potential energy terms: MS data derived experimental constraints (long and short range constants, site specific secondary structure elements, hydrogen exchange and charging patterns), globally applicable covalent restraints for proteins (bond length and angles, improper angles, van-der-Waals-terms, etc.), and data-base potentials (any type of constraining knowledge: function, tendency of a given stretch of amino acids to form a specific secondary structure derived from the large body of available high resolution 3D structures in protein databases, phylogenic information). Templates for homologous structures available in 3D databases are used for threading of unknown structures.

Intact Peptides and Proteins

MW determinations can be necessary to confirm the purity of a bioanalyte or for its further characterization as was shown for a number of earthworm peptides such as those from *E. fetida* coelomic fluid[34], lumbricin I of *L. rebellus*[35], and GGNG peptides of *E. fetida* and *Pheretima vittata*[36].

In cases, it can be advisable to measure the uncleaved protein, because information might gained, which is lost in digests. The MS spectrum allows conclusions, if a protein is pure or partially modified. The number of attached groups can be determined (i.e. the number of phosphorylated sites), if the resolution of the instrument is sufficient and the protein is not too large or too inhomogeneous. MALDI-MS can be used to measure within a wide mass range up to 200 kDa and more depending on the instrument. However, ESI analyses sometimes deliver better masses for proteins. Up to ~70 kDa (albumin), ESI spectra can still be measured, for higher-MW proteins this appears to be difficult. The reason is not MS, but the incapability to purify higher mass proteins in such a way that the charge state distribution can be resolved. MALDI is more tolerant, but mass accuracy also suffers from that effect. Generally, MS experiment are carried out at acidic conditions to support ionization. Using volatile buffers such as ammoniumbicarbonat also non-covalent protein complexes can be measured. Although questions remain, correlations of the solution structure and the gas phase structure have been shown and MS can be a valuable tool to study protein interaction[37].

Intact annelida hemoglobins have been the main object of ESI-MS measurements where the technique was very useful in determining accurate MWs of globin subunits and their modifications[38]. Examples include the globin of the nerve cord of the polychaete annelid *Aphrodite aculeate*[39], three major monomeric hemoglobins (Hbs) from the marine annelid *Glycera dibranchiate*[40], and the ~3.5-MDa hexagonal bilayer Hb from the pond leech *Macrobdella decora*[41]. Maximum entropy analysis of the ESI mass spectra of the native, carbamidomethylated, and reduced forms of the extracellular ca. 400 kDa Hb of the pogonophoran *Oligobrachia mashikoi* has shown it to consist of eight globin chains of ~15 kDa forming disulfide-bonded dimers and trimers[42]. In a similar way *Alvinella pompejana* Hb was studied[43-45]. ESI-MS of the subassemblies of hexagonal bilayer Hbs obtained by gel filtration of partially dissociated *L. terrestris* and *Arenicola marina* Hbs showed the presence of noncovalent complexes of monomer and trimer subunits with masses in the 213.3-215.4 and 204.6-205.6 kDa ranges[46]. Such multidomain arrangements were also detected for scaleworms *B. symmytilida* and *B. seepensis*[47]. In addition to Hbs, also other proteins such as fibrinolytic enzymes *Lumbricus rubellus*[48] and hemolysins[15] have been studied.

Affinity MS

Interfacing biospecific interaction analysis based on surface plasmon resonance (SPR) and MS allows the combination of information on binding events and the determination of the MW of the interacting molecules. Thereby, chromatographic or bioaffinity surfaces are used on chips. MS serves to measure the MW of (in the SPR experiment) isolated biospecific markers and identify them.[49, 50]

Oligonucleotides

The soft ionization methods permit the near routine analysis of oligonucleotides and intact nucleic acids. Among the various uses of MS in genomics, applications focused on the characterization of single nucleotide polymorphisms (SNPs) and short tandem repeats (STRs) are particularly important[51]. For instance, genomic DNA methylation was demonstrated in the polychaete annelid worm *Chaetopterus variopedatus*[52].

Small Analytes – Fatty Acids, Hydrocarbons, Lipids, Sterols, Sugars, Toxins

Although low-MW biomolecules also can be investigated using soft ionisation methods, gas chromatography (GC)/MS is still the method of choice combining valuable separation with MS detection[53]. There, volatile analytes are separated at high temperatures (~300 °C) on a chromatographic column using a carrier gas (usually Helium). The most frequently used method for ion production in toxicology labs is EI, the occasionally used alternative is chemical ionization (CI). For EI a collimated beam of electrons impact the sample molecules causing the loss of an electron from the molecule (molecular ion M^+). CI begins with ionization of a gas such as methane, creating a radical which in turn will impact the sample molecule to produce MH^+ molecular ions. Some of the molecular ions fragment into smaller daughter ions and neutral fragments. Less fragmentation occurs with CI than with EI.

The giant nephridia in *S. pavonina* contains cholesterol and its long-chain fatty acid esters (mainly palmitoleate and palmitate), which were identified by thin-layer chromatography and EI- and CI-MS[54]. Structures of a novel series of fucolipids obtained from whole tissues of the marine annelid, *Pseudopotamilla occelata* could be elucidated by compositional and methylation analyses, gas-liquid chromatography, GC- and MALDI-MS, and proton NMR spectroscopy demonstrating that a synergistic combination of analytical techniques is often necessary[55]. Similarly, glycosphingolipids require extensive analytics involving exoglycosidase cleavage and multistage chromatography as well as a variety of MS methods by which sugar, sphingoid and fatty acid components can be clarified. Examples include galactose-containing glycosphingilipids from earthworm, *Pheretima hilgendorfi*[56] as such containing choline phosphate[57] and neogala series glycosphingolipids containing mannose[58], neutral glycosphingolipids isolated from the leech, *Hirudo nipponica*[59], acidic, neutral and amphoteric glycosphingolipids extracted from the lugworm, *Tylorhynchus heterochaetus*[60, 61], and the marine annelid, *P. occelata*[62], and digalactosylceramides carrying a choline phosphate group in the outer galactose moiety isolated from the earthworm, *Pheretima asiatica*[63].

GC/MS is a common method in toxicology, drug surveillance as well as for the study of organic acids. Sterol compounds of oceanic and intertidal annelida species were compared using GC/MS as early as 1978[64]. Later on, the sterol mixtures of two marine sedentary annelides, *Chaetopterus variopedatus* and *Spirographis spallanzani* were fractionated by argentation chromatogography and were GC/MS analyzed[65]. Specimens of the ribbon worm, "himomushi" *Cephalothrix sp.* (Nemertean) adherent to the shells of cultured oysters were found to contain tetrodotoxin-derived compounds which showed strong paralytic action in mice[66]. Organic acid analysis analysis of a sperm maturation factor in the intertidal polychaete *A. marina* revealed that arachidonic acid and 8,11,14-eicosatrienoic acid cause sperm activation[67]. The metabolism of ecdysone and 20-hydroxyecdysone was investigated in various animal species, including annelids. Some of the major metabolites were isolated and characterized by MS and 2D 1H-NMR[68]. Other applications include the determination of the bioconcentration factors of parent and alkylated polycyclic aromatic hydrocarbons isolated from polychaete worms collected in coastal sediments with GC/MS employing a flame-ionization detector[69], and the structural analysis of purified hallachrome from the marine annelid, *H. parthenopeia*, which was shown to be the first anthraquinone pigment unsubstituted at positions 9 and 10[70].

Metals

Inductively-Coupled Plasma-Ms

Inorganic MS differs from organic MS particularly in the ionisation sources, but also in the detection range[71]. A recent work on the detection of heavy metals (Cr, Mn, Cu, Zn, Cd, Hg, As) in the freshwater leech *Erpobdella octoculata* demonstrates the complex procedure involving inductively coupled plasma (ICP)-MS and absorption spectrometry[72].

ICP-MS is a very powerful tool for trace (ppb-ppm) and ultra-trace (ppq-ppb) elemental analysis. The ICP source operates at temperatures of ~5000 °C so that virtually all molecules in a sample are broken up into their component atoms. The plasma is generated in argon gas with a spark from a Tesla unit. The sample which consists of a fine aerosol from nebulized liquids or ablated solids is directly injected into the heart of the plasma. Any molecules initially present in the aerosol are quickly and completely broken down to charged atoms (M^+, M^{++}) some of which will recombine with other species in the plasma to create both stable and meta-stable molecular species (*e.g.* MAr^+, M_2^+, MO^+, etc.).[73]

Secondary Ion Ms

Secondary ion MS (SIMS)[74] is an measurement technique that is being used for the compositional analysis of small samples. In a SIMS instrument (or "ion microprobe") a high energy primary ion beam interacts with the sample surface causing mixing of the upper layers of the sample, resulting in an amorphization of the surface. Atoms from the primary ion beam are implanted in the sample and some secondary particles (atoms and small molecules) are ejected from sample. Charged particles of one polarity ("secondary ions") can then be extracted from the sputtering area with the help of an electrical field between the sample and an extraction lens. Since the size of the sputtered area is determined by the primary ion beam diameter, which typically is in the order of micro-meter, a SIMS analysis has a relatively high lateral resolution. SIMS can be used for practically all elements of the periodic table, including hydrogen. Only the noble gases are difficult to measure because they do not ionize easily. SIMS allows the routine measurement of many trace elements at very low concentration (ppb-range). During a measurement, the sample is slowly sputtered (eroded) away and, therefore, SIMS is capable of measuring depth profiles and perform 3D measurements. In this way, Li was investigated in 37 species of aquatic mammals, fish, crustacea, annelids, molluscs belonging to cephalopods, gasteropods and lamellibranches[75].

Isotope-Ratio MS

Measuring isotopic ratios on light elements such as C, H, O, N or S has been an important tool in the biological sciences[76]. Originally developed by geochemists, stable isotope techniques are now widely used in many different fields like food control, medical diagnostic, environment, forensic sciences, doping control, or climate reconstruction. Radioisotope labelling studies have traditionally been carried out by marking chemicals of interest with specific rare radioisotopes of elements typically found in organics (^{3}H, ^{11}C, ^{14}C, ^{32}P, and ^{35}S) because they can usually be incorporated into biomolecules without modifying their natural properties and they have low natural abundances. Isotope quantification is also advantageous since it is independent of the physical properties of the labeled chemical and is distinctive in a complex biological matrix[76]. The approach has long been used in small molecule MS, but was recently proposed for the quantification of proteins in a high-throughput environment. With isotope-coded affinity tags (ICAT[77]), the relative protein abundance between two samples (for example, healthy and diseased tissues) can be compared using multi-dimensional liquid chromatography followed by MS.

Detection of radioisotopes can be performed by decay-counting, which is often limited by high backgrounds, as well as low specificity and low decay counting efficiency. The possible advantages of MS for long-lived radioisotope detection where individual nuclei are counted independent of decay have long been recognized. However, until the advent of accelerator MS (AMS)[78], measurement of ^{14}C as an important example for radioisotope measurement has been fraught with difficulty, mostly due to problems in resolution and isobaric interference. AMS is 1,000 to 10,000 times more sensitive than decay counting methods. The enhanced sensitivity is achieved by accelerating sample ions to high energies using a particle accelerator and nuclear particle detection techniques. An example is the determination of ^{14}C concentrations in total organic carbon, pore-water dissolved inorganic carbon, infauna, and bulk of sediments by AMS in samples taken at and near the Isla Vista hydrocarbon seep off southern California to estimate the extent of fossil carbon cycling.[79]

Conclusion and Outlook

Both organic and inorganic MS may be of interest in the biosciences and the paper tried to give a general, although not complete, overview covering important applications in proteomics, small molecule and metal analysis. MS can give very detailed answers such as accurate masses, sequences, and protein assignments. The availability of the technique increases although the local experimental focus and knowledge may differ and especially extensive functional studies warrant the effort to find specialists for collaborators.

References

[1]   Vetvicka, V. S., P.; Cooper, E.L.; Bilej, M.; Roch, P. *Immunology of Annelids*; CRC Press: Boca Raton, 1993.

[2]   Cooper Edwin, L.; Kauschke, E.; Cossarizza, A. *BioEssays : news and reviews in molecular, cellular and developmental biology 2002*, 24, 319-333.

[3]   Bilej, M.; De Baetselier, P.; Beschin, A. *Folia microbiologica 2000*, 45, 283-300.

[4]   Warr, G. W.; Marchalonis, J. J. *Quarterly review of biology 1978*, 53, 225-241.

[5]   Downard, K. M. *Journal of Mass Spectrometry 2000*, 35, 493-503.

[6]   *ASMS Measuring Mass. From Positive Rays to Proteins*; Chemical Heritage Press: Philadelphia, 2002.

[7]   Chapman, J. R. *Protein and Peptide Analysis by Mass Spectrometry*; Humana Press: Totowa, NJ, 1996.

[8]   Mills, I., Cvitaš, T., Homann, K., Kallay, N. , Kuchitsu, K. *Quantities, Units and Symbols in Physical Chemistry., 2nd ed.*; Blackwell Science: London, 1993.

[9]   DeLaeter, J. R. B., J.K.; DeBievre, P.; Hidaka, H.; Peiser, H.S.; Rosman, K.J.R.; Taylor, *P.D.P. Pure Appl. Chem. 2003*, 75, 683-800.

[10]   Krajniak, K. G.; Price, D. A. *Peptides* (New York, NY, United States) 1990, 11, 75-77.

[11]   Kellner, R. M., J.-M.; Otto, M.; Widmer, H.M. *Analytical Chemistry;* Wiley-VCH: Weinheim, 1998.

[12]   Gross, J. H. *Mass Spectrometry*; Springer: Berlin, 2004.

[13]   Godovac-Zimmermann, J., Brown, L.R. *Mass Spectrom. Rev. 2001*, 20, 1-57.

[14]   Koenig, S. G., J.; Zeller, M. *Detection and Identification of Proteins by Mass Spectrometry;* Wiley-VCH: Weinheim, 2003.

[15]   Koenig, S.; Wagner, F.; Kauschke, E.; Peter-Katalinic, J.; Cooper, E. L.; Eue, I. *Developmental & Comparative Immunology 2003*, 27, 513-520.

[16]   Koenig, S.; Wagner, F.; Kauschke, E.; Eue, I. *Spectroscopy* (Amsterdam, Netherlands) 2004, 18, 347-353.

[17]   De Cian, M.-C.; Bailly, X.; Morales, J.; Strub, J.-M.; Van Dorsselaer, A.; Lallier Francois, H. *Proteins 2003*, 51, 327-339.

[18]   Shevchenko, A., Wilm, M., Vorm, O., Mann, M. *Anal. Chem. 1996*, 68, 850-858.

[19]   Shevchenko, A., Jennsen, O.N., Podtelejnikov, A.V., Sagliocco, F., Wilm, M., Vorm, O., Mortensen, P. et al. *Proc. Natl. Acad. Sci. 1996*, 93, 14440-14445.

[20]   Fernandez, J., Andrews, L. , Mische, S.M. *Anal. Biochem. 1994*, 218, 112-117.

[21]   Stults, J. T., Henzel, W.J., Wong, S.C. , Watanabe, C. In *Mass spectrometry in the biological sciences,*; A.L. Burlingame, S. A. C., Ed.; Humana Press Inc.,: Totowa, 1996.

[22]   Binz, P.-A., Müller, M., Walther, D., Bienvenut, W.V., Gras, R., Hoogland, C., Bouchet, G. et al. *Anal. Chem. 1999*, 71, 4981-4988.

[23]   Veelaert, D.; Baggerman, G.; Derua, R.; Waelkens, E.; Meeusen, T.; Vande Water, G.; De Loof, A.; Schoofs, L. *Biochemical and Biophysical Research Communications 1999*, 266, 237-242.

[24]   Salzet, M.; Bulet, P.; Verger-Bocquet, M.; Malecha, J. *FEBS Letters 1995*, 357, 187-191.

[25] Salzet, M.; Verger-Bocquet, M.; Bulet, P.; Beauvillain, J.-C.; Malecha, J. *Journal of Biological Chemistry 1996*, 271, 13191-13196.

[26] Huang, R.-H., Wang, D.-C. *Journal of Chromatography B 2004*, 803, 167-172.

[27] Patterson, D. H., Tarr, G.E., Regnier, F.E., Martin, S.A. *Anal Chem. 1995*, 67, 3971-3978.

[28] Link, A. J., Eng, J., Schieltz, D.M., Carmack, E., Mize, G.J., Morris, D.R., Garvik, B.M. et al. *Nature Biotechn. 1999*, 17, 676-682.

[29] Peng, J., Gygi, S.P. *J. Mass Spectrom. 2001*, 36, 1083-1091.

[30] Zeller, M., Koenig, S. *Analytical and Bioanalytical Chemistry 2004*, 378, 898-909.

[31] Bill, R. M., Revers, L., Wilson, *I.B.H. Protein Glycosylation*; Kluwer Academic Pub, 1998.

[32] Rademaker, G. J., Thomas-Oates, J. In *Methods in Molecular Biology;* Chapman, J. R., Ed.; Humana Press Inc: Totowa, 1996; Vol. 61.

[33] Young, M., Tang, N., Hempel, J.C., Oshiro, C.M., Taylor, E.W., Kuntz, I.D., Gibson, B.W. et al. *PNAS 2000*, 97, 5802-5806.

[34] Hanusova, R.; Tuckova, L.; Halada, P.; Bezouska, K.; Bilej, M. *Developmental and comparative immunology 1999*, 23, 113-121.

[35] Cho, J. H.; Park, C. B.; Yoon, Y. G.; Kim, S. C. *Biochimica et biophysica acta 1998*, 1408, 67-76.

[36] Oumi, T.; Ukena, K.; Matsushima, O.; Ikeda, T.; Fujita, T.; Minakata, H.; Nomoto, K. *Biochemical and biophysical research communications 1995*, 216, 1072-1078.

[37] Hernandez, H., Robinson, C.V. *JBC 2001*, 276, 46685-46688.

[38] Green, B. N.; Hutton, T.; Vinogradov, S. N. *Methods in Molecular Biology* (Totowa, New Jersey) 1996, 61, 279-294.

[39] Dewilde, S.; Blaxter, M.; Van Hauwaert, M.-L.; Vanfleteren, J.; Esmans, E. L.; Marden, M.; Griffon, N.; Moens, L. *Journal of Biological Chemistry 1996*, 271, 19865-19870.

[40] Teske, J. G.; Edmonds, C. G.; Deckert, G.; Satterlee, J. D. *Journal of Protein Chemistry 1997*, 16, 139-150.

[41] Suzuki, T.; Vinogradov, S. N. *Journal of Protein Chemistry 2003*, 22, 231-242.

[42] Yuasa, H. J.; Green, B. N.; Takagi, T.; Suzuki, N.; Vinogradov, S. N.; Suzuki, T. *Biochimica et Biophysica Acta 1996*, 1296, 235-244.

[43] Zal, F.; Green, B. N.; Martineu, P.; Lallier, F. H.; Toulmond, A.; Vinogradov, S. N.; Childress, J. J. *European Journal of Biochemistry 2000*, 267, 5227-5236.

[44] Zal, F.; Green, B. N.; Lallier, F. H.; Toulmond, A. *Biochemistry 1997*, 36, 11777-11786.

[45] Hourdez, S.; Lallier, F. H.; De Cian, M.-C.; Green, B. N.; Weber, R. E.; Toulmond, A. *Physiological and Biochemical Zoology 2000*, 73, 365-373.

[46] Green, B. N.; Bordoli, R. S.; Hanin, L. G.; Lallier, F. H.; Toulmond, A.; Vinogradov, S. N. *Journal of Biological Chemistry 1999*, 274, 28206-28212.

[47] Hourdez, S.; Lallier, F. H.; Green, B. N.; Toulmond, A. *Proteins: Structure, Function, and Genetics 1999*, 34, 427-434.

[48] Nakajima, N.; Mihara, H.; Sumi, H. Bioscience, *Biotechnology, and Biochemistry 1993*, 57, 1726-1730.

[49] Nedelkov, D., Nelson, R.W. *Analytica Chimica Acta 2000*, 423, 1-7.

[50] Grote, J.; Dankbar, N.; Gedig, E.; Koenig, S. Analytical Chemistry 2005, 77, 1157-1162.

[51] Meng, Z., Simmons-Willis, T.A., Limbach, P.A. *Biomolecular Engineering 2004*, 21, 1-13.

[52] del Gaudio, R.; Di Giaimo, R.; Geraci, G. *FEBS Letters 1997*, 417, 48-52.

[53] Huebschmann, H.-J. *Handbook of GC/MS. Fundamentals and Applications*; Wiley-VCH: Weinheim, 2001.

[54] Koechlin, N.; Polonsky, J.; Varenne, J. *Comparative Biochemistry and Physiology, Part A: Molecular & Integrative Physiology 1981*, 68A, 391-397.

[55] Itonori, S.; Hamana, H.; Hada, N.; Takeda, T.; Dulaney, J. T.; Sugita, M. *Journal of Oleo Science 2001*, 50, 537-544.

[56] Sugita, M.; Hayata, C.; Suzuki, M.; Takeda, T.; Mizunoma, T.; Makino, T.; Narushima, K.; Hori, T. *Yukagaku 1992*, 41, 568-573.

[57] Sugita, M.; Fujii, H.; Inagaki, F.; Suzuki, M.; Hayata, C.; Hori, T. *Journal of Biological Chemistry 1992*, 267, 22595-22598.

[58] Sugita, M.; Mizunoma, T.; Inagaki, F.; Suzuki, M.; Suzuki, A.; Hori, T.; Nakatani, F.; Ohta, S. *Yukagaku 1994*, 43, 1052-1061.

[59] Noda, N.; Tanaka, R.; Miyahara, K.; Sukamoto, T. *Chemical & pharmaceutical bulletin 1996*, 44, 895-899.

[60] Sugita, M.; Makino, T.; Hayata, C.; Suzuki, M.; Suzuki, A.; Hori, T.; Miwa, S.; Nakatani, F. *Yukagaku 1993*, 42, 935-941.

[61] Sugita, M.; Miwa, S.; Aoki, K.; Dulaney, O. T.; Ichikawa, S.; Inagaki, F.; Suzuki, M. *Nihon Yukagakkaishi 2000*, 49, 33-44.

[62] Sugita, M.; Yamake, N.; Hamana, H.; Sasaki, K.; Dulaney, J. T. *Nihon Yukagakkaishi 1999*, 48, 671-679.

[63] Tanaka, R.; Miyahara, K.; Noda, N. *Chemical & pharmaceutical bulletin 1996*, 44, 1152-1156.

[64] Ballantine, J. A.; Lavis, A.; Roberts, J. C.; Morris, R. J.; Elsworth, J. F.; Cragg, G. M. L. *Comparative Biochemistry and Physiology, Part B: Biochemistry & Molecular Biology 1978*, 61B, 43-47.

[65] Sica, D.; Di Giacomo, G. *Comparative Biochemistry and Physiology, Part B: Biochemistry & Molecular Biology 1981*, 70B, 719-723.

[66] Asakawa, M.; Toyoshima, T.; Shida, Y.; Noguchi, T.; Miyazawa, K. *Toxicon 2000*, 38, 763-773.

[67] Bentley, M. G.; Clark, S.; Pacey, A. A. *Biological Bulletin* (Woods Hole, MA, United States) 1990, 178, 1-9.

[68] Lafont, R.; Beydon, P.; Blais, C.; Garcia, M.; Lachaise, F.; Riera, F.; Somme, G.; Girault, J. P. *Insect Biochemistry 1986*, 16, 11-16.

[69] Fernandez, P.; Bayona, J. M. *Journal of High Resolution Chromatography 1989*, 12, 802-806.

[70] Prota, G.; D'Agostino, M.; Misuraca, G. *Experientia 1971*, 27, 15-16.

[71] de Laeter, J. R. *Applications of Inorganic Mass Spectrometry*; Wiley Interscience, 2001.

[72] Friese, K.; Froemmichen, R.; Witter, B.; Mueller, H. *Acta Hydrochimica et Hydrobiologica 2004,* 31, 346-355.

[73] Hill, S. J. *ICP Spectrometry and its Applications*; CRC Press, 1998.

[74] Wilson, R. G. *Secondary Ion Mass Spectrometry: A Practical Handbook for Depth Profiling and Bulk Impurity Analysis;* John Wiley & Sons, 1989.

[75] Chassard-Bouchaud, C.; Galle, P.; Escaig, F.; Miyawaki, M. *Comptes Rendus de l'Academie des Sciences, Serie III: Sciences de la Vie 1984,* 299, 719-724.

[76] Platzner, I. T. *Modern Isotope Ratio Mass Spectrometry*; John Wiley & Sons, 1997.

[77] Gygi, S. P., Rist, B., Gerber, S.A., Turecek, F., Gelb, M.H., Aebersold, R. *Nature Biotechn. 1999,* 17, 994-999.

[78] Tuniz, J. R., Fink Bird, D., Tuniz, C. *Accelerator Mass Spectrometry*; CRC Press, 1998.

[79] Bauer, J. E.; Spies, R. B.; Vogel, J. S.; Nelson, D. E.; Southon, J. R. *Nature* (London, United Kingdom) 1990, 348, 230-232.

[80] Rosman, K. J. R. T., P.D.P. *Pure Appl. Chem. 1998,* 70, 217-236.

In: Progress in Immunology Research
Editor: Barbara A. Veskler, pp. 69-81

ISBN 1-59454-380-1
©2005 Nova Science Publishers, Inc.

Chapter IV

Cathelicidins: Antimicrobial Peptides with Nonmicrobicidal Activities

Balaji Ramanathan and Frank Blecha

Department of Anatomy and Physiology, College of Veterinary Medicine, Kansas State University, Manhattan, Kansas 66506

Abstract

Antimicrobial peptides are ancient host defense molecules widely represented across the animal and plant kingdoms. Animals possess numerous antimicrobial peptides, which are expressed in many tissues and cells, including neutrophils, macrophages and mucosal epithelial cells. A prominent family of mammalian antimicrobial peptides is the cathelicidins. These peptides have broad-spectrum antimicrobial activity and are an important innate immune defense. However, in addition to their well-recognized antimicrobial activity, cathelicidins function in initiating and amplifying host innate and adaptive immune responses. This chapter reviews recent data illustrating the nonmicrobicidal activities of cathelicidins.

Introduction

In mammals innate immunity is an essential first line of defense against pathogens. A key component of innate immunity is the production of small, cationic antimicrobial peptides, a protection strategy that is conserved from insects through humans. Animals possess numerous antimicrobial peptides, which are expressed in many tissues and cells, including neutrophils, macrophages and mucosal epithelial cells [1]. Most of these compounds are relatively simple structures such as short cationic peptides and many are being evaluated as potential novel antibiotics to combat emerging drug-resistant bacteria. One family of antimicrobial peptides is the cathelicidins, which are characterized by conserved pro-peptide sequences that have been identified in several mammalian species [2]. In addition to their direct antimicrobial function, cathelicidins have multiple roles as mediators of inflammation influencing diverse processes such as cell proliferation and migration, immune modulation, wound healing, angiogenesis and the release of cytokines. This review provides an overview of nonmicrobicidal properties of the cathelicidin family of antimicrobial peptides and discusses their biological functions (Table 1).

Table 1. Nonmicrobicidal activities of cathelicidins

Cathelicidin	Activity	References
PR-39	Anti-tumor	54
	Inhibition of reactive oxygen species	56-58
	Promotes angiogenesis	67-69
	Wound repair	59
	Chemoattractant	64
	Inhibition of apoptosis	76, 77
Probac 5	Prevention of tissue injury	4
LL-37	Chemoattractant	65, 73
	Gene regulation	70
	Degranulation of mast cells	72
	Binds endotoxin	31
	Reproduction	78
BMAP-28	Chemoattractant	66
Indolicidin	Binds endotoxin	74

Cathelicidins

Cathelicidins are a class of gene-encoded antibiotics found in humans and animals. The term cathelicidins was introduced in 1995 to encompass bipartite molecules containing both a cathelin domain and a C-terminal antimicrobial peptide domain. Cathelin, an acronym for cathepsin L inhibitor, was first isolated as a 96-residue porcine peptide [3]. The key event in the discovery of cathelicidins was the cloning of the cDNA for bactenecin (Bac) 5, a 43-residue antimicrobial peptide that had previously been isolated from bovine neutrophils [4]. Cathelin domains display considerable interspecies conservation of their primary sequences. For example, more than 60% of residues in the cathelin domain of human cationic antimicrobial peptide-18 (hCAP-18) and porcine cathelin are identical. Cathelicidins range from 12 to 97 amino acids. The most widespread are linear peptides of 23 to 37 amino acid residues that fold into amphipathic α-helices mimicking biological membranes. Other members of this family include a number of small-sized (12-18 residues) molecules with beta-hairpin structures stabilized by one or two disulphide bonds, and a 13-residue linear peptide characterized by a high proportion of tryptophan. Each family member has been named individually using acronyms (e.g., PMAPs, CRAMP) or one-letter symbols of key amino acid residues present in the antimicrobial sequence, followed by the number of residues (e.g., LL-37, PR-39). In other instances, specific features of the antimicrobial domains are designated (e.g., protegrin, dodecapeptide, indolicidin). The general structural and biological properties of the peptides were recently reviewed [5-7]. An interpretative summary of procedures used to purify, identify, and test antimicrobial peptides was also recently published [8].

To date, cathelicidins have been described only in mammals, including guinea pigs [9], rats [10], rabbits [11], mice [12, 13], goats [14], sheep [15, 16], pigs [17-23], cows [4, 24-28], horses [29], nonhuman primates [30] and humans [31-33]. Each species shows a varied number of congeners. Cathelicidin genes are organized into four exons, with the coding sequence for the preproregion contained in exons 1-3, and the region encoding the varied antimicrobial domain in exon 4. This structural organization is conserved among species and several potential consensus sites for nuclear factors involved in hematopoiesis and inflammation are present in the 5' flanking sequences of all the genes sequenced, suggesting coordinated control of their expression [34-38]. Circulating neutrophils are a major source of cathelicidins, and myeloid bone marrow cells are their primary site of expression. Cathelicidins are synthesized at a myelocyte and metamyelocyte stage of neutrophil maturation and the propeptides are then targeted to the secretory granules, i.e., the specific granules [9, 39, 40]. Processing of the storage forms to active peptides has been shown to take place in activated neutrophils following exocytosis of the granules [37, 41, 42]. This activation mechanism is mediated by elastase, or proteinase 3 [43-45]. These results were confirmed by analysis of the sequences showing that the propiece and the antimicrobial domain of cathelicidins are joined by cleavage sites for elastase (Val, Ala, or Thr) [46]. In addition to myeloid-derived cells, some cathelicidins, have been found to be expressed in other tissues. The human cathelicidin is present in testis and is expressed in a variety of squamous epithelia where synthesis is up-regulated in inflammatory states [32, 47, 48]. In addition to myeloid-derived cells, porcine cathelicidins have been detected in myeloid/lymphoid organs such as the spleen, thymus, and lymph nodes [49]. Studies have also shown developmentally regulated gene expression of porcine cathelicidins [49].

A wealth of published data indicates that cathelicidin antimicrobial peptides kill a variety of bacteria, fungi, and enveloped viruses with a broad overlap in specificity and significant difference in potency. A limited number of papers have reported on the function of the cathelin propiece [50, 51], and the biological role of this evolutionarily conserved domain is still a matter of debate. According to a recent report, the cathelin domain of the human cathelicidin hCAP-18 is endowed with antimicrobial properties after being released by proteolytic processing of the holoprotein [50].

Nonmicrobicidal Activities

Anti-Tumor Activity

The cationic or amphiphilic feature of cathelicidins determines the mode of their antimicrobial action. Cationic parts of the peptides are capable of interacting with negatively charged structures of the microbial membrane and lead to its permeabilization. Several models of membrane permeabilization have been suggested [52]. Selectivity of their actions stems from differences in membrane composition of higher eukaryotes and microbes. The outer membrane of higher eukaryotes is made of electrically neutral phospholipids like phosphatidylcholine and sphingomyelin, whereas bacterial membranes have exposed negatively charged phosphatidylglycerol and cardiolipin. Another difference is the lack of cholesterol in bacterial membranes. These parameters were found to be important for selectivity of antimicrobial peptides. This property may also prove useful in cancer therapy. Tumor cells can differ in membrane composition from nontransformed cells. For example, phosphatidylserine content is increased in the membrane of melanoma and carcinoma cells in comparison to that in normal human keratinocytes [53]. Such differences can result in higher susceptibility of tumor cells to membrane-permeabilizing peptides. Recently, it was shown that PR-39 gene transduction can alter invasive activity in human hepatocarcinoma cells. These cells have a high metastatic potential as a consequence of a reduced expression of syndecan-1 [54]. Cells transfected with the PR-39 gene showed a highly increased expression of syndecan-1 accompanied by suppression of invasive activity. Moreover, the transfectants showed disorganization of actin filaments with suppression of motile activity; a novel effect of PR-39 [54].

Inhibition of Tissue Injury

Reactive oxygen species (ROS) generated by phagocyte NADPH oxidase are important components of host defense. However, these highly toxic oxidants can cause significant tissue injury during inflammation; thus, it is essential that ROS are tightly regulated [55]. One such mechanism of regulation involves attenuation of NADPH oxidase activity. Inhibition of phagocyte NADPH oxidase activity by PR-39 has been described at concentrations that are slightly higher than those required to induce inflammatory cell recruitment (chemotaxis) [56]. The inhibition of NADPH activity has the direct effect of reducing local tissue injury, since upregulated superoxide anion formation by NADPH oxidase contributes to ongoing tissue damage during inflammation. Cytosolic binding of PR-39 involves interaction with cytosolic SH3 domains of $p47^{phox}$, thereby resulting in attenuation of superoxide anion development. These effects were also demonstrated in rat lung reperfusion studies indicating that PR-39 has therapeutic potential in post-ischemic inflammation [4, 57, 58]. Alternative mechanisms of protection are provided by other peptides. For example, probac5, a proform of Bac5, attenuates tissue injury by inhibiting the actions of cathepsin L protease, a predominant enzyme involved in cell damage [4].

Wound Repair

The initial discovery that mammalian proline-rich cathelicidins had biological activities beyond microbicidal properties was identified when PR-39 present in wound fluid induced the synthesis of syndecan-1 and -4 in mesenchymal cells [59]. The syndecan family of cell surface heparan sulfate proteoglycans has been implicated in a number of biological processes including regulation of blood coagulation, cell adhesion, and wound repair [60]. The expression of syndecan-1 on the surface of mesenchymal cells is increased during wound repair, enabling cells to become more responsive to a variety of effector molecules present in the wound environment. These effectors include heparin-binding growth factors such as fibroblast growth factor (FGF), vascular endothelial growth factor (VEGF), and transforming growth factor β (TGF-β) [61]. Syndecans are thought to play a role in wound repair by regulating cell proliferation and migration in response to these effectors.

Syndecan-inducing activity of PR-39 suggested that the peptide might function as a signaling molecule in wound repair. To investigate the mechanism by which it could induce gene expression, its biologically active N-terminal fragment 1-15 was used to identify potential binding targets. In these experiments it was first found that the peptide bound cells in a saturable manner, consistent with the presence of a 'receptor' molecule, and that it rapidly entered cells without membrane permeabilization and localized to the cytoplasm. The peptide was also shown to selectively bind cytosolic proteins with Src homology 3 (SH3) domains [61, 62]. This was not unexpected as PR-39 contains several SH3 consensus binding motifs, including the RxxPxxP sequence at the N-terminus.

To characterize the structure-activity relationship of PR-39 with respect to syndecan induction and SH3 domain binding activity, analogues of this peptide were tested. In brief, it was found that the N-terminal arginines are crucial for binding to SH3-containing targets and for syndecan induction in NIH 3T3 cultures, as well as for antibacterial activity [62].

Chemotactic Activity

Chemoattractants and adhesion molecules coordinate the recruitment of circulating leukocytes to an area of inflammation that allow inflammatory cell interaction with the endothelium and extracellular matrix [63]. Among cathelicidin peptides, PR-39 has been identified as a potential chemoattractant present at the site of inflammation [64]. Chemoattractant activity for neutrophils was apparent at sub-antimicrobial concentrations. Acting through formyl peptide-like receptor-1, a Gi protein-coupled receptor, LL-37 was also found to be chemotactic for human neutrophils, monocytes and T cells, all of which express this receptor. LL-37 induced Ca^{2+} mobilization both in monocytes and in human embryonic kidney cells transfected with formyl peptide-like receptor-1 [65]. A potential role as a chemoattractant exists for BMAP-28 as it has been shown to be secreted at higher concentrations at inflammatory sites with a subsequent increase in the number of neutrophils [66].

Angiogenesis

PR-39 exerts cardioprotective effects in mouse and rat myocardial ischemia-reperfusion models. In mice, myocardial infarct per risk area was significantly reduced in PR-39-treated mice, as well as the accumulation of neutrophils within the ischemic reperfused myocardium [67]. Similar results were obtained with ischemic and reperfused hearts administered neutrophils at the onset of reperfusion, that were isolated from PR-39-treated or untreated rats [68]. These peptide effects were likely due to reduction of neutrophil adherence to coronary vascular endothelium that caused a significant reduction in neutrophil infiltration, and inhibition of NADPH oxidase of both endothelial cells and neutrophils. In line with the cardioprotective effect, PR-39 also induced angiogenesis in cell culture and *in vivo* in mouse myocardium. Angiogenesis is a complex process that can be stimulated by inflammation and hypoxia, which determines the release of several pro-angiogenic factors such as growth factors and cytokines. Pro-angiogenic activity of PR-39 was mediated by the inhibition of ubiquitine-proteasome-dependent degradation of the hypoxia-inducible factor (HIF)-1α, which upregulates the expression of angiogenesis-related genes, including vascular endothelial growth factor (VEGF) [69]. This finding provides a link between the two major stimulants of angiogenesis, inflammation and hypoxia, in that a peptide released during the inflammatory response promotes an increased concentration of HIF-1α that mediates the hypoxia effects.

Regulation of Inflammatory Mediators

Gene expression of inflammatory mediators is influenced by LL-37 [70]. Of particular interest is the upregulation of the chemokines CXCL8 and CCL2 and their corresponding receptors, CXCR2 and CCR2, by LL-37 [70]. Neutrophils, monocytes and T cells express CXCR2, whereas CCR2 is expressed by monocytes and interstitial dendritic cells (iDCs) [63, 71, 72]. Therefore, LL-37 may indirectly facilitate the recruitment of phagocytes, iDCs, and T cells to inflammatory sites through the induction of chemokines and their corresponding receptors. LL-37 has also been shown to degranulate mast cells, leading to the release of proinflammatory mediators such as histamine and prostaglandins, which enhances the inflammatory reaction [72]. The ability of LL-37 to chemoattract human peripheral blood T cells indicates that it can participate in the recruitment of effector T cells to sites of microbial infection, thereby contributing to adaptive antimicrobial immunity [65, 73].

Antisepsis

Killing bacteria by antimicrobial peptides, phagocytes, and complement releases bacterial components, such as endotoxin from Gram-negative bacteria and lipotechoic acid from Gram-positive bacteria. These bacterial components, if allowed to enter the circulation, may cause septic shock by inducing the production of high levels of systemic proinflammatory cytokines. LL-37, SMAP-29, and indolicidin, bind LPS with high affinity and neutralize its biological activities [31, 70, 74, 75]. In addition, the activity of lipotechoic acid can also be inhibited by cathelicidins [70].

Apoptosis

Apoptosis is a physiologically, tightly regulated mode of cell death that plays an important role in development and tissue homeostasis. In a recent study, a novel mechanism for the inhibition of apoptosis in macrophages by the antimicrobial peptide, PR-39 was discovered. Apoptosis, induced by several stimuli including, nutrient depletion, LPS stimulation and camptothecin treatment, was significantly reduced when cells were treated with PR-39 at bactericidal and physiologically relevant concentrations [76]. The anti-apoptotic effect of PR-39 was associated with a decrease in caspase-3 activity. These findings were confirmed in a study showing that PR-39 inhibits apoptosis in endothelial cells [77] by down regulating caspase-3 and inducing the inhibitor of apoptosis protein (IAP-2).

Reproduction

Recent evidence suggests that LL-37 may participate in reproduction. Seminal plasma from healthy donors contains 40 to 140 μg/ml hCAP-18, levels that are 70-fold higher than those in blood plasma [78]. Most of this hCAP-18 is associated with spermatozoa. Independently, a high level of expression of the hCAP-18 gene was found in the body and tail of the epididymis, and hCAP-18 mRNA could be detected by reverse transcription polymerase chain reaction in round spermatids. Perhaps hCAP-18 coating allows sperm to eliminate microbial pathogens when they enter the sterile upper female genital tract. Alternatively, the peptide could participate in later events directly associated with fertilization.

Cross-Linking Innate and Adaptive Immunity

In addition to direct antimicrobial properties, cathelicidins modulate other components of innate immunity. They are generally present at the sites of microbial entry due to extracellular release or secretion and form a chemotactic gradient, which results in the recruitment and activation of various subsets of leukocytes. This contributes to the elimination of invading pathogens, thereby contributing to innate host defense. The recruitment to inflammatory sites and activation of in situ effector T cells would enhance the effector phase of host adaptive immunity against infection. Although LL-37 does not seem to activate dendritic cells, it is reported to be capable of degranulating mast cells and enhance HLA-DLR expression by human dendritic cells, suggesting that it may have an enhancing effect on the induction phase of adaptive immunity. The participation of cathelicidins in host innate and adaptive immunity against microbial invasion has been shown in an *in vivo* study showing that adenoviral vector targeted systemic overexpression of cathelicidins/LL-37 in vivo results in decreased bacterial load and mortality of experimental mice following challenge with *Pseudomonas aeruginosa* or *Escherichia coli* [79].

Future Prospects

Collectively, data on cathelicidins support a role for these natural peptides in innate immunity and in epithelial defense. These peptides possess multiple biological properties, including the ability to kill microorganisms that have developed resistance to conventional antibiotics. Despite recent progress, research on this class of antimicrobials is still in its infancy. As we learn more about cathelicidins, additional diversification and functions will be discovered. Therefore, several issues remain to be addressed, and additional studies are required to elucidate aspects of their biology. (i) Identification of novel antimicrobial peptides -- More members of cathelicidin peptides are likely to exist in mammalian species.

Progress in the genome projects will also reveal ways to shortcut conventional bioscreening procedures for the identification of these family members. (ii) Analysis of the biologically relevant functions of the antimicrobial peptides -- Aside from *in vitro* experiments that give the first insight into the function of peptide antibiotics, a broader approach involving genetic animal models such as knock-out animals is necessary to interpret *in vitro* data in the context of whole organisms. Structure-function studies of these antimicrobial peptides may increase understanding of mechanisms of action and provide guidance for therapeutic design. (iii) Identifying molecular factors involved in cathelicidin regulation will also be useful therapeutically by stimulating the natural expression of the innate immune response. (iv) Evaluation of the function of antimicrobial peptides in diseases might provide information about the corresponding pathogenesis. (v) Development of antimicrobial peptides as drugs -- Studying the biology of antimicrobial peptides may allow the development of novel therapeutics including anti-infectious, anti-inflammatory, or pro-angiogenic drugs. Several naturally occurring peptides have been used to develop prototypical antibiotic drugs that are currently being evaluated in phase I-III clinical trials [80]. Thus, critical information will come from animal model systems, and the role of cathelicidins and their synergism with other defense molecules will be better understood.

References

[1]   Lehrer, R. I., and T. Ganz. 1999. Antimicrobial peptides in mammalian and insect host defence. *Curr. Opin. Immunol.* 11:23-27.

[2]   Lehrer, R. I., and T. Ganz. 2002. Cathelicidins: a family of endogenous antimicrobial peptides. *Curr. Opin. Hematol.* 9:18-22.

[3]   Ritonja, A., M. Kopitar, R. Jerala, and V. Turk. 1989. Primary structure of a new cysteine proteinase inhibitor from pig leucocytes. *FEBS Lett.* 255:211-214.

[4]   Zanetti, M., G. Del Sal, P. Storici, C. Schneider, and D. Romeo. 1993. The cDNA of the neutrophil antibiotic Bac5 predicts a pro-sequence homologous to a cysteine proteinase inhibitor that is common to other neutrophil antibiotics. *J. Biol. Chem.* 268:1:522-526.

[5]   Gennaro, R., and M. Zanetti. 2000. Structural features and biological activities of the cathelicidin-derived antimicrobial peptides. *Biopolymers* 55:31-49.

[6]   Zanetti, M., R. Gennaro, M. Scocchi, and B. Skerlavaj. 2000. Structure and biology of cathelicidins. *Adv. Exp. Med. Biol.* 479:203-218.

[7]   Ramanathan, B., E. G. Davis, C. R. Ross, and F. Blecha. 2002. Cathelicidins: microbicidal activity, mechanisms of action, and roles in innate immunity. *Microbes. Infect.* 4:361-372.

[8]   Cole, A. M., and T. Ganz. 2000. Human antimicrobial peptides: analysis and application. *Biotechniques* 29:822-831.

[9]   Nagaoka, I., Y. Tsutsumi-Ishii, S. Yomogida, and T. Yamashita. 1997. Isolation of cDNA encoding guinea pig neutrophil cationic antibacterial polypeptide of 11 kDa (CAP11) and evaluation of CAP11 mRNA expression during neutrophil maturation. *J. Biol. Chem.* 272:36:22742-22750.

[10] Termen, S., M. Tollin, B. Olsson, T. Svenberg, B. Agerberth, and G. H. Gudmundsson. 2003. Phylogeny, processing and expression of the rat cathelicidin rCRAMP: a model for innate antimicrobial peptides. *Cell Mol. Life Sci.* 60:536-549.

[11] Larrick, J. W., J. G. Morgan, I. Palings, M. Hirata, and M. H. Yen. 1991. Complementary DNA sequence of rabbit CAP18--a unique lipopolysaccharide binding protein. *Biochem. Biophys. Res. Commun.* 179:170-175.

[12] Gallo, R. L., K. J. Kim, M. Bernfield, C. A. Kozak, M. Zanetti, L. Merluzzi, and R. Gennaro. 1997. Identification of CRAMP, a cathelin-related antimicrobial peptide expressed in the embryonic and adult mouse. *Biol. Chem.* 272:20:13088-13093.

[13] Popsueva, A. E., M. V. Zinovjeva, J. W. Visser, J. M. Zijlmans, W. E. Fibbe, and A. V. Belyavsky. 1996. A novel murine cathelin-like protein expressed in bone marrow. *FEBS Lett.* 391:5-8.

[14] Shamova, O., K. A. Brogden, C. Zhao, T. Nguyen, V. N. Kokryakov, and R. I. Lehrer. 1999. Purification and properties of proline-rich antimicrobial peptides from sheep and goat leukocytes. *Infect. Immun.* 67:8:4106-4111.

[15] Bagella, L., M. Scocchi, and M. Zanetti. 1995. cDNA sequences of three sheep myeloid cathelicidins. *FEBS Lett.* 376:225-228.

[16] Mahoney, M. M., A. Y. Lee, D. J. Brezinski-Caliguri, and K. M. Huttner. 1995. Molecular analysis of the sheep cathelin family reveals a novel antimicrobial peptide. *FEBS Lett.* 377:519-522.

[17] Zanetti, M., P. Storici, A. Tossi, M. Scocchi, and R. Gennaro. 1994. Molecular cloning and chemical synthesis of a novel antibacterial peptide derived from pig myeloid cells. *J. Biol. Chem.* 269:11:7855-7858.

[18] Storici, P., and M. Zanetti. 1993. A cDNA derived from pig bone marrow cells predicts a sequence identical to the intestinal antibacterial peptide PR-39. *Biochem. Biophys. Res. Commun.* 196:3:1058-1065.

[19] Storici, P., and M. Zanetti. 1993. A novel cDNA sequence encoding a pig leukocyte antimicrobial peptide with a cathelin-like pro-sequence. *Biochem. Biophys. Res. Commun.* 196:3:1363-1368.

[20] Storici, P., M. Scocchi, A. Tossi, R. Gennaro, and M. Zanetti. 1994. Chemical synthesis and biological activity of a novel antibacterial peptide deduced from a pig myeloid cDNA. *FEBS Lett.* 337:303-307.

[21] Tossi, A., M. Scocchi, M. Zanetti, P. Storici, and R. Gennaro. 1995. PMAP-37, a novel antibacterial peptide from pig myeloid cells. cDNA cloning, chemical synthesis and activity. *Eur. J. Biochem.* 228:941-946.

[22] Agerberth, B., J. Y. Lee, T. Bergman, M. Carlquist, H. G. Boman, V. Mutt, and H. Jornvall. 1991. Amino acid sequence of PR-39. Isolation from pig intestine of a new member of the family of proline-arginine-rich antibacterial peptides. *Eur. J. Biochem.* 202:849-854.

[23] Kokryakov, V. N., S. S. L. Harwig, E. A. Panyutich, A. A. Schevchenko, G. M. Aleshina, O. V. Shamova, H. A. Korneva, and R. I. Lehrer. 1993. Protegrins: leukocyte antimicrobial peptides that combine features of corticostatic defensins and tachyplesins. *FEBS Lett.* 327:2:231-236.

[24] Gennaro, R., B. Skerlavaj, and D. Romeo. 1989. Purification, composition, and activity of two bactenecins, antibacterial peptides of bovine neutrophils. *Infect. Immun.* 57:10:3142-3146.

[25] Frank, R. W., R. Gennaro, K. Schneider, M. Przybylski, and D. Romeo. 1990. Amino acid sequences of two proline-rich bactenecins. Antimicrobial peptides of bovine neutrophils. *J. Biol. Chem.* 265:31:18871-18874.

[26] Del Sal, G., P. Storici, C. Schneider, D. Romeo, and M. Zanetti. 1992. cDNA cloning of the neutrophil bactericidal peptide indolicidin. *Biochem. Biophys. Res. Commun.* 187:1:467-472.

[27] Storici, P., G. Del Sal, C. Schneider, and M. Zanetti. 1992. cDNA sequence analysis of an antibiotic dodecapeptide from neutrophils. *FEBS Lett.* 314:2:187-190.

[28] Selsted, M. E., M. J. Novotny, W. L. Morris, Y. Q. Tang, W. Smith, and J. S. Cullor. 1992. Indolicidin, a novel bactericidal tridecapeptide amide from neutrophils. *J. Biol. Chem.* 267:7:4292-4295.

[29] Scocchi, M., D. Bontempo, S. Boscolo, L. Tomasinsig, E. Giulotto, and M. Zanetti. 1999. Novel cathelicidins in horse leukocytes(1). *FEBS Lett.* 457:459-464.

[30] Bals, R., C. Lang, D. J. Weiner, C. Vogelmeier, U. Welsch, and J. M. Wilson. 2001. Rhesus monkey (*Macaca mulatta*) mucosal antimicrobial peptides are close homologues of human molecules. *Clin. Diagn. Lab Immunol.* 8:370-375.

[31] Larrick, J. W., M. Hirata, R. F. Balint, J. Lee, J. Zhong, and S. C. Wright. 1995. Human CAP18: a novel antimicrobial lipopolysaccharide-binding protein. *Infect. Immun.* 63:4:1291-1297.

[32] Agerberth, B., H. Gunne, H. Odeberg, P. Kogner, H. G. Boman, and G. H. Gudmundsson. 1995. FALL-39, a putative human peptide antibiotic, is cysteine-free and expressed in bone marrow and testis. *Proc. Natl. Acad. Sci. USA* 92:195-199.

[33] Cowland, J. B., A. H. Johnsen, and N. Borregaard. 1995. hCAP-18, a cathelin/pro-bactenecin-like protein of human neutrophil specific granules. *FEBS Lett.* 368:173-176.

[34] Scocchi, M., S. Wang, and M. Zanetti. 1997. Structural organization of the bovine cathelicidin gene family and identification of a novel member. *FEBS Lett.* 417:311-315.

[35] Gudmundsson, G. H., K. P. Magnusson, B. P. Chowdhary, M. Johansson, L. Andersson, and H. G. Boman. 1995. Structure of the gene for porcine peptide antibiotic PR-39, a cathelin gene family member: comparative mapping of the locus for the human peptide antibiotic FALL-39. *Proc. Natl. Acad. Sci. U. S. A* 92:7085-7089.

[36] Zhao, C., T. Ganz, and R. I. Lehrer. 1995. The structure of porcine protegrin genes. *FEBS Lett.* 368:197-202.

[37] Gudmundsson, G. H., B. Agerberth, H. Odeberg, T. Bergman, B. Olsson, and R. Salcedo. 1996. The human gene FALL39 and processing of the cathelin precursor to the antibacterial peptide LL-37 in granulocytes. *Eur. J. Biochem.* 238:325-332.

[38] Zhao, C., T. Ganz, and R. I. Lehrer. 1995. Structures of genes for two cathelin-associated antimicrobial peptides: prophenin-2 and PR-39. *FEBS Lett.* 376:130-134.

[39] Zanetti, M., L. Litteri, R. Gennaro, H. Horstmann, and D. Romeo. 1990. Bactenecins, defense polypeptides of bovine neutrophils, are generated from precursor molecules stored in the large granules. *J. Cell Biol.* 111:1363-1371.

[40] Sorensen, O., K. Arnljots, J. B. Cowland, D. F. Bainton, and N. Borregaard. 1997. The human antibacterial cathelicidin, hCAP-18, is synthesized in myelocytes and metamyelocytes and localized to specific granules in neutrophils. *Blood* 90:7:2796-2803.

[41] Zanetti, M., L. Litteri, G. Griffiths, R. Gennaro, and D. Romeo. 1991. Stimulus-induced maturation of probactenecins, precursors of neutrophil antimicrobial polypeptides. *J. Immunol.* 146:12:4295-4300.

[42] Gennaro, R., L. Dolzani, and D. Romeo. 1983. Potency of bactericidal proteins purified from the large granules of bovine neutrophils. *Infect. Immun.* 40:2:684-690.

[43] Panyutich, A., J. Shi, P. L. Boutz, C. Zhao, and T. Ganz. 1997. Porcine polymorphonuclear leukocytes generate extracellular microbicidal activity by elastase-mediated activation of secreted proprotegrins. *Infect. Immun.* 65:978-985.

[44] Scocchi, M., B. Skerlavaj, D. Romeo, and R. Gennaro. 1992. Proteolytic cleavage by neutrophil elastase converts inactive storage proforms to antibacterial bactenecins. *Eur. J. Biochem.* 209:589-595.

[45] Sorensen, O. E., P. Follin, A. H. Johnsen, J. Calafat, G. S. Tjabringa, P. S. Hiemstra, and N. Borregaard. 2001. Human cathelicidin, hCAP-18, is processed to the antimicrobial peptide LL-37 by extracellular cleavage with proteinase 3. *Blood* 97:3951-3959.

[46] Zanetti, M., R. Gennaro, and D. Romeo. 1995. Cathelicidins: a novel protein family with a common proregion and a variable C-terminal antimicrobial domain. *FEBS Lett.* 374:1-5.

[47] Frohm, M., B. Agerberth, G. Ahangari, M. Ståhle-Bläckdahl, S. Lidén, H. Wigzell, and G. H. Gudmundsson. 1997. The expression of the gene coding for the antibacterial peptide LL-37 is induced in human keratinocytes during inflammatory disorders. *J. Biol. Chem.* 272:24:15258-15263.

[48] Frohm, N. M., B. Sandstedt, O. Sorensen, G. Weber, N. Borregaard, and M. Stahle-Backdahl. 1999. The human cationic antimicrobial protein (hCAP18), a peptide antibiotic, is widely expressed in human squamous epithelia and colocalizes with interleukin-6. *Infect. Immun.* 67:2561-2566.

[49] Wu, H., G. Zhang, C. R. Ross, and F. Blecha. 1999. Cathelicidin gene expression in porcine tissues: roles in ontogeny and tissue specificity. *Infect. Immun.* 67:1:439-442.

[50] Zaiou, M., V. Nizet, and R. L. Gallo. 2003. Antimicrobial and protease inhibitory functions of the human cathelicidin (hCAP18/LL-37) prosequence. *J. Invest Dermatol.* 120:810-816.

[51] Shinnar, A. E., K. L. Butler, and H. J. Park. 2003. Cathelicidin family of antimicrobial peptides: proteolytic processing and protease resistance. *Bioorg. Chem.* 31:425-436.

[52] Yeaman, M. R., and N. Y. Yount. 2003. Mechanisms of antimicrobial peptide action and resistance. *Pharmacol. Rev.* 55:27-55.

[53] Utsugi, T., A. J. Schroit, J. Connor, C. D. Bucana, and I. J. Fidler. 1991. Elevated expression of phosphatidylserine in the outer membrane leaflet of human tumor cells and recognition by activated human blood monocytes. *Cancer Res.* 51:3062-3066.

[54] Ohtake, T., Y. Fujimoto, K. Ikuta, H. Saito, M. Ohhira, M. Ono, and Y. Kohgo. 1999. Proline-rich antimicrobial peptide, PR-39 gene transduction altered invasive activity and actin structure in human hepatocellular carcinoma cells. *Br. J. Cancer* 81:393-403.

[55] Korthuis, R. J., and D. N. Granger. 1993. Reactive oxygen metabolites, neutrophils, and the pathogenesis of ischemic-tissue/reperfusion. *Clin. Cardiol.* 16:I19-I26.

[56] Shi, J., C. R. Ross, T. L. Leto, and F. Blecha. 1996. PR-39, a proline antibacterial peptide that inhibits phagocyte NADPH oxidase activity by binding to Src homology 3 domains of $p47^{phox}$. *Proc. Natl. Acad. Sci. USA* 93:6014-6018.

[57] Al Mehdi, A. B., G. Zhao, C. Dodia, K. Tozawa, K. Costa, V. Muzykantov, C. Ross, F. Blecha, M. Dinauer, and A. B. Fisher. 1998. Endothelial NADPH oxidase as the source of oxidants in lungs exposed to ischemia or high K+. *Circ. Res.* 83:730-737.

[58] Korthuis, R. J., D. C. Gute, F. Blecha, and C. R. Ross. 1999. PR-39, a proline/arginine-rich antimicrobial peptide, prevents postischemic microvascular dysfunction. *Am. J. Physiol* 277:H1007-H1013.

[59] Gallo, R. L., M. Ono, T. Povsic, C. Page, E. Eriksson, M. Klagsbrun, and M. Bernfield. 1994. Syndecans, cell surface heparan sulfate proteoglycans, are induced by a proline-rich antimicrobial peptide from wounds. *Proc. Natl. Acad. Sci. U. S. A* 91:11035-11039.

[60] Bernfield, M., R. Kokenyesi, M. Kato, M. T. Hinkes, J. Spring, R. L. Gallo, and E. J. Lose. 1992. Biology of the syndecans: a family of transmembrane heparan sulfate proteoglycans. *Annu. Rev. Cell Biol.* 8:365-393.

[61] Chan, Y. R., and R. L. Gallo. 1998. PR-39, a syndecan-inducing antimicrobial peptide, binds and affects $p130^{Cas}*$. *J. Biol. Chem.* 273:28978-28985.

[62] Chan, Y. R., M. Zanetti, R. Gennaro, and R. L. Gallo. 2001. Anti-microbial activity and cell binding are controlled by sequence determinants in the anti-microbial peptide PR-39. *J. Invest Dermatol.* 116:230-235.

[63] Kunkel, S. L., N. Lukacs, and R. M. Strieter. 1995. Expression and biology of neutrophil and endothelial cell-derived chemokines. *Semin. Cell Biol.* 6:327-336.

[64] Huang, H. J., C. R. Ross, and F. Blecha. 1997. Chemoattractant properties of PR-39, a neutrophil antibacterial peptide. *J. Leukoc. Biol.* 61:624-629.

[65] De, Y., Q. Chen, A. P. Schmidt, G. M. Anderson, J. M. Wang, J. Wooters, J. J. Oppenheim, and O. Chertov. 2000. LL-37, the neutrophil granule- and epithelial cell-derived cathelicidin, utilizes formyl peptide receptor-like 1 (FPRL1) as a receptor to chemoattract human peripheral blood neutrophils, monocytes, and T cells. *J. Exp. Med.* 192:1069-1074.

[66] Gennaro, R., M. Scocchi, L. Merluzzi, and M. Zanetti. 1998. Biological characterization of a novel mammalian antimicrobial peptide. *Biochim Biophys Acta.* 1425:361-368.

[67] Hoffmeyer, M. R., R. Scalia, C. R. Ross, S. P. Jones, and D. J. Lefer. 2000. PR-39, a potent neutrophil inhibitor, attenuates myocardial ischemia- reperfusion injury in mice. *Am. J. Physiol Heart Circ. Physiol* 279:H2824-H2828.

[68] Ikeda, Y., L. H. Young, R. Scalia, C. R. Ross, and A. M. Lefer. 2001. PR-39, a proline/arginine-rich antimicrobial peptide, exerts cardioprotective effects in myocardial ischemia-reperfusion. *Cardiovasc. Res.* 49:69-77.

[69] Li, J., M. Post, R. Volk, Y. Gao, M. Simons, C. Metais, K. Sato, J. Tsai, W. Aird, R. D. Rosenberg, T. G. Hampton, J. Li, F. Sellke, P. Carmeliet, and M. Simons. 2000. PR39, a peptide regulator of angiogenesis. *Nat Med* 6:1:49-55.

[70] Scott, M. G., D. J. Davidson, M. R. Gold, D. Bowdish, and R. E. Hancock. 2002. The human antimicrobial peptide LL-37 is a multifunctional modulator of innate immune responses. *J. Immunol.* 169:3883-3891.

[71] Baggiolini, M. 1998. Chemokines and leukocyte traffic. *Nature* 392:565-568.

[72] Niyonsaba, F., A. Someya, M. Hirata, H. Ogawa, and I. Nagaoka. 2001. Evaluation of the effects of peptide antibiotics human beta-defensins-1/-2 and LL-37 on histamine release and prostaglandin D(2) production from mast cells. *Eur. J. Immunol.* 31:1066-1075.

[73] Agerberth, B., J. Charo, J. Werr, B. Olsson, F. Idali, L. Lindbom, R. Kiessling, H. Jornvall, H. Wigzell, and G. H. Gudmundsson. 2000. The human antimicrobial and chemotactic peptides LL-37 and alpha-defensins are expressed by specific lymphocyte and monocyte populations. *Blood* 96:3086-3093.

[74] Larrick, J. W., M. Hirata, H. Zheng, J. Zhong, D. Bolin, J. M. Cavaillon, H. S. Warren, and S. C. Wright. 1994. A novel granulocyte-derived peptide with lipopolysaccharide-neutralizing activity. *J. Immunol.* 152:231-240.

[75] Hirata, M., Y. Shimomura, M. Yoshida, J. G. Morgan, I. Palings, D. Wilson, M. H. Yen, S. C. Wright, and J. W. Larrick. 1994. Characterization of a rabbit cationic protein (CAP18) with lipopolysaccharide-inhibitory activity. *Infect. Immun.* 62:1421-1426.

[76] Ramanathan, B., H. Wu, C. R. Ross, and F. Blecha. 2004. PR-39, a porcine antimicrobial peptide, inhibits apoptosis: involvement of caspase-3. *Dev. Comp Immunol.* 28:163-169.

[77] Wu, J., C. Parungo, G. Wu, P. M. Kang, R. J. Laham, F. W. Sellke, M. Simons, and J. Li. 2004. PR39 inhibits apoptosis in hypoxic endothelial cells: role of inhibitor apoptosis protein-2. *Circulation* 109:1660-1667.

[78] Malm, J., O. Sorensen, T. Persson, M. Frohm-Nilsson, B. Johansson, A. Bjartell, H. Lilja, M. Stahle-Backdahl, N. Borregaard, and A. Egesten. 2000. The human cationic antimicrobial protein (hCAP-18) is expressed in the epithelium of human epididymis, is present in seminal plasma at high concentrations, and is attached to spermatozoa. *Infect. Immun.* 68:4297-4302.

[79] Bals, R., X. Wang, M. Zasloff, J. M. Wilson. 1998. The peptide antibiotic LL-37/hCAP-18 is expressed in epithelia of the human lung where it has broad antimicrobial activity at the airway surface. *Proc. Natl. Acad. Sci. USA* 95:9541-9546.

[80] Giles, F. J., C. B. Miller, D. D. Hurd, J. R. Wingard, T. R. Fleming, S. T. Sonis, W. Z. Bradford, J. G. Pulliam, E. J. Anaissie, R. A. Beveridge, M. M. Brunvand, and P. J. Martin. 2003. A phase III, randomized, double-blind, placebo-controlled, multinational trial of iseganan for the prevention of oral mucositis in patients receiving stomatotoxic chemotherapy (PROMPT-CT trial). *Leuk. Lymphoma* 44:1165-1172.

In: Progress in Immunology Research
Editor: Barbara A. Veskler, pp. 83-103

ISBN 1-59454-380-1
©2005 Nova Science Publishers, Inc.

Chapter V

Expression of Membrane Peptidases on Cultured Human Keratinocytes

Jelka Gabrilovac[1]*, Barbara Čupić[1], Davorka Breljak[1], Ognjen Kraus[2], Jasminka Jakić-Razumović[3]

[1] Ruđer Bošković Institute, Division of Molecular Medicine, Zagreb, Croatia
[2] Clinical Hospital Sestre milosrdnice, Zagreb, Croatia
[3] Clinical Hospital Center Rebro, Zagreb, Croatia

Abstract

Keratinocytes participate in immune response and inflammation by secreting cytokines and chemokines. Membrane-bound peptidases control local concentrations of signalling peptides and recently have been proposed as an additional mechanism of cell-to-cell interaction and signal transmission. We examined expression of three membrane-bound peptidases: aminopeptidase N (APN; EC 3.4.11.2; CD13), neutral endopeptidase (NEP; EC 3.4.24.11; CD10) and dipeptidyl-peptidase IV (DPPIV; EC 3.4.14.5; CD26) on cultured keratinocytes obtained from normal human skin. Membrane expression of peptidase markers was assessed by means of monoclonal antibodies and FACS and their enzyme activity by measuring hydrolysis of selective substrates. Cultured human skin fibroblasts served as positive controls. CD13 was expressed on $36.0 \pm 19.4\%$ of cultured keratinocytes (n = 25) as compared to fibroblasts which are 99% CD13$^+$ (n = 6). Density of CD13 on keratinocytes was several times lower than on fibroblasts. Expression of CD13 on keratinocytes was associated with significant APN enzyme activity. Inhibitors of APN, actinonin, and substance-P, as well as the APN blocking antibody WM-15, decreased keratinocyte proliferation. The ability of actinonin to inhibit APN enzyme activity was stronger than its ability to suppress keratinocyte proliferation. APN activity was inhibited up to 80% ($IC_{50} = 0.18$ μM) and the cell proliferation maximally for 25%. The growth inhibition by actinonin occurred in the range of high (50 to 100 μM) or low

* Dr. Jelka Gabrilovac; Ruđer Bošković Institute; Division of Molecular Medicine; Laboratory of Experimental Haematology, Immunology and Oncology; Bijenička c. 54; HR-10000 Zagreb; Croatia; Tel.: (385 1) 45 61 082; Fax: (385 1) 456 10 10; E-mail address: gabril@irb.hr

concentrations (0.2 to 1.5 μM). That bimodal dose-response probably reflects inhibitory action on cytosolic aminopeptidases (at high concentrations) and on membrane APN (at low concentrations). CD10 and CD26 membrane expression on cultured keratinocytes was negligible (1.6 ± 1.7%; n=10 and 2.5 ± 2.0%; n=16) as well as their corresponding enzyme activities (NEP and DPPIV). Expression of membrane peptidases was also meassured on freshly isolated epidermal cell samples before culturing. CD13 was found in 5/5 samples tested, but at a lower level (18.9 ± 10.2%) than on cultured keratinocytes. CD10 was practically absent in two epidermal cell samples tested before culturing (2.2 ± 1.6%) but surprisingly a high level of CD26 (61.6 ± 10.5%) was detected. Its expression decreased to negligible values in subsequent passages. In conclusion, functional CD13 with measurable APN activity was found in 36% of cultured, non-stimulated keratinocytes. Its inhibition interfered with keratinocyte proliferation. No significant expression of other two membrane peptidases, CD10/NEP and CD26/DPPIV was found in the same keratinocyte samples. The data suggest a role of APN in regulation of keratinocyte growth.

Key words: human keratinocytes; CD13, aminopeptidase N/APN; CD10, neutral endopeptidase/NEP; CD26, dipeptidyl-peptidase IV; human skin fibroblasts; membrane expression; substance-P; growth regulation

Introduction

Skin represents an important part of the immune system. Langerhans cells serve as professional antigen-presenting cells (APC), which upon activation migrate into proximal lymph node(s). Keratinocytes also participate actively in immune response and inflammation by secreting variety of cytokines (IL-1, IL-10, IL-15, IL-17, IL-18, IL-20), chemokines (RANTES, TARC, MIP-3a), growth factors (GM-CSF) [1], and neuropeptides (NPY, met-enkephalin) [2] in response to antigens, allergens, UV-radiation. By means of cytokines and growth factors, keratinocytes influence growth and differentiation of adjacent cells and promote migration of Langerhans cells into lymph nodes; and by means of chemokines, keratinocytes promote accumulation of T lymphocytes and granulocytes at the sites of immune/inflammatory reactions [3]. Keratinocytes participate in the immune response also by changing their membrane profile and expressing molecules necessary for cell-to-cell interactions, such as CD11a and CD11c [4], CD13 and CD14 [5], CD36 [6; 7], CD54 [4], CD68 [5] and MHC-class II molecules [6]. Non-stimulated keratinocytes either lack those markers or express them at very low density [1, 7].

An important arc in the regulation of immune/inflammatory responses are ectopeptidases. These membrane-bound peptidases cleave bioactive peptides, like chemokines, cytokines and neuropeptides and thus affect their activity and specificity. There are three main membrane-bound peptidases: neutral endopeptidase (NEP; EC 3.4.24.11; CD10), aminopeptidase N (APN; EC 3.4.11.2; CD13) and dipeptidyl-peptidase IV (DPPIV, EC 3.4.14.5; CD26). Their expression has been documented on cells of various tissue origins – epithelial, endothelial, nervous and hematopoietic. Function of these membrane-bound peptidases depends on the cell type and tissue. Membrane peptidases expressed on cells of hematopoietic origin are involved in specific immune response and inflammation [8]. They process peptide antigens protruding from the MHC-II groove [9], interfere with signalling molecules, perform signalling function themselves (after interaction with ligands not yet identified) [10], and cleave some cytokines, chemokines and growth factors [11].

Thus, keratinocytes participate in skin immune responses and inflammation in several ways: (a) secrete soluble pro- or anti-inflammatory mediators, (b) up-regulate membrane receptors for them, and (c) up-regulate accessory and adhesive molecules [1].

Membrane-bound peptidases that control local concentrations of peptide mediators might play important role in immune and inflammatory responses in skin. Fibroblasts constitutively express APN [12], NEP [13, 14] and DPPIV [15, 16]. Endothelial cells of skin vessels constitutively express APN [8] and NEP [14]. Dipeptidyl peptidase IV (DPPIV) has been found on keratinocytes and shown to take part in their growth regulation [17]. In sections of inflamed skin, keratinocytes express NEP [14]. It was suggested that NEP may interfere with inflammatory response in two ways: (a) by degrading proinflammatory neuropeptide substance-P, and (b) by degrading growth factors required for wound healing [14, 18, 19]. Aminopeptidase N (APN) has been found in basal and suprabasal epidermal layers in biopsies of tuberculin-positive skin reactions and in skin samples from patients with various skin diseases such as psoriasis, mycosis fungoides and urticaria [5]. Thus, skin cells of various types and origin express ectopeptidases, either constitutively or upon activation, and these membrane molecules apparently participate in cell-to-cell communication.

We have studied expression of CD13/APN, CD10/NEP and CD26/DPPIV on human keratinocytes prepared from biopsies of normal skin or cultured. Membrane expression of CD10, CD13 and CD26 at the population level was analysed by means of a cell sorter (FACS) and at the single cell level by fluorescent microscope. Functional properties of CD13, CD10 and CD26 were examined by testing their enzyme activity towards selective substrates. The same parameters were analyzed in cultured human skin fibroblasts expressing CD10/NEP, CD13/APN and CD26/DPPIV. The role of APN in regulation of keratinocyte growth was examined by comparing the ability of APN inhibitors, actinonin and substance-P, as well as of APN blocking WM-15 antibody, to inhibit the enzyme activity and the cell growth.

Materials and Methods

Chemicals

FITC-labelled anti-CD13 (clone WM-47; Sigma, cat. no. F-5671), anti-CD10 (clone SS2/36; DAKO, cat. no. F0826), anti-CD26 (clone M-A261; Pharmingen, cat. no. 30814X), and isotype control (mouse IgG$_1$ kappa; Sigma, cat. no. F-6397) were used. Suc-Ala-Ala-Phe-pNA (cat. no. 158000), Gly-Pro-pNA (cat. no. L-1880), substance-P (cat. no. H-1890) and diprotin-A (cat. no. H-3825) were purchased from Bachem. L-Ala-βNA (cat. no. I-2628), bestatin (cat. no. B-8385), actinonin (cat. no. A-6671), DL-thiorphan (cat. no. T-6031), and porcine microsomal leucine aminopeptidase (cat. no. L-5006) were purchased from Sigma. L-Ala-βNA was dissolved in methanol, Suc-L-Ala-Ala-Phe-pNA and thiorphan in dimethyl formamide, actinonin, bestatin and diprotin-A in phosphate buffered salt solution (PBS; pH = 7.4). All stock solutions were stored at -20 °C until use. Blocking (WM-15) and non-blocking

(WM-47) anti-CD13 antibodies, were generous gift from dr. E.J. Favaloro (Institute of Clinical Pathology and Medical Research, New South Wales, Australia).

Cell Cultures

Keratinocytes were obtained from foreskins of boys 3 to 8 years of age who have had phymosis. The skin was not inflamed and the children were free of any local or systemic treatment for at least one month before surgery. Keratinocytes were isolated according to the procedure described by Boyce and Ham [20], and cultured in serum-free medium with selective growth supplements (DKSFM = defined keratinocyte-serum-free medium; Gibco, cat. no. 10744). The seeding density was approximately 20.000 cells per cm^2 and confluence was reached after 5 to 7 days. Adherent cells were detached by means of trypsin-EDTA (Gibco, cat. no. 25200-056) and split 1: 5 for the next passage. Keratinocytes were used between the 1st and 5th passages. Purity of cultured populations used for analysis was above 95% as determined by immunocytochemical examination of cells labelled with antibodies to cytokeratin and vimentin. Human foreskin fibroblasts were grown in RPMI-1640 supplemented with 10% FCS and used between the 1st and 4th passage.

Determination of Membrane CD13, CD10 and CD26

Keratinocytes or fibroblasts (0.5×10^6 per sample) were incubated with FITC-labelled anti-CD13, anti-CD10, anti-CD26, or with the isotype control for 30 min at 4 °C according to recommendation of the provider. After 2 washings the cells were resuspended in PBS with 0.1% BSA and analysed either on a FACScan (Becton-Dickinson), or by means of a fluorescent microscope.

APN Enzyme Activity

APN enzyme activity was determined on cultured keratinocytes or skin fibroblasts cultured in 96-well flat-bottomed microtiter plates (Sarstedt) to approximately 80% confluence, as described before [21]. Briefly, the cells were washed twice in PBS and the selective substrate (L-Ala-βNA) was added in various concentrations. The enzyme reaction was carried out in water bath at 37 °C for 60 min. At the end of incubation, the reaction was terminated by immersing the plates into ice-bath and by the addition of ice cold PBS (200 µl per well). The cells were spun down by centrifugation for 5 min at 200 x g and the supernatants (200 µl per well) were transferred into another 96-well plate. βNA was visualized by the addition of 40 µl/well of 0.2 mM Fast Blue B (FBB) solution in 2 mM sodium acetate with 10% TWEEN; pH = 4.2). Rates of βNA hydrolysis were determined by measuring optical density (OD) at 540 nm on an ELISA-reader. The samples were run in pentaplicates.

NEP Enzyme Activity

NEP enzyme activity on cultured keratinocytes or skin fibroblasts was determined according to the method described by Mari et al. [22]. The cells were grown in 96-well flat-bottomed microtiter plates to approximately 80% confluence, harvested and washed twice in PBS, and then incubated for 30 min at 37 °C with selective substrate (Suc-Ala-Ala-Phe-pNA) in various concentrations. Leucine-aminopeptidase was added (5 U/ml) and the incubation continued for additional 30 min at 37 °C. The reaction was terminated by immersing the plates into ice-bath and adding ice cold PBS (200 µl per well). After centrifugation for 5 min at 200 x g the supernatants (200 µl per well) were transferred into another 96-well plate. Rates of pNA hydrolysis were determined by measuring optical density (OD) of cell-free supernatants at 405 nm on an ELISA-reader. The samples were run in pentaplicates.

DPPIV Enzyme Activity

DPPIV enzyme activity was determined on keratinocytes or skin fibroblasts cultured in 96-well flat-bottomed microtiter plates to approximately 80% confluence. The procedure described previously by us [23] was used. Briefly, the cells were washed twice in PBS and the selective substrate (Gly-Pro-pNA) was added in various concentrations. The enzyme reaction was carried out in a water bath at 37 °C for 60 min. Further steps were as for NEP. The samples were also run in pentaplicates.

Inhibition Studies

In order to test specificity of the enzyme activities, the cells were incubated with selective inhibitors (bestatin, actinonin, thiorphan or diprotin-A) for 30 min at room temperature, except of the inhibition of APN with substance-P or with the blocking antibody WM-15 which was assayed after 60 min. The excess of WM-15 was removed by two washings. Selective substrates were added and the enzyme activities were determined as above. Control cells were incubated in PBS.

WST-Assay

The number of keratinocytes per culture at the end of treatment with APN inhibitors bestatin or substance-P was estimated by using WST reagent (Roche Diagnostic, cat. no. 1644807; 10 µl per well). After 3 h of incubation at 37 °C the absorbance was measured at 450 nm by using an ELISA reader.

Statistical Analysis

The differences between groups were analysed by Student's t-test. The level of significance was set at $p \leq 0.05$.

Results

Membrane Expression of CD10, CD13 and CD26 on Keratinocytes and Fibroblasts

Membrane expression of CD10, CD13 and CD26 on cultured human keratinocytes was examined by means of FACS or fluorescent microscope using anti-CD10, anti-CD13 or anti-CD26 monoclonal antibodies labelled with FITC. CD10 and CD26 were not found on keratinocytes by means of FACS (Fig. 1A, E; Table 1) but were amply expressed on skin fibroblasts (Fig. 1B, F; Table 1). CD13 was found on 36.0% of human keratinocytes (Fig. 1C; Table 1) and practically on all skin fibroblasts (Fig. 1D; Table 1). Average intensity of CD13 expression on keratinocytes was low as compared to strong expression on fibroblasts (FITC-CD13/FITC-isotype control: 2.8 to 11.2 for keratinocytes *vs* 43 to 110 for fibroblasts). CD13 distribution on the *per* cell level was examined by fluorescent microscope. Moderate, homogenous, ring-like CD13 distribution was observed on keratinocytes (Fig. 2A), in contrast to strong, predominantly patchy or polar CD13 expression on fibroblasts (Fig. 2B).

Possible effect of keratinocyte cultivation on the expression of peptidase membrane markers was examined by comparing freshly prepared keratinocytes from skin biopsies with the cultured ones. CD13 expression was comparable (Fig. 3B; Table 2) and CD10 was absent in both cell types (Fig. 3A; Table 2). A significant expression of CD26 was found on freshly prepared epidermal skin cells (Fig. 3C; Table 2) but decreased to negligible levels in culture (Fig. 4).

NEP, DPPIV and APN Enzyme Activity of Cultured Human Skin Cells

The observed pattern of CD10, CD13 and CD26 expression on keratinocytes and fibroblasts was compared with the corresponding enzyme activity of those membrane markers. Keratinocytes or fibroblasts were cultured in 96-well flat-bottomed plates and their NEP or APN activities were tested when cells reached approximately 80% confluence. Keratinocytes had negligible NEP and DPPIV activities (data not shown). Fibroblasts, however, exerted significant NEP and DPPIV activities which could be inhibited by respective inhibitors thiorphan or diprotin-A (Fig. 5A, B). APN activity towards L-Ala-βNA as the substrate was obtained with keratinocytes (Fig. 6A), but compared to fibroblasts (Fig. 6B) it was significantly lower. APN of fibroblasts and keratinocytes could be blocked by actinonin to similar levels (Fig. 6A, B), suggesting that the same enzyme was detected on both types of cells.

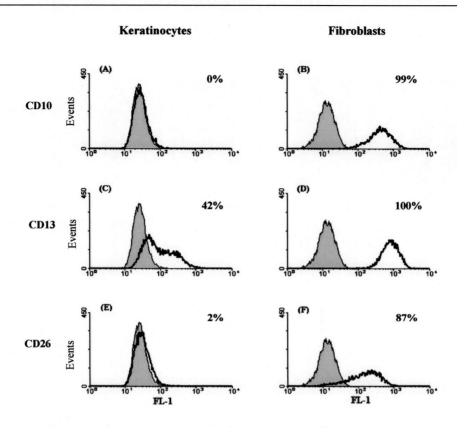

Figure 1. Membrane expression of CD10, CD13 and CD26 on cultured human keratinocytes and skin fibroblasts. Membrane expression of CD10 (A), CD13 (C) and CD26 (E) on cultured human keratinocytes was compared to cultured skin fibroblasts (B, D, F). Shaded histograms represent isotype controls, bold lines show distribution of CD10-, CD13-, or CD26-labelled cells. Data of one out of 5 experiments with similar results were presented.

Table 1. Membrane expression of CD10, CD13 and CD26 on cultured human keratinocytes and skin fibroblasts

Cell type	Membrane markers		
	CD10	CD13	CD26
Keratinocytes	1.6 ± 1.7 (n = 10)	36.0 ± 19.4 (n = 25)	2.5 ± 2.0 (n = 16)
Fibroblasts	93.9 ± 5.9 (n = 6)	98.9 ± 7.7 (n = 6)	63.7 ± 19.6 (n = 6)

Data are presented as percentage of positive cells (mean values ± s.d. and sample sizes in brackets).

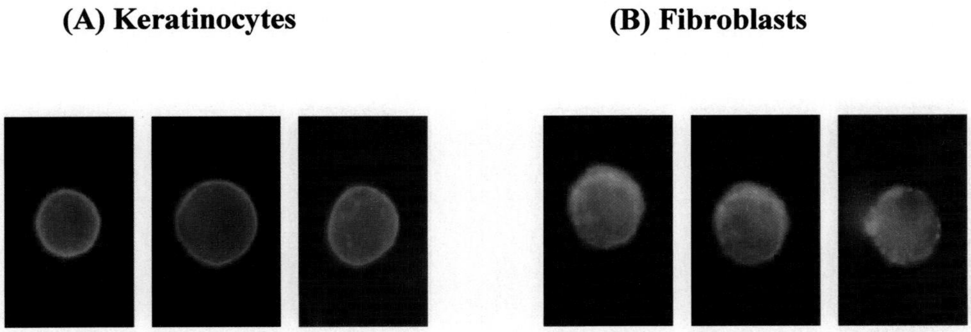

(A) Keratinocytes   **(B) Fibroblasts**

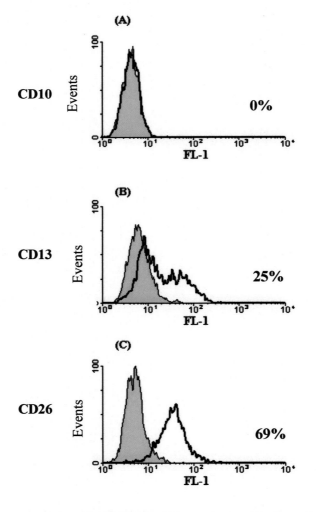

Figure 2. Distribution of CD13 on cultured keratinocytes and skin fibroblasts. Note moderate, homogenous, ring-like CD13 distribution on keratinocytes as compared with strong, patchy or polar CD13 expression on fibroblasts (from reference [21]).

Figure 3. Membrane expression of CD10 CD13 and CD26 on freshly prepared epidermal skin cells. Shaded histograms represent isotype control, bold lines show distribution of CD10-, CD13- and CD26-labelled cells.

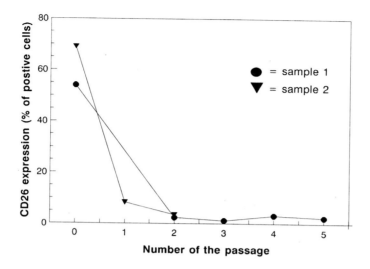

Figure 4. Membrane expression of CD26 on epidermal cells before and during culturing. CD26 expression was followed on two samples of epidermal cells before culturing and in subsequent passages. The cells were labelled with FITC-labelled anti-CD26 antibody and analysed by means of a cell sorter (FACS). Data are expressed as percentages of positive cells.

Table 2. Membrane expression of CD10, CD13
and CD26 on freshly prepared epidermal cells

Membrane marker		
CD10	CD13	CD26
2.2 ± 1.6	18.9 ± 10.2	61.6 ± 10.5
(n = 2)	(n = 5)	(n = 2)

Data are presented as percentages of positive cells (mean values ± s.d. and sample sizes in brackets).

The Effect of APN Inhibitors on Keratinocyte Proliferation

If the keratinocyte APN was involved in growth regulation, APN inhibition would result in decreased cell proliferation. In order to examine that idea, we determined the capacity of substance-P, an endogenous APN inhibitor [24], as well as of two potent exogenous APN inhibitors, actinonin and bestatin, to suppress keratinocyte APN and keratinocyte growth. Substance-P decreased keratinocyte APN activity by 50% (Fig. 7A), with $IC_{50} = 30$ μM. Bestatin and actinonin were more potent inhibitors. Keratinocyte APN activity was decreased by 90% with bestatin ($IC_{50} = 0.75$ μM) and by 80% with actinonin ($IC_{50} = 0.18$ μM) (Fig. 7B, C). Although actinonin appeared to be the most potent inhibitor of keratinocyte APN, its ability to inhibit keratinocyte proliferation was much weaker, up to 25%. The inhibition was detected in the range of high (50 and 100 μM; Fig. 7D) and low concentrations (0.2 to 1.5 μM; Fig. 7D). That biphasic mode suggests that actinonin affected other target enzymes in addition to APN, possibly cytosolic aminopeptidases. In order to distinguish between the membrane-bound (APN) and cytosolic aminopeptidase activity on cell proliferation, we used two extracellular inhibitors, substance-P and an APN blocking antibody WM-15. Their abilities to decrease keratinocyte proliferation were compared to their inhibition of APN enzyme activity. Substance-P diminished keratinocyte proliferation for 50% (Fig. 8B) and APN activity for 47% (Fig. 8A) and WM-15 did so for 21% (Fig. 8D) and 24% (Fig. 8C) respectively. Thus, growth inhibition with both substances paralleled the APN inhibition, suggesting a causal relationship.

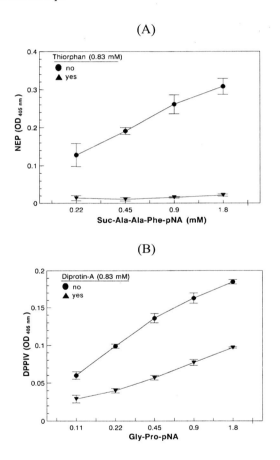

Figure 5. NEP and DPPIV enzyme activity of human skin fibroblasts. Human skin fibroblasts were seeded in 96-well plates (20.000 cells per well). After culturing the cells to approximately 80% confluence, enzyme activities of NEP (A) and DPPIV (B) were measured. Cells were washed and preincubated for 30 min at room temperature with a selective inhibitor (thiorphan or diprotin-A, respectively). Control cultures were incubated in PBS. Selective substrates (Suc-L-Ala-Ala-Phe-pNA or Gly-Pro-pNA) were added in various concentrations and the incubation continued for 60 min. For NEP determination, microsomal APN was added simultaneously with the substrate. The data are expressed as means ± s.d. of 5 parallel samples. One of two experiments with similar results is presented.

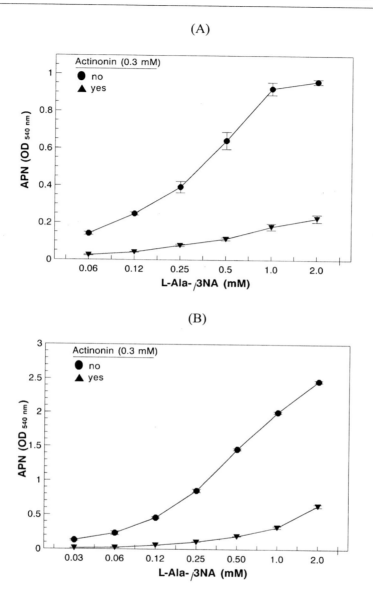

Figure 6. The abilities of cultured human keratinocytes and skin fibroblasts to hydrolyse L-Ala-βNA as a measure of APN activity. Keratinocytes (A) or fibroblasts (B) were seeded in 96-well plates (20.000 cells per well). After culturing the cells to approximately 80% confluence, APN enzyme activity was determined. Cells were washed and preincubated for 30 min at room temperature with a selective inhibitor actinonin (or in PBS in control cultures). Selective substrate L-Ala-βNA was added in various concentrations and the incubation continued for 60 min. The data are expressed as means ± s.d. of 5 parallel samples.

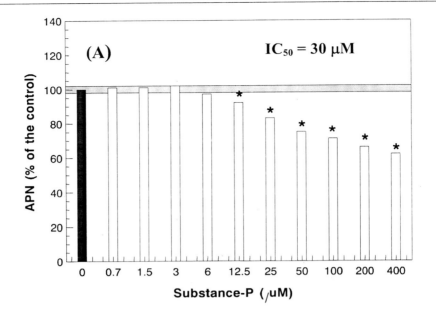

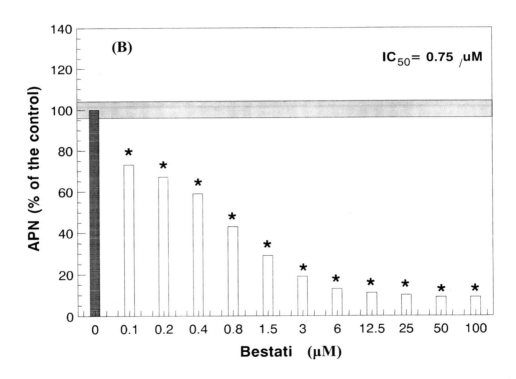

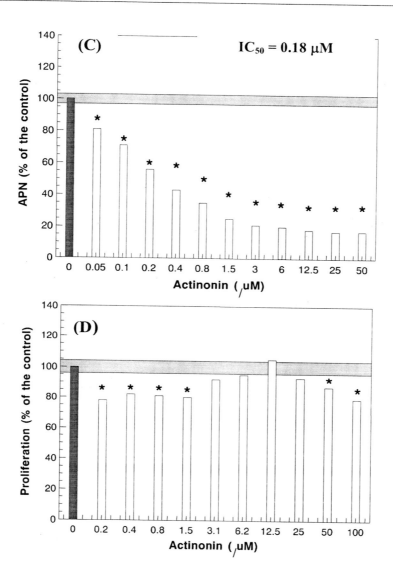

Figure 7. The ability of APN inhibitors to block APN enzyme activity and diminish keratinocyte proliferation. *APN inhibition (A, B, C)*: Keratinocytes (40.000 per sample) were preincubated at room temperature with the inhibitors at indicated concentrations for 30 min (actinonin and bestatin) or 60 min (substance-P). APN activity was determined by using L-Ala-βNA (0.3 mM) as a substrate. The results are presented as percentage of the respective control. *Growth inhibition (D)*: Keratinocytes were seeded in 96-well plates, 20.000 cells per well. Upon semiconfluence, actinonin was added at indicated concentrations and the incubation continued for 24 hrs. Proliferation was estimated by the WST-assay. Results are presented as percentage of the respective control. Data of two representative experiments are presented. Each experiment was run in 5 parallel samples. Asterisks (*) indicate significant differences from the respective control.

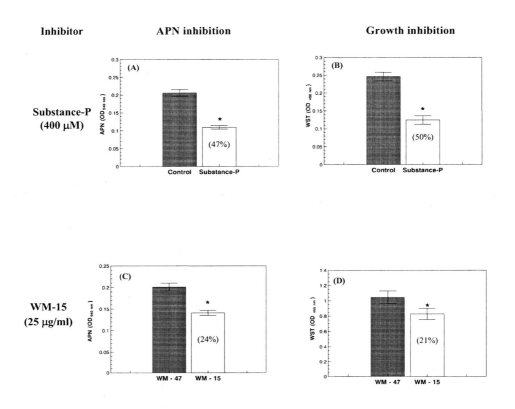

Figure 8. The abilities of substance-P and the APN blocking antibody WM-15 to inhibit APN enzyme activity (left) and keratinocyte growth (right). *APN inhibition (left)*: Keratinocytes (50.000 per sample) were preincubated at room temperature with the inhibitors at indicated concentrations for 60 min. Unbound WM-15 was removed by two washings. Peptide inhibitor, substance-P, was not removed. APN activity was determined by using L-Ala-βNA (0.2 mM) as a substrate. The results are presented as means ± s.d. (n = 5 parallel samples). *Growth inhibition (right)*: Keratinocytes were seeded in 96-well plates (15.000 or 20.000 cells per well). Upon semiconfluence, APN inhibitors were added at indicated concentrations and the incubation continued for 24 hrs. Proliferation was measured by the WST-assay. The results are presented as means ± s.d. (n = 5 parallel samples) and asterisks (*) indicate significant differences from the respective controls. Numbers in brackets show percentages of inhibition in comparison to the respective control.

Conclusion

This study shows that approximately one third (36%) of cultured non-stimulated human keratinocytes isolated from normal skin express membrane marker CD13 and that the marker expression is associated with corresponding APN enzyme activity. CD13 expression and APN enzyme activity of the keratinocytes are both lower than in cultured human skin fibroblasts. APN inhibitors - actinonin and substance-P, as well as an APN-blocking antibody (WM-15), decreased keratinocyte growth, suggesting a physiological role of CD13/APN in the regulation of keratinocyte growth. In contrast to CD13/APN, neither CD10/NEP nor CD26/DPPIV could be detected on cultured human keratinocytes, whereas both CD10 and CD26 with the respective NEP and DPPIV enzyme activity were detected on human skin fibroblasts. CD13 was also found on freshly prepared epidermal cells before culturing, which suggests its constitutive expression under physiological conditions. No CD10 was found on epidermal cells before culturing, whereas CD26 was amply expressed but disappeared during cultivation. Thus, culturing conditions differently affect the expression of membrane peptidases.

Our finding of CD13/APN on non-stimulated cultured keratinocytes is in contrast to the data of Hunyadi et al. [5] who did not find CD13 in normal epidermis and of Moehrle et al. [25] who did not detect APN enzyme activity. Discrepancy might be due to different sensitivities of the methods. Hunyadi et al. [5] and Moehrle et al. [25] applied immunohistochemistry on tissue sections whereas we analysed keratinocyte suspensions by means of FACS. However, Hunyadi et al. [5] did find CD13 on keratinocytes in skin samples of patients with various inflammatory skin diseases. APN activity was also found in epithelial skin tumours [25]. Strong CD13 expression associated with APN activity was described by Menrad et al. [26] and Fujii et al. [27] on melanoma cells. That feature was proposed as an important property associated with tumour spread [26, 27]. Cells of human keratinocyte line HaCaT were found by Steinstrasser et al. [28] to exert enzyme activities corresponding to aminopeptidase B, aminopeptidase N and Leu-aminopeptidase.

Our study has also shown membrane expression of functionally active CD13 on freshly harvested, non-stimulated keratinocytes from normal human skin. Kinetic parameters of the APN expressed on cultured keratinocyte and on skin fibroblasts were comparable (K_m = 0.307 vs 0.766 mM, respectively) [21]. Those K_m values are in good agreement with those described by Stefanović et al. [29], Amoscato et al. [30] and Xu et al. [24], respectively, for APN on human dermal fibroblasts (0.630 mM), rat lymphocytes (0.490 mM) and human myelo-monocytic cell line HL-60 (0.402 mM).

Sorrell et al. [31] have recently described APN expression on normal skin cells located at the mesenchymal-epithelial interface and proposed that APN might play a role in epidermal-dermal communication. In their view, APN positive cells were fibroblasts, since antibody (DF-5) used for APN detection had been raised against fibroblast membranes. If so, freshly prepared epidermal cells expressing APN in our study would represent cells of the mesenchymal-epithelial interface, but our data indicate that APN positive cells in normal skin samples are of epithelial and not of mesenchymal origin. APN was detected in selective cultures of keratinocytes.

In order to see whether keratinocyte APN participated in regulation of keratinocyte growth, we compared the ability of several APN inhibitors to abrogate APN enzyme activity and to affect cell proliferation. APN inhibitors bestatin, actinonin and substance-P blocked APN activity in the following decreasing order: actinonin at $IC_{50} = 0.18$ µM, bestatin at $IC_{50} = 0.75$ µM and substance-P at $IC_{50} = 30$ µM. Similar IC_{50} values for APN inhibition by actinonin and bestatin were found by Grujić and Renko [32] and by Xu et al. [24] who used monocytic or myelo-monocytic cell lines U937 and HL-60, respectively. However, the potency (IC_{50}) of substance-P to inhibit keratinocyte APN was about two orders of magnitude below that reported by Xu et al. [24] for HL-60 cells (30 µM vs 0.44 µM). The discrepancy may be ascribed to the expression of neurokinin receptors on keratinocytes [33] which may bind substance-P and thus decrease its availability for APN inhibition. Since actinonin was, in our hands, the most effective inhibitor of keratinocyte APN, it was further used in experiments addressing keratinocyte proliferation. Actinonin inhibited it moderately (up to 25%) and did so in the range of high and low concentrations. Grujić and Renko [32] report that actinonin/bestatin in high concentrations enter the cells, inhibit cytosolic aminopeptidases and diminish cell proliferation in that manner. Inhibition of cytosolic aminopeptidases by bestatin was shown to trigger apoptosis [34]. In our hands, actinonin decreased keratinocyte proliferation even at low concentrations (0.2 to 1.5 µM) that are probably below the level required for accessing the cell interior. Nevertheless, those concentrations were still sufficient to inhibit APN and to cause a moderate but significant decrease of cell proliferation (up to 25%). That decrease could be attributed to the inhibition of membrane APN.

Substance-P, a potent endogenous APN inhibitor [24], also decreased keratinocyte proliferation. Substance-P is an undecapeptide and probably does not enter cells by diffusion. Therefore it could be considered an inhibitor of APN expressed at the cell membrane. Substance-P inhibited APN activity and keratinocyte proliferation to similar degrees (about 50%).

APN blocking antibody WM-15 has been proposed as a convenient tool for investigations of membrane APN as it acts exclusively at the outer side of cell membrane [35]. Being an immunoglobulin, WM-15 is much larger in comparison with actinonin and bestatin, which are small dipeptides, and therefore may have limited access to the enzymatically active site of membrane APN. That may explain its relatively poor inhibition of the enzyme activity of APN (24%). The inhibition of keratinocyte growth was comparable (for 21%).

In short, inhibition studies have shown that low concentrations of actinonin, substance-P and the APN blocking antibody WM-15 all decrease keratinocyte proliferation. Keratinocytes express CD13/APN and APN seems to be involved in the control of proliferation.

The enzymatically active site of APN seems be needed for described effects on cell proliferation. Enkephalins, natural substrates of APN, regulate growth and differentiation of keratinocytes in an autocrine manner [36]. In addition, keratinocyte APN may regulate concentrations of other peptide mediators of inflammation and immune response. For example, degrade chemokines fMLP [37] and MCP-1 [38]. Synthetic peptides corresponding to N-terminal sequence of IL-6 [39] are also susceptible to degradation by APN, whereas the susceptibility of IL-8 to APN has been questioned [40, 41]. By modifying local concentrations of regulatory peptides, keratinocyte APN may affect immune responses and inflammation in the skin. Its regulatory role could be affected by interaction with endogenous inhibitors such as substance-P and bradikinin [24].

Keratinocyte APN might also serve as a molecule for interkeratinocyte communication, as reported for melanoma cells [26], and for communication of keratinocytes with other cell types in the skin. It may function as an adhesive molecule and/or chemoatractant involved in recruitment of T-lymphocytes at the sites of inflammation, as reported for APN of synovial fibroblasts [42] and APN on alveolar macrophages of patients with pulmonary sarcoidosis [43]. Finally, keratinocyte APN might be a viral receptor in skin infections, as reported for epithelial cells of the gut in gastrointestinal viral infections [44].

Cultured, non-stimulated keratinocytes did not express membrane CD10 and did not exert NEP enzyme activity as judged by substrate and inhibitor selectivity. Absence of CD10/NEP on keratinocytes is in contrast to the data of Olerud et al. [14] who found CD10 on cells of the epidermal basal layer in sections of normal human skin and detected mRNA for NEP in skin homogenates. The reason for this discrepancy is not clear. It might be due to different locations from which skin samples were taken. Olerud et al. [14] studied skin samples from the leg, arm and sacrum, and we used foreskin of young boys. Similar discrepancy between *in vivo* and *in vitro* expression of Toll-like receptor 4 (TLR4) on keratinocytes has recently been reported by Kawai et al. [45]. It was explained as a consequence of continuous contact of keratinocytes with resident skin flora *in vivo*. Another explanation for the discrepancy between our findings and the data of Olerud et al. [14] might be the presence of other cells expressing high levels of CD10, like fibroblasts and endothelial cells, in their skin homogenates. Purity of keratinocyte suspensions in our study was above 95% as checked by microscopic examination of cells stained with antibodies to cytokeratin and vimentin (data not shown).

Data on keratinocyte expression of CD26 are contradictory. Novelli et al. [46] described only sparse foci of $CD26^+$ cells in sections of normal skin, whereas Reinhold et al. [17] found significant expression of CD26 associated with DPPIV enzyme activity in cultured human keratinocytes. Inhibition of CD26/DPPIV resulted in decreased keratinocyte proliferation. A growth regulatory role was therefore attributed to CD26/DPPIV [17]. We have found significant CD26 expression in freshly prepared epidermal cell samples (cca 60% of the cells), but the expression faded out in culture. The meaning of that finding is not clear at present. No CD26 was found on a human keratinocyte cell line, HaCaT (data not shown).

In conclusion, this study shows that about one third of cultured, non-stimulated keratinocytes isolated from normal human skin express CD13, and that the CD13 expression was associated with APN enzyme activity. APN inhibition decreased keratinocyte growth. CD13 was also present in native keratinocytes obtained from skin biopsies. The data suggest a role of CD13/APN in regulation of keratinocyte growth. CD10/NEP and CD26/DPPIV were not detected in the same keratinocyte samples.

Acknowledgements

We thank Professor Milivoj Boranić, MD, PhD, for critical reading of the manuscript and Marija Zekušić, BSc, for the help in preparing primary cultures of epidermal skin cells. Excellent technical assistance of Mrs. Margareta Cvetkovski is greatly appreciated. This work was financially supported by project No. 0098094 of the Croatian Ministry for Science, Education and Sport.

References

[1] Steinhoff, M; Brzoska, T; Luger, TA. Keratinocytes in epidermal immune responses. Curr Opin Allergy Clin Immunol (2001) 1:469-476.

[2] Slominski, A; Wortsman, J. Neuroendocrinology of the skin. Endocr Rev (2000) 21:457-487.

[3] Uchi, H; Terao, H; Koga, T; Furue, M. Cytokines and chemokines in the epidermis. J Dermatol Sci (2000) 24 (Supll 1):S29-S38.

[4] Hunyadi, J; Simon, MJr; Dobozy, A. Immune-associated surface markers of human keratinocytes. Immunol Lett (1992) 31:209-216.

[5] Hunyadi, J; Simon, MJr; Kenderessy, AS; Dobozy, A. Expression of monocyte/macrophage markers (CD13, CD14, CD68) on human keratinocytes in healthy and diseased skin. J Dermatol (1993) 20:341-345.

[6] Hunyadi, J; Simon, MJr. Expression of OKM5 antigen on human keratinocytes in vitro upon stimulation with gamma-interferon. Acta Derm Venereol (1986) 66:527-530.

[7] Castagnoli, C; Stella, M; Menegatti, E; Trombotto, C; Calcagni, M; Magliacani, G, Alasia, ST; Alessio, M. CD36 is one of the immunological markers expressed by keratinocytes in active hypertrophic scars. Annals of Burns and Fire Disasters VIII (1995) 4:1-7.

[8] Lendeckel, U; Kähne, T; Riemann, D; Neubert, K; Arndt, M; Reinhold, D. Review: The role of membrane peptidases in immune functions. Adv Exp Med Biol (2000) 477:1-24.

[9] Larsen, SL; Pedersen, LO; Buus, S; Stryhn, A. T cell responses affected by aminopeptidase N (CD13)-mediated trimming of major histocompatibility complex class II-bound peptides. J Exp Med (1996) 184:183-189.

[10] Santos, AN; Langer, J; Herrmann, M; Riemann, D. Aminopeptidase N/CD13 is directly linked to signal transduction pathways in monocytes. Cell Immunol (2000) 201:22-32.

[11] Riemann, D; Kehlen, A; Langner, J. CD13 – not just a marker in leukemia typing. Immunol Today (1999) 20:83-88.

[12] Piela-Smith, TH; Korn, JH. Aminopeptidase N: A constitutive cell-surface protein on human dermal fibroblasts. Cell Immunol (1995) 162:42-48.

[13] Lorkowski, G; Zijderhand-Bleekemolen, JE; Erdos, EG; von Figura, K; Hasilik, A. Neutral endopeptidase-24.11 (enkephalinase). Biosynthesis and localization in human fibroblasts. Biochem J (1987) 248:345-350.

[14] Olerud, JE; Usui, ML; Seckin, D; Chiu, DS; Haycox, CL; Song I-S; Ansel, JC; Bunnett, NW. Neutral endopeptidase expression and distribution in human skin and wounds. J Invest Dermatol (1999) 112:873-881.

[15] Bauvois, B. A collagen-binding glycoprotein on the surface of mouse fibroblasts is identified as dipeptidyl peptidase IV. Biochem J (1988) 252:723-731.

[16] Atherton, AJ; Monaghan, P; Warburton, MJ; Robertson, D; Kenny, AJ; Gusterson, BA. Dipeptidyl peptidase IV expression identifies a functional sub-population of breast fibroblasts. Int J Cancer (1992) 50:15-19.

[17] Reinhold, D; Vetter, RW; Mnich, K; Buchling, F; Lendeckel, U; Born, I; Faust, J; Neubert, K; Gollnick, H; Ansorge, S. Dipeptidyl peptidase IV (DP IV, CD26) is involved in regulation of DNA synthesis in human keratinocytes. FEBS Lett (1998) 428:100-104.

[18] Antezana, MA; Sullivan, SR; Usui, ML; Gibran, NS; Spenny, ML; Larsen, JA; Ansel, JC; Bunnett, NW; Olerud, JE. Neutral endopeptidase activity is increased in the skin of subjects with diabetic ulcers. J Invest Dermatol (2002) 119:1400-1404.

[19] Luger, TA. Neuromediators – a crutial component of the skin immune system. J Dermatol Sci (2002) 30:87-93.

[20] Boyce, ST; Ham, RG. Calcium-regulated differentiation of normal human epidermal keratinocytes in chemically defined clonal culture and serum-free serial culture. J Invest Dermatol (1983) 81 (1 Suppl):33s-40s.

[21] Gabrilovac, J; Čupić, B; Breljak, D; Zekušić, M; Boranić, M. Expression of CD13/aminopeptidase N and CD10/neutral endopeptidase on cultured human keratinocytes. Immunol Lett (2004) 91:39-47.

[22] Mari, B; Checler, F; Ponzio, G; Peyron, JF; Manie, S; Farahifar, D; Rossi, B; Auberger, P. Jurkat T cells express a functional neutral endopeptidase activity (CALLA) involved in T cell activation. EMBO J (1992) 11:3875-3885.

[23] Gabrilovac, J; Abramić, M; Užarević, B; Andreis, A; Poljak, Lj. Dipeptidyl peptidase IV (DPPIV) enzyme activity on immature T-cell line R1.1 is down-regulated by dynorphin-A(1-17) as a non-substrate inhibitor. Life Sci (2003) 73:151-166.

[24] Xu, Y; Wellner, D; Scheinberg, DA. Substance-P and bradykinin are natural inhibitors of CD13 aminopeptidase N. Biochem Biophys Res Commun (1995) 208:664-674.

[25] Moehrle, MC; Schlagenhauff, BE; Klessen, C; Rassner, G. Aminopeptidase M and dipeptidyl peptidase IV activity in epithelial skin tumors: a histochemical study. J Cutan Pathol (1995) 22:241-247.

[26] Menrad, A; Speicher, D; Wacker, J; Herlyn, M. Biochemical and functional characterization of aminopeptidase N expressed by human melanoma cells. Cancer Res (1993) 53:1450-1455.

[27] Fujii, H; Nakajima, M; Saiki, I; Yoneda, J; Azuma, I; Tsuruo, T. Human melanoma invasion and metastasis enhancement by high expression of aminopeptidase N/CD13. Clin Exp Metastasis (1995) 13:337-344.

[28] Steinstrasser, I; Koopmann, K; Merkle, HP. Epidermal aminopeptidase activity and metabolism as observed in an organized HaCaT cell sheet model. J Pharm Sci (1997) 86:378-383.

[29] Stefanović, V; Vlahović, P; Mitić-Zlatković, M. Receptor-mediated induction of human dermal fibroblast ectoaminopetidase N by glucocorticoids. Cell Mol Life Sci (1998) 54:614-617.

[30] Amoscato, AA; Spiess, RR; Brumfield, AM; Herberman, RB; Chambers, WH. Surface aminopeptidase activity of rat natural killer cells. Biochemicals and biological properties. 1. Biochim Biophys Acta - Mol Cell Research (1994) 1221:221-232.

[31] Sorrell, JM; Baber, MA; Brinon, L; Carrino, DA; Seavolt, M; Asselineau, D; Caplan, AI. Production of a monoclonal antibody, DF-5, that identifies cells at the epithelial-mesenchymal interface in normal human skin. APN/CD13 is an epithelial-mesenchymal marker in skin. Exp Dermatol (2003) 12:315-323.

[32] Grujić, M; Renko, M. Aminopeptidase inhibitors bestatin and actinonin inhibit cell proliferation of myeloma cells predominantly by intracellular interactions. Cancer Lett (2002) 182:113-119.

[33] Staniek, V; Misery, L. Substance-P and human skin. Pathol Biol (Paris) (1996) 44:860-866.

[34] Sekine, K; Fujii, H; Abe, F. Induction of apoptosis by bestatin (ubenimex) in human leukemic cell lines. Leukemia (1999) 13:729-734.

[35] vanderVelden, VHJ; Naber, BAE; vanderSpoel, P; Hoogsteden HC; Versnel MA. Cytokines and glucocorticoids modulate human bronchial epithelial cell peptidases. Cytokine (1998) 10:55-65.

[36] Nissen, JB; Kragballe, K. Enkephalins modulate differentiation of normal human keratinocytes in vitro. Exp Dermatol (1997) 6:222-229.

[37] Shipp, MA; Look, AT. Hematopoietic differentiation antigens that are membrane-associated enzymes: Cutting is the key! Blood (1993) 82:1052-1070.

[38] Weber, M; Uguccioni, M; Baggiolini, M; ClarkLewis, I; Dahinden, CA. Deletion of the NH2-terminal residue converts monocyte chemotactic protein 1 from an activator of basophil mediator release to an eosinophil chemoattractant. J Exp Med (1996) 183:681-685.

[39] Hoffmann, T; Faust, J; Neubert, K; Ansorge, S. Dipeptidyl peptidase IV (CD26) and aminopeptidase N (CD13) catalyzed hydrolysis of cytokines and peptides with N-terminal cytokine sequences. FEBS Lett (1993) 336:61-64.

[40] Kanayama, N; Kajiwara, Y; Goto, J; el Maradny, EE; Maehara, K; Andou, K; Terao, T. Inactivation of interleukin-8 by aminopeptidase N (CD13). J Leukoc Biol (1995) 57:129-134.

[41] Kehlen, A; Egbert, I; Thiele, K; Fischer, K; Riemann, D; Langner, J. Increased expression of interleukin-8 and aminopeptidase N by cell-cell contact: interleukin-8 is resistant to degradation by aminopeptidase N/CD13. Eur Cytokine Netw (2001) 12:316-324.

[42] Riemann, D; Kehlen, A; Thiele, K; Löhn, M; Langner, J. Induction of aminopeptidase N/CD13 on human lymphocytes after adhesion to fibroblast-like synoviocytes, endothelial cells, epithelial cells, and monocytes/machrophages. J Immunol (1997) 158:3425-3432.

[43] Tani, K; Ogushi, F; Huang, L; Kawano, T; Tada, H; Hariguchi, N; Sone, S. CD13/aminopeptidase N, a novel chemoattractant for T lymphocytes in pulmonary sarcoidosis. Am J Respir Crit Care Med (2000) 161:1636-1642.

[44] Delmas, B; Gelfi, J; L'Haridon, R; Vogel, LK; Sjostrom, H; Noren, O; Laude, H. Aminopeptidase N is a major receptor for the entero-pathogenic coronavirus TGEV. Nature (1992) 357:417-420.

[45] Kawai, K; Shimura, H; Minagawa, M; Ito, A; Tomiyama, K; Ito, M. Expression of functional Toll-like receptor 2 on human epidermal keratinocytes. J Dermatol Sci (2002) 30:185-194.

[46] Novelli, M; Savoia, P; Fierro, MT; Verrone, A; Quaglino, P; Bernengo, MG. Keratinocytes express dipeptidyl-peptidase IV (CD26) in benign and malignant skin diseases. Br J Dermatol (1996) 134:1052-1056.

In: Progress in Immunology Research
Editor: Barbara A. Veskler, pp. 105-121

ISBN 1-59454-380-1
©2005 Nova Science Publishers, Inc.

Chapter VI

Secretory Lysosomes of Hematopoietic Cells as Vehicles for Therapeutically Active Proteins

Markus Hansson[*] *and Inge Olsson*

Department of Hematology, C14, BMC, SE-221 84 Lund, Sweden

Abstract

We suggest a concept for local deposition of therapeutically active proteins using secretory lysosomes of hematopoietic cells as delivery vehicles. During hematopoietic differentiation secretory proteins are stored in granules such as secretory lysosomes for release upon activation e.g. in areas of inflammation and malignancy. The protein of interest is expressed in progenitor cells, granule targeted, and deposited into a tumor or site of inflammation, to which the mature cells are migrating. The concept is supported by gene transfer and granule loading of soluble TNF-α receptor (sTNFR1) in hematopoietic cell lines and progenitor cells. Endoplasmic reticulum export was made easier by the addition of a transmembrane domain, and secretory lysosome targeting was achieved by incorporating a tyrosine sorting signal. The sTNFR1 was cleaved off from the transmembrane domain and finally regulated sTNFR1-secretion was triggered by activation. Results from *in vivo* investigations will determine the suitability of this local protein delivery principle in inflammation and tumors.

Introduction

Secretory lysosomes are lysosome-related organelles, combing storage of cell specific proteins together with lysosome hydrolases, present in most hematopoietic cells (Dell'Angelica, Mullins et al. 2000; Blott and Griffiths 2002). If an exogenous protein were

[*] Department of Hematology, C14, BMC, SE-221 84 Lund, Sweden. Telephone: +46-46-2220737; Telefax: +46-46-184493; E-mail: Markus.Hansson@med.lu.se

sorted into secretory lysosomes of hematopoietic cells and remained stable during cell maturation, the exogenous protein could be deposited at a site of inflammation or tumor by degranulation after cell migration. Thereby using the secretory lysosomes as vehicles for targeting sites of inflammation or tumors with exogenous protein. The cellular protein sorting mechanisms allowed sorting of exogenous proteins, expressed by cDNA transfection (Bulow, Gullberg et al. 2000). Moreover, a mixture of proteases and other proteins could co-exist in secretory lysosomes without degradation. Accordingly, we have applied sorting signals for secretory lysosomes in order to target a protein of interest in hematopoietic precursor cells. The goal was to accomplish local release of this protein e.g. an anti-cytokine with a therapeutic potential in dampening an inflammatory reaction or cytokines, stimulating an immune response in tumors. Thus, a local delivery and local action of cytokines and soluble cytokine receptors would be combined with low systemic effects.

Leukocytes such as, neutrophils, mast cells, natural killer (NK)-cells, cytotoxic T-lymphocytes (CTLs) and other hematopoietic cells have a critical role in host defense. Neutrophils eliminate microorganisms by phagocytosis and phagosomal killing (Boxer and Dale 2002). Mast cells and eosinophils have a primary role in parasitic defense but are more known for their role in allergic disorders with release of stored histamine and serotonin after IgE stimulation (Benoist and Mathis 2002). NK-cells and CTLs destroy virus-infected and malignant cells by the action of cytolytic proteins, including perforin and the serine proteases, released from cytotoxic granules by regulated secretion (Dell'Angelica, Mullins et al. 2000; Blott and Griffiths 2002). Platelets contribute to blood clotting by releasing serotonin and P-selectin from dense granules (McNicol and Israels 1999). These are all examples of cells using secretory lysosomes for storage of bioactive proteins, some produced only during certain windows of differentiation, others produced in mature cells. The content of secretory lysosomes are ready to be released by degraulation upon certain cell-specific exogenous stimulation (Borregaard and Cowland 1997; Gullberg, Bengtsson et al. 1999). Secretory lysosomes are not only found in hematopoietic cell but also in other cells such as neural crest derived melanocytes that store pigment proteins called melanins in secretory lysosomes (Setaluri 2000).

Neutrophils have many subsets of storage granules. Secretory lysosomes, also called azurophil granules are the first produced during neutrophil differentiation. They occur during the promyelocyte stage of maturation and contain anti-microbial proteins and cationic serine proteases as well as lysosome hydrolases (Borregaard and Cowland 1997; Gullberg, Bengtsson et al. 1999). Secondary or specific neutrophil granules are manufactured after the azurophil granules during the myelocyte stage of maturation. These granules are equipped with anti-microbial proteins, matrix metalloproteinases and NADPH-oxidase. (Borregaard and Cowland 1997). Later during differentiation, neutrophils also produce gelatinase granules, mostly thought to be released during migration and secretory granules containing alkaline phosphatase and plasma proteins.

Secretory Lysosomes

As the name implies, the secretory lysosome has two main functions, regulated secretion and lysosomal activity (Blott and Griffiths 2002). The morphology and biochemical content differs from that of the conventional lysosome. Secretory lysosomes have an external limiting membrane and a matrix compartment but are more diverse than conventional lysosomes. Some have dense cores of tightly packed proteins and internal membrane-bound vesicles, other have a more pronounced multi-laminar structure depending on cell type (Blott and Griffiths 2002). Both conventional and secretory lysosomes share the degradative machinery with acid hydrolases and members of the LAMP and heat shock protein 70 families. In addition, secretory lysosomes contain, depending on cell type, specific secretory components. The internal vesicles are formed by inward budding towards the compartment lumen, similar to how viruses are formed. The vesicles may serve a lytic function while the dense core and matrix compartments serve a storage function (Blott and Griffiths 2002).

Several human autosomal genetic disorders illustrate the importance of functional secretory lysosomes. Common for all are immunodeficiency and skin involvement such as hypopigmentation or albinism. Genes of importance in secretory lysosome function have been identified during the study of these diseases. The Chediak-Higashi syndrome (CHS) was described first, characterized by hypopigmentation and defective NK-cell and CTLs function (Stinchcombe, Bossi et al. 2004). Although other cell types have functional conventional lysosomes, giant lysosomes are found in most cells from these patients. The gene affected codes for the LYST protein that has a role in membrane fusion. Consequently, the enlarged lysosomes are formed because of defective membrane fusion, e.g NK-cell and CTL dysfunction results from a loss of secretion to the immunological synapse (Ward, Shiflett et al. 2003). The Griscelli syndrome shows a phenotype similar to that of CHS, but enlarged conventional lysosomes are lacking. The Rab27a gene is affected in this disorder (Menasche, Pastural et al. 2000) facilitates the dissociation between the secretory lysosome and the microtuble cytoskeleton and the fusion with the plasma membrane during secretion. The Hermansky-Pudlak syndrome (HSP) shows a bleeding tendency and partial albinism as a result of platelet and melanocyte deficiency. Seven separate genes (HPS1-HPS7) have been identified to cause the disease. These encode proteins termed biogenesis of lysosomal–related organelle complexes (BLOCs), including the lysosomal adaptor proteins (AP) (Stinchcombe, Bossi et al. 2004).

Secretory Lysosome Targeting

Newly synthesized proteins have to be rescued from constitutive secretion to allow granule targeting. In addition, the targeting requires sorting determinants allowing entry into the cargo route. We try to employ these sorting determinants for exogenous proteins not normally produced in hematopoietic cells. Trans-membrane proteins require a cytosol-sorting signal with a tyrosine (Y) motif, conforming to the lysosome sorting sequence YXXØ (Blott and Griffiths 2002; Bonifacino and Traub 2003). The X position accommodates an unspecified amino acid, while Ø is an amino acid with bulky hydrophobic side chains. To act as a secretory lysosome-targeting signal, the YXXØ sequence should be at the carboxy-terminus close to the trans-membrane domain (Bonifacino and Traub 2003). In contrast to transmembrane protein targeting, luminal protein targeting to the secretory lysosome is due to unknown mechanims (Gullberg, Bengtsson et al. 1999). In many cells, lysosome hydrolases are however targeted by the mannose-6-phosphate receptor (MPR) that takes cargo from the TGN to secretory lysosomes by the endosome route. The MPR system is also responsible for granzyme targeting to secretory lysosomes in NK cells and CTLs (Griffiths and Isaaz 1993), but neutrophil hydrolase targeting seems to be independent of this system (Gullberg, Bengtsson et al. 1999). Transport to the sub-compartments of secretory lysosomes may require separate transport routes to avoid mixing of incompatible proteins (Fig 1A).

Secretory Lysosome Degranulation

Degranulation of secretory lysosomes is a multi-step event starting with an exogenous signal such as cell-membrane binding of a ligand to an activating receptor leading to intracellular calcium mobilization. An increase in calcium is the signal to secretory lysosoms for starting the mobilization towards degranulation (Lyubchenko, Wurth et al. 2001). Synaptotagmins is thought to act as calcium-sensors on secretory lysosomes and regulate the mobilization (Baram, Adachi et al. 1999). The mobilized secretory lysosomes are transported along microtubules by the kinesin familiy as motor against the cell membrane (Burkhardt, McIlvain et al. 1993). Near the cell periphery, actin participates in the movement of secretory lysosome docking into the cell membrane (Langford 1995). Extracellular release of granule matrix and dense core proteins occurs after fusion of the secretory lysosome limiting membrane with the plasma membrane. Membrane fusion is mediated by vesicle (v)- and target (t)-SNARES (Brumell, Volchuk et al. 1995; Guo, Turner et al. 1998) and by isoforms of GTPases of the Rab protein family (Tardieux, Webster et al. 1992). Internal vesicles can be released intact and designated exosomes (Raposo, Nijman et al. 1996). In NK-cells and cytotoxic T-cells, dense core granzymes and perforins are packed in complexes with ser-gly proteoglycans (Bleackley, Lobe et al. 1988) (Fig 1B).

Tissue Damage – Inflammation

Normally, inflammation occurs as a local, beneficial, vigilantly controlled response to injury or infection. However, if the regulation fails, the inflammatory response can give systemic complications, such as sepsis (Cohen 2002) or persistent inflammation or even malignant transformation.

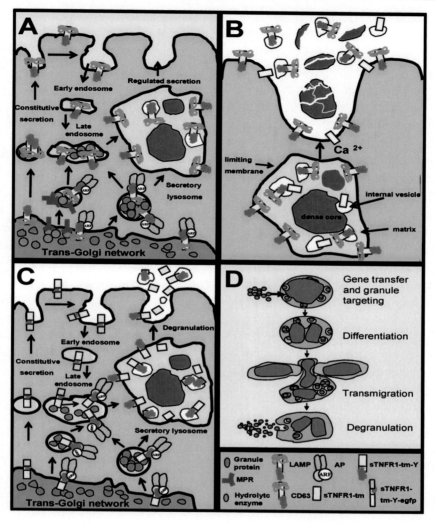

Figure 1. Secretory lysosome targeting in hematopoietic cells as a principle for targeting of selected proteins to areas of inflammation and tumours. (A) Secretory lysosome targeting of lysosomal hydrolases, other soluble granule proteins and transmembrane proteins such as LAMPs and CD63. (B) Secretory lysosome degranulation triggered by a calcium signal. Extracellular degranulation is characteristic for cytotoxic T-cells, NK-cells, DC-cells and mast cells, while neutrophil degranulation occurs into phagocytic vacuoles. Dense cores can be released intact and dissolved extracellularly. Internal vesicles can also be released intact as exosomes. (C) Secretory lysosome targeting and regulated secretion of sTNFR1-tm-Y and sTNFR1-tm-Y-egfp. sTNFR1-tm-Y was targeted to secretory lysosomes through a tyrosine-based cytosol-sorting motif (Y). Delivery to the internal vesicle was combined with release of sTNFR1 into the matrix and discharge to the exterior during degranulation. Some sTNFR1-tm-Y might remain intact and become released with exosomes. In contrast, newly synthesized sTNFR1-tm-Y-egfp was directed to the plasma membrane, subjected to endosomal uptake, proteolytic release of egfp, and secretory lysosome targeting of generated sTNFR1-tm-Y. (D) Secretory lysosome as vehicle for delivery of exogenous therapeutic molecules at inflamed sites. Gene transfer and granule loading of encoded protein is achieved in hematopoietic progenitor cells, differentiation is induced with cytokines, and mature cells are directed to inflamed or tumour sites by chemotactic gradients. During degranulation, the targeted protein becomes delivered at the inflamed focus. AP; adaptor protein. MPR; mannose-6-phosphate receptor.

As a result of tissue injury or infection, pro-inflammatory cytokine-production is triggered from tissue resident mast cells and macrophages. These cytokines affect the endothelium to further neutrophil adhesion and emigration. Later, other hematopoietic cells are recruited. A lack of leukocyte recruitment contributes to serious infection. But also excess leukocyte mobilization may contribute to prolonged inflammation. Selectin facilitates rolling along the vascular endothelium, integrin activation facilitates neutrophil sticking, and chemokine gradients direct neutrophils to the injury site. Moreover, downstream secondary responses are generated by cytokines that enlarge the inflammatory reaction. Then, stop signals turn the process towards healing and repair at which stage TNF-α, IFN-γ, TGF-β and others become anti-inflammatory instead of proinflammatory (Nathan 2002). Apoptotic neutrophils are removed by macrophage ingestion to prevent local tissue damage. A continuous active suppression of inflammation is probably necessary as disruption of many genes predisposes to inflammatory disease. Very often the gene products of importance for inflammation suppression have a role in clearance of immune complexes, activation, proliferation and apoptosis of inflammatory cells, and inhibition of oxidative injury (Nathan 2002). New ideas for understanding inflammatory disorders and for the development of new therapeutic modalities may come forward from knowledge of the function of these gene products.

A novel finding indicates that the nervous system too could influence the inflammatory process. Thus, vagus nerve stimulation has been shown to inhibit lipopolysaccharide-induced release of TNF-α in peripheral macrophages (Bernik, Friedman et al. 2002; Wang, Yu et al. 2003). This is mediated by acetycholin that acts via nicotin receptors on macrophages, emphasizing a controlling mechanism between the nervous system and the inflammatory response, where new therapeutic target could be found.

TNF-α in Chronic Inflammatory Disorders – Anti-Cytokine Therapy

TNF-α has a beneficial and protective effect in homeostasis. Among the pro-inflammatory cytokines it (Carswell, Old et al. 1975; Kawakami and Cerami 1981; Pennica, Nedwin et al. 1984; Beutler, Greenwald et al. 1985) is a central player in inflammation initiation and by inducing many other cytokines and chemokines (Marucha, Zeff et al. 1991; Polentarutti, Introna et al. 1997). However, overproduction of TNF-α can cause deleterious systemic effects and give rise to chronic inflammation such as in inflammatory arthritis and Crohn's inflammatory bowel disease (Kollias, Douni et al. 1999). Hence, TNF-α has become a target for new therapeutic modalities (Feldmann and Maini 2001). Cytokine inhibitors such as soluble TNF-α receptors identified as TNF-α-binding proteins in biological fluids (Peetre, Thysell et al. 1988; Seckinger, Isaaz et al. 1988; Engelmann, Aderka et al. 1989), and anti-TNF-α antibodies (Siegel, Shealy et al. 1995) are now in the therapeutic arsenal of anti-inflammatory agents.

Although TNF-α can have harmful pro-inflammatory activities, it can also be anti-inflammatory (Kollias and Kontoyiannis 2002). Blocking TNF-α in multiple sclerosis patients furthered disease exacerbation suggesting a need for this cytokine in suppressing autoimmunity (van Oosten, Barkhof et al. 1996). Moreover, TNF-α prevents autoimmune diabetes (Kollias, Douni et al. 1999). But pro-inflammatory and immunosuppressive properties of TNF-α can be dissociated. The p55 TNF-α receptor (TNFR1) was required for detrimental TNF-α effects during inflammation, but was not required for TNF-α-mediated immunosuppression and protection against autoimmune encephalitis (Kassiotis and Kollias 2001). Beneficial and detrimental TNF-α effects depend on concentration and cell specificity.

Signs and symptoms can be diminished in patients with rheumatoid arthritis by systemic administration of anti-TNF-α therapy that reduces both inflammation and joint destruction (Feldmann and Maini 2001). Clear responses have been achieved with both an IgG1 murine-human chimeric antibody (infliximab) (Elliott, Maini et al. 1994), and a fusion protein between recombinant soluble human TNF-α receptor and human IgG1 (etanercept) (Moreland, Schiff et al. 1999; Weinblatt, Kremer et al. 1999; Bathon, Martin et al. 2000). However, the disease is not cured by anti-TNF-α therapy. The cytokine cascade is downregulated; the IL-6 concentration was normalized, IL-8, MCP-1 and VEGF decreased. A reduction in leukocyte trafficking into the inflamed joint was seen, which is probably important for the clinical response (Feldmann, Brennan et al. 1996). Not all patients with rheumatoid arthritis patients respond and around one-third remains as non-responders. TNF-α promoter gene polymorphism (Suryaprasad and Prindiville 2003) or disease phenotype heterogeneity might explain the response pattern. Furthermore, in some patients TNF-α may not be the central mediator of the disease. Also patients with Crohn's disease respond to the monoclonal anti-TNF-α antibody (infliximab) (D'Haens 2003), but not to soluble TNF-α-receptor (D'Haens, Swijsen et al. 2001; Sandborn, Hanauer et al. 2001). In addition, one-third of patients were non-responders to the monoclonal antibody. The TNF-α source in Crohn's disease is thought to be the CD4$^+$ T cell (Suryaprasad and Prindiville 2003). The anti-TNF-α antibody may induce effector T cell apoptosis by TNF-α activation supported by the lack of clinical effect by soluble TNF-α receptor (etanercept). The variable therapy response in Crohn's disease may also depend on TNF-α receptor polymorphisms (Mascheretti, Hampe et al. 2002).

It is not unexpected to find anti-TNF-α side effects such as infections, autoimmune reactions, lymphomas and neurological complications. Considerable adverse effects are expected upon inhibition of this cytokine as it has an important role in host defense. The most serious infectious complication has been reactivation of tuberculosis (Gardam, Keystone et al. 2003). The side effects could probably be reduced by the principle of local delivery if feasible.

Inflammatory Response in Tumors

In the 19th century Virchow postulated that cancer originates from chronic inflammation. A relationship between inflammation and cancer is indeed observed frequently, and new knowledge about the interplay between inflammation and transformation has implications for prevention and treatment.

Hepatitis C infections and development of hepatocellular carcinoma (Marotta, Vangieri et al. 2004), ulcerative colitis and colon adenocarcinoma (Munkholm 2003) are both examples of cancers developing in an inflammatory milieu. Further supported by the protective effect of non-steroidal anti-inflammatory drugs (NSAIDs) is shown in gastrointestinal cancer (Baron and Sandler 2000).

The Tumor Microenviroment

Many tumor cells manufacture cytokines and chemokines (Negus, Stamp et al. 1997) that can stimulate inflammatory cell production and migration to tumor environments (Coussens and Werb 2002). The tumor therefore becomes surrounded by macrophages, neutrophils, dendritic cells, eosinophils, mast cells and other cells. The inflammatory infiltrate has a dual role. First, it can give rise to anti-tumor effects by adaptive immune responses. Second, it can do the opposite: promote tumor growth and dissemination. Thus, tumor-infiltrating inflammatory cells may aid in early neoplastic progression. A major part of the inflammatory infiltrate in malignant disease consists of tumor-associated macrophages (TAM) (Eccles and Alexander 1974). These cells are derived from monocytes recruited from peripheral blood possibly attracted by the monocyte chemotactic protein (Daly and Rollins 2003). TAMs have diverse effects on malignant disease and can both kill tumor cells and promote tumor progression. The latter effect may be mediated by TAMs release of angiogenetic growth factors – necessary for tumor growth; cytokines (Sica, Saccani et al. 2000) and radicals (Hellstrand, Brune et al. 2000) – decreasing CTL and NK-cell responses and proteases – promoting tumor spreading (review in (Coussens and Werb 2002)). Clinically, this could be observed as a relation between high TAM density within malignant tumors and disease severity and prognosis: in patients with colorectal carcinomas (Allen and Hogg 1985), malignant melanoma (Brocker, Zwadlo et al. 1987; Brocker, Zwadlo et al. 1988), gastric carcinoma (Heidl, Davaris et al. 1987) and breast cancer (Leek, Lewis et al. 1996).

Dendritic cells are required for the antigen specific immune-response. However, most dendritic cells in tumors have an immature phenotype and thereby a defective ability to stimulate T-cells (Katsenelson, Shurin et al. 2001): a phenomenon possibly mediated by tumor cell produced cytokine (IL-6) by inducing the maturation of functionally imparied dendritic cells with defect antigen presentation and cytokine production (Hegde, Pahne et al. 2004).

It seems that TNF-α is a central regulator in early malignancy by enhancing tumor development and spread (Balkwill 2002). This has been shown on a molecular level, by a TNF-α mediated NF-κB activation that protects pre-malignant hepatocytes apoptosis (Pikarsky, Porat et al. 2004), linking inflammation with malignant disease (Greten and Karin 2004). Anti-TNF-α therapy could therefore have a potential role in cancer prevention and or early stage therapy. Local anti-TNF-α- and pro-lymphocyte response-therapy could be valid approaches provided that tumor targeting can be achieved. Our local delivery concept could use neutrophils as well as T- and NK-cells as vehicles for the transport of therapeutic proteins into tumors. Local anti-TNF-α deposition by granulocytes might reduce inflammatory cell activity in tumors. NK-cells, that patrol tissues and destroy tumor cells by release of lytic granules, are of particular interest as potential vehicles for the local delivery of therapeutic molecules against tumor cells. The therapeutic success with improving NK-cell tumor cell toxicity has been limited (see review (Smyth, Hayakawa et al. 2002)). Local delivery of chemokines with NK-cells as vehicles may amplify NK-cell accumulation and promote and sustain an anti-tumor effect.

Secretory Lysosome Targeting of Non-Hematopoietic Protein

Secretory lysosome sorting and secretion have been investigated in hematopoietic cell lines after cDNA transfection and gene expression. The human NK-cell lines, YT-Indy (Yodoi, Teshigawara et al. 1985) and NK-92 (Gong, Maki et al. 1994), derived from patients with large granulated lymphocyte leukemia, were used. Both show normal NK-cell functions including regulated cytotoxicity and cytokine production. In addition, a number of murine cell lines have been used such as the myeloblast-like leukemia 32D (Liu, Oren et al. 1995) and the rat basophilic leukaemia (RBL) (Seldin, Adelman et al. 1985) cell line. Protein targeting to secretory lysosomes took place in these cell lines upon expression but regulated secretion was impaired in the YT-indy and RBL-1 cells (Hansson, Jonsson et al. 2003; Gao, Hansson et al. 2004). But the RBL-2H3 (Gao, Hansson et al. 2004) and NK-92 (Hansson unpublished) cells had capacity for regulated secretion.

Proteins receive native conformation in the endoplasmic reticulum (ER) and are exported for Golgi sorting. Proteins with abnormal fold are directed by a quality control system to proteasomes for degradation (Ellgaard, Molinari et al. 1999). The gene product of the sTNFR1 cDNA failed the quality control and was to a large extent retained in the ER and degraded. Furthermore, the small amount of gene product that was exported was constitutively secreted without retention (Gao, Rosen et al. 2003). Other truncated proteins behave in a similar manner e.g. propeptide-deleted myeloperoxidase (Andersson, Hellman et al. 1998) and chimeric proteins (Bulow, Nauseef et al. 2002). However, anchoring sTNFR1 by a transmembrane domain (tm) overcame ER retention (Gao, Rosen et al. 2003). The tm domain enhanced ER export of sTNFR1-tm to the Golgi but did not allow granule targeting (Gao, Rosen et al. 2003). Consequently, clearence by the ER quality control followed by Golgi trafficking does not assure granule targeting as constitutive secretion occurred in this case (Bulow, Nauseef et al. 2002). Actually, sTNFR1-tm was translocated to the plasma membrane, where sTNFR1 was shed by proteolytic cleavage. Clearly, any protein that arrives in the Golgi compartment of hematopoietic cells is not granule targeted, some are constitutively secreted. This indicates a need for cell- and/or protein-specific mechanisms to avoid constitutive secretion. That chimeras and their multimers are prevented from targeting suggests protein conformation to be important (Rosen, Gao et al. 2003). Truncated proteins expressed in hematopoietic cells are secreted and not targeted.

The sTNFR1-tm obviously lacked a retention signal for granule targeting (Fig 4A). A carboxy-terminal cytosol sorting sequence SIRSGYEVM (Y), corresponding to the common secretory lysosome sorting sequence YXXØ, was therefore incorporated to rescue sTNFR1-tm-Y from constitutive secretion (Gao, Rosen et al. 2003). This sequence is identical with the tyrosine-based sorting motif of CD63, a secretory lysosome membrane protein (Fukuda 1991; Cham, Gerrard et al. 1994). The assumption was that this motif should direct trafficking of not only CD63 but also sTNFR1-tm-Y to the secretory lysosome. Efficient targeting of sTNFR1-tm-Y to this organelle was in fact achieved, even more efficient than the endogenous secretory lysosome targeting of granzyme B that uses the MPR transport system (Fig 4A).

Addition of enhanced green fluorescent protein (egfp) to the carboxy-terminus for construction of sTNFR1-tm-Y-egfp resulted in decreased targeting efficiency (Gao, Hansson et al. 2004). In this construct the YXXØ motif (Y) is not any longer exposed at the carboxy-terminus but folded within the structure of the protein. Hiding the YXXØ motif obviously prevented the sorting machinery at the TGN from interacting with it but allowing cell surface trafficking by default (Gao, Hansson et al. 2004). However, endosomal reuptake of the cell membrane-located sTNFR1-tm-Y-egfp was observed followed by proteolytic release of egfp and generation of sTNFR1-tm-Y. The YXXØ motif would now again be positioned at the carboxy-terminus making secretory lysosome targeting possible. Consequently, sTNFR1-tm-Y was trafficking to the secretory lysosome by the biosynthetic pathway, while sTNFR1-tm-Y-egfp seemed to make a detour to the plasma membrane before secretory lysosome trafficking via the endocytic pathway.

In secretory lysosomes processing of newly targeted sTNFR1-tm-Y gave rise to a lower molecular weight form consisting of sTNFR1 cleaved off by limited proteolysis. Co-localization of sTNFR1-tm-Y and its processed form with endogenous secretory lysosome constituents was observed by immunoelectron-microscopy. Co-localization was observed of both rat mast cell protease–II (RMCP-II) (Seldin, Adelman et al. 1985) and lysosome-associated membrane protein 1 (LAMP-1) (Granger, Green et al. 1990) with the sTNFR1 signal. In contrast to LAMP that is localized along the outer limiting membrane, the sTNFR1 signal was observed in the interior of the secretory lysosome (Fig. 4B). This is consistent with sTNFR1-tm-Y translocation into the internal vesicles of secretory lysosomes (Jiang, Erickson et al. 2002). Such transfer has recently been reported to require a mono-ubiquitin tag and binding by the escorting complex ESCRT-1 (Katzmann, Babst et al. 2001). The CD63 transport to internal vesicles does, however, not require ubiquitination (Stoorvogel, Kleijmeer et al. 2002). The carboxy-terminal cytosol-sorting motif (Y) of TNFR1-tm-Y will be placed inside the internal vesicle that is formed by inward budding, and the amino-terminal part of the protein will be in the luminal compartment outside the internal vesicle membrane. We assume that delivery to internal vesicles was concomitant with hydrolysis for generation of sTNFR1 that was released into the secretory lysosome lumen. As an effect the sTNFR1 would be deposited in the matrix and/or dense core. Hydrolytic cleavage with release of sTNFR1 into the lumen might also occur on the outer limiting membrane without transfer to internal vesicles (Fig 1C).

Secrtory lysosomes also called cytotoxic granules of NK-cells could be one target for a therapeutic invention. These cells play a role in destruction of virus-infected and tumour cells. Therefore, anti-tumour protein targeting to the secretory lysosomes of NK-cells and delivery at the tumour site could be therapeutically beneficial. It may be feasible to achieve secretory lysosome targeting of a gene product in NK-cell precursors. After maturation the transduced NK-cells migrating into the tissue could encounter a tumour that would result in gene product delivery. Several non-NK-cell proteins were expressed in NK-cell lines (Hansson, Jonsson et al. 2003). For instance, the liver secretory protein, α_1-antitrypsin, showed defective retention and was constitutively secreted. Another liver secretory protein, α_1-microglobulin showed some cellular retention. Even newly synthesized endogenous granzyme B was constitutively secreted in the NK-cell line used indicating a rather low sorting capacity for soluble proteins.

However, strong intracellular retention was accomplished for sTNFR1-tm-Y when expressed in an NK-cell line. The secretory lysosome sorting signal (Y) obviously furthered cell retention also in NK-cells. Golgi import was observed for newly synthesized sTNFR1-tm-Y as well as targeting to the densest organelles with concomitant release of sTNFR1. The sTNFR1-tm-Y targeting was more efficient than that of endogeneous granzyme B. Results from both immunofluorescence and immunoelectron microscopy verified the co-localization of sTNFR1 with endogenous CD63 and granzyme B of NK-cell lines. The NK-cells showed granule profiles with small internal vesicles and dense cores. The CD63 labeling was observed on the outer membrane and on the parallel tubular arrays, and the granzyme B labelling was observed on the dense core and the tubular arrays. Accordingly, sTNFR1-tm-Y and processed forms were targeted to the same granules as granzyme B and CD63, consistent with secretory lysosome trafficking (Hansson, Jonsson et al. 2003) (Fig 1C).

Regulated Secretion

Combined storage, hydrolytic activity and secretory capacity is characteristic of the secretory lysosome. Compartmentalization of storage and lysosomal functions might be required to prevent direct interactions. However, interactions between components from separate compartments can be necessary for activation of proteins stored as inactive pro-proteins. This occurs during degranulation when compartment constituents become mixed. Exocytosis is critical for hematopoietic cell function (Blott and Griffiths 2002). Unlike constitutive secretion, regulated secretion is provoked in a calcium-dependent manner by extracellular signals. For instance, CTLs degranulate when targets are recognized. NK-cells degranulate upon formation of an intercellular cleft after recognition and binding to a tumour cell. IgE receptor cross-linking triggers mast cell degranulation. The sTNFR1 secretion as well as endogenous secretory lysosome constituent secretion was achieved in RBL-2H3 cells expressing sTNFR1-tm-Y. Both calcium-ionophore and IgE receptor cross-linking independently stimulated secretion of endogeneous RMCP-II and the sTNFR1.

Concluding Remarks

We suggest that storage organelles of hematopoietic cells might be used as vehicles for transport to inflammatory sites. Deposition of therapeutically active agents carried by these cells should prevent systemic effects. The concept is summarized in figure 1D. Homeostasis and inflammation are regulated by pro-inflammatory and anti-inflammatory cytokines. A TNF-α excess can further chronic inflammatory disorders as well as cell transformation. Therefore, TNF-α inhibition by systemic anti-TNF-α antibodies or soluble TNF-α receptor has been successful in these disorders. This therapy, however, goes with a risk of systemic side effects such as infectious complications as TNF-α has an important role in host defense.

Our research has been focused on the expression of a soluble TNF-α receptor form in hematopoietic cell lines. Targeting the secretory lysosomes for storage and regulated secretion has been accomplished. ER retention was overcome by incorporation of a transmembrane domain that facilitated ER-export. Constitutive secretion was prevented by a cytosol-sorting signal that led to retention and secretory lysosome trafficking in transformed myeloblastic, basophilic and NK-cells. Our results should be applicable to non-transformed hematopoietic precursor cells as well since hematopoietic cells in particular have lysosome-like organelles with secretory capacity.

For storage stability of the sorted protein resistance to proteolysis is required. The presence of active hydrolytic enzymes in the secretory lysosome may therefore be a problem. While endogenous degradative enzymes, antibiotic proteins and other granule proteins can co-exist, it is a risk that heterologously expressed non-hematopoietic proteins may be unstable as they are not adopted to their new environment. The membrane-free form of sTNFR1 that was generated remained intact and seemed to be relatively stable in secretory lysosomes.

The functions of the secretory lysosome differ among hematopoietic cells. The neutrophil secretory lysosome (azurophil granule) degranulates primarily to the phagosome when formed during phagocytosis. NK-cell and CTL secretory lysosomes degranulate into the extracellular space upon stimulation. Accordingly, the azurophil granule could serve as a potential vehicle for protein deposition into the phagosome and modification of antimicrobial defense. In contrast, NK-cell and CTL secretory lysosomes would be potential vehicles for stimulation of an anti-tumour effect upon degranulation. Future studies will show whether secretory lysosome targeting is also feasible in normal hematopoietic cells. The granules of these cells are final vehicles for delivering agents at an inflamed site. It is possible that targeting and secretion are more efficient in normal hematopoietic cells than in transformed cell lines.

Experiments in animal models of inflammatory and malignant disease are now possible to test our principle *in vivo*. This will require gene transfer in hematopoietic progenitor cells combined with animal cell transfer. Finally, exploration of this concept could also shed new light on hematopoietic granule formation, secretion and granule protein deposition in the inflammatory process.

References

[1]   Allen, C. and N. Hogg (1985). "Monocytes and other infiltrating cells in human colorectal tumours identified by monoclonal antibodies." *Immunology* 55(2): 289-99.

[2]   Andersson, E., L. Hellman, et al. (1998). "The role of the propeptide for processing and sorting of human myeloperoxidase." *J Biol Chem* 273(8): 4747-53.

[3]   Balkwill, F. (2002). "Tumor necrosis factor or tumor promoting factor?" *Cytokine Growth Factor Rev* 13(2): 135-41.

[4]   Baram, D., R. Adachi, et al. (1999). "Synaptotagmin II negatively regulates Ca2+-triggered exocytosis of lysosomes in mast cells." *J Exp Med* 189(10): 1649-58.

[5]   Baron, J. A. and R. S. Sandler (2000). "Nonsteroidal anti-inflammatory drugs and cancer prevention." *Annu Rev Med* 51: 511-23.

[6]   Bathon, J. M., R. W. Martin, et al. (2000). "A comparison of etanercept and methotrexate in patients with early rheumatoid arthritis." *N Engl J Med* 343(22): 1586-93.

[7]   Benoist, C. and D. Mathis (2002). "Mast cells in autoimmune disease." *Nature* 420(6917): 875-8.

[8]   Bernik, T. R., S. G. Friedman, et al. (2002). "Pharmacological stimulation of the cholinergic antiinflammatory pathway." *J Exp Med* 195(6): 781-8.

[9]   Beutler, B., D. Greenwald, et al. (1985). "Identity of tumour necrosis factor and the macrophage-secreted factor cachectin." *Nature* 316(6028): 552-4.

[10] Bleackley, R. C., C. G. Lobe, et al. (1988). "The isolation and characterization of a family of serine protease genes expressed in activated cytotoxic T lymphocytes." *Immunol Rev* 103: 5-19.

[11] Blott, E. J. and G. M. Griffiths (2002). "Secretory lysosomes." *Nat Rev Mol Cell Biol* 3(2): 122-31.

[12] Bonifacino, J. S. and L. M. Traub (2003). "Signals for sorting of transmembrane proteins to endosomes and lysosomes." *Annu Rev Biochem* 72: 395-447.

[13] Borregaard, N. and J. B. Cowland (1997). "Granules of the human neutrophilic polymorphonuclear leukocyte." *Blood* 89(10): 3503-21.

[14] Boxer, L. and D. C. Dale (2002). "Neutropenia: causes and consequences." *Semin Hematol* 39(2): 75-81.

[15] Brocker, E. B., G. Zwadlo, et al. (1988). "Inflammatory cell infiltrates in human melanoma at different stages of tumor progression." *Int J Cancer* 41(4): 562-7.

[16] Brocker, E. B., G. Zwadlo, et al. (1987). "Infiltration of primary and metastatic melanomas with macrophages of the 25F9-positive phenotype." *Cancer Immunol Immunother* 25(2): 81-6.

[17] Brumell, J. H., A. Volchuk, et al. (1995). "Subcellular distribution of docking/fusion proteins in neutrophils, secretory cells with multiple exocytic compartments." *J Immunol* 155(12): 5750-9.

[18] Bulow, E., U. Gullberg, et al. (2000). "Structural requirements for intracellular processing and sorting of bactericidal/permeability-increasing protein (BPI): comparison with lipopolysaccharide-binding protein." *J Leukoc Biol* 68(5): 669-78.

[19] Bulow, E., W. M. Nauseef, et al. (2002). "Sorting for storage in myeloid cells of nonmyeloid proteins and chimeras with the propeptide of myeloperoxidase precursor." *J Leukoc Biol* 71(2): 279-88.

[20] Burkhardt, J. K., J. M. McIlvain, Jr., et al. (1993). "Lytic granules from cytotoxic T cells exhibit kinesin-dependent motility on microtubules in vitro." *J Cell Sci* 104 (Pt 1): 151-62.

[21] Carswell, E. A., L. J. Old, et al. (1975). "An endotoxin-induced serum factor that causes necrosis of tumors." *Proc Natl Acad Sci U S A* 72(9): 3666-70.

[22] Cham, B. P., J. M. Gerrard, et al. (1994). "Granulophysin is located in the membrane of azurophilic granules in human neutrophils and mobilizes to the plasma membrane following cell stimulation." *Am J Pathol* 144(6): 1369-80.

[23] Cohen, J. (2002). "The immunopathogenesis of sepsis." *Nature* 420(6917): 885-91.

[24] Coussens, L. M. and Z. Werb (2002). "Inflammation and cancer." *Nature* 420(6917): 860-7.

[25] D'Haens, G. (2003). "Anti-TNF therapy for Crohn's disease." *Curr Pharm Des* 9(4): 289-94.

[26] D'Haens, G., C. Swijsen, et al. (2001). "Etanercept in the treatment of active refractory Crohn's disease: a single-center pilot trial." *Am J Gastroenterol* 96(9): 2564-8.

[27] Daly, C. and B. J. Rollins (2003). "Monocyte chemoattractant protein-1 (CCL2) in inflammatory disease and adaptive immunity: therapeutic opportunities and controversies." *Microcirculation* 10(3-4): 247-57.

[28] Dell'Angelica, E. C., C. Mullins, et al. (2000). "Lysosome-related organelles." *Faseb J* 14(10): 1265-78.

[29] Eccles, S. A. and P. Alexander (1974). "Macrophage content of tumours in relation to metastatic spread and host immune reaction." *Nature* 250(468): 667-9.

[30] Ellgaard, L., M. Molinari, et al. (1999). "Setting the standards: quality control in the secretory pathway." *Science* 286(5446): 1882-8.

[31] Elliott, M. J., R. N. Maini, et al. (1994). "Randomised double-blind comparison of chimeric monoclonal antibody to tumour necrosis factor alpha (cA2) versus placebo in rheumatoid arthritis." *Lancet* 344(8930): 1105-10.

[32] Engelmann, H., D. Aderka, et al. (1989). "A tumor necrosis factor-binding protein purified to homogeneity from human urine protects cells from tumor necrosis factor toxicity." *J Biol Chem* 264(20): 11974-80.

[33] Feldmann, M., F. M. Brennan, et al. (1996). "Role of cytokines in rheumatoid arthritis." *Annu Rev Immunol* 14: 397-440.

[34] Feldmann, M. and R. N. Maini (2001). "Anti-TNF alpha therapy of rheumatoid arthritis: what have we learned?" *Annu Rev Immunol* 19: 163-96.

[35] Fukuda, M. (1991). "Lysosomal membrane glycoproteins. Structure, biosynthesis, and intracellular trafficking." *J Biol Chem* 266(32): 21327-30.

[36] Gao, Y., M. Hansson, et al. (2004). "Sorting soluble tumor necrosis factor (TNF) receptor for storage and regulated secretion in hematopoietic cells." (2004) *J Leukoc Biol.* 76(4): 876-85.

[37] Gao, Y., H. Rosen, et al. (2003). "Sorting of soluble TNF-receptor for granule storage in hematopoietic cells as a principle for targeting of selected proteins to inflamed sites." *Blood* 102(2): 682-8.

[38] Gardam, M. A., E. C. Keystone, et al. (2003). "Anti-tumour necrosis factor agents and tuberculosis risk: mechanisms of action and clinical management." *Lancet Infect Dis* 3(3): 148-55.

[39] Gong, J. H., G. Maki, et al. (1994). "Characterization of a human cell line (NK-92) with phenotypical and functional characteristics of activated natural killer cells." *Leukemia* 8(4): 652-8.

[40] Granger, B. L., S. A. Green, et al. (1990). "Characterization and cloning of lgp110, a lysosomal membrane glycoprotein from mouse and rat cells." *J Biol Chem* 265(20): 12036-43.

[41] Greten, F. R. and M. Karin (2004). "The IKK/NF-kappaB activation pathway-a target for prevention and treatment of cancer." *Cancer Lett* 206(2): 193-9.

[42] Griffiths, G. M. and S. Isaaz (1993). "Granzymes A and B are targeted to the lytic granules of lymphocytes by the mannose-6-phosphate receptor." *J Cell Biol* 120(4): 885-96.

[43] Gullberg, U., N. Bengtsson, et al. (1999). "Processing and targeting of granule proteins in human neutrophils." *J Immunol Methods* 232(1-2): 201-10.

[44] Guo, Z., C. Turner, et al. (1998). "Relocation of the t-SNARE SNAP-23 from lamellipodia-like cell surface projections regulates compound exocytosis in mast cells." *Cell* 94(4): 537-48.

[45] Hansson, M., S. Jonsson, et al. (2003). "Targeting proteins to secretory lysosomes of natural killer cells as a principle for immunoregulation." *Mol Immunol* 40(6): 363-72.

[46] Hegde, S., J. Pahne, et al. (2004). "Novel immunosuppressive properties of interleukin-6 in dendritic cells: inhibition of NF-kappaB binding activity and CCR7 expression." *Faseb J* 18(12): 1439-41.

[47] Heidl, G., P. Davaris, et al. (1987). "Association of macrophages detected with monoclonal antibody 25 F 9 with progression and pathobiological classification of gastric carcinoma." *J Cancer Res Clin Oncol* 113(6): 567-72.

[48] Hellstrand, K., M. Brune, et al. (2000). "Alleviating oxidative stress in cancer immunotherapy: a role for histamine?" *Med Oncol* 17(4): 258-69.

[49] Jiang, L., A. Erickson, et al. (2002). "Multivesicular bodies: a mechanism to package lytic and storage functions in one organelle?" *Trends Cell Biol* 12(8): 362-7.

[50] Kassiotis, G. and G. Kollias (2001). "Uncoupling the proinflammatory from the immunosuppressive properties of tumor necrosis factor (TNF) at the p55 TNF receptor level: implications for pathogenesis and therapy of autoimmune demyelination." *J Exp Med* 193(4): 427-34.

[51] Katsenelson, N. S., G. V. Shurin, et al. (2001). "Human small cell lung carcinoma and carcinoid tumor regulate dendritic cell maturation and function." *Mod Pathol* 14(1): 40-5.

[52] Katzmann, D. J., M. Babst, et al. (2001). "Ubiquitin-dependent sorting into the multivesicular body pathway requires the function of a conserved endosomal protein sorting complex, ESCRT-I." *Cell* 106(2): 145-55.

[53] Kawakami, M. and A. Cerami (1981). "Studies of endotoxin-induced decrease in lipoprotein lipase activity." *J Exp Med* 154(3): 631-9.

[54] Kollias, G., E. Douni, et al. (1999). "The function of tumour necrosis factor and receptors in models of multi-organ inflammation, rheumatoid arthritis, multiple sclerosis and inflammatory bowel disease." *Ann Rheum Dis* 58 Suppl 1: I32-9.

[55] Kollias, G., E. Douni, et al. (1999). "On the role of tumor necrosis factor and receptors in models of multiorgan failure, rheumatoid arthritis, multiple sclerosis and inflammatory bowel disease." *Immunol Rev* 169: 175-94.

[56] Kollias, G. and D. Kontoyiannis (2002). "Role of TNF/TNFR in autoimmunity: specific TNF receptor blockade may be advantageous to anti-TNF treatments." *Cytokine Growth Factor Rev* 13(4-5): 315-21.

[57] Langford, G. M. (1995). "Actin- and microtubule-dependent organelle motors: interrelationships between the two motility systems." *Curr Opin Cell Biol* 7(1): 82-8.

[58] Leek, R. D., C. E. Lewis, et al. (1996). "Association of macrophage infiltration with angiogenesis and prognosis in invasive breast carcinoma." *Cancer Res* 56(20): 4625-9.

[59] Liu, L., A. Oren, et al. (1995). "Murine 32D c13 cells--a transfectable model of phagocyte granule formation." *J Immunol Methods* 181(2): 253-8.

[60] Lyubchenko, T. A., G. A. Wurth, et al. (2001). "Role of calcium influx in cytotoxic T lymphocyte lytic granule exocytosis during target cell killing." *Immunity* 15(5): 847-59.

[61] Marotta, F., B. Vangieri, et al. (2004). "The pathogenesis of hepatocellular carcinoma is multifactorial event. Novel immunological treatment in prospect." *Clin Ter* 155(5): 187-99.

[62] Marucha, P. T., R. A. Zeff, et al. (1991). "Cytokine-induced IL-1 beta gene expression in the human polymorphonuclear leukocyte: transcriptional and post-transcriptional regulation by tumor necrosis factor and IL-1." *J Immunol* 147(8): 2603-8.

[63] Mascheretti, S., J. Hampe, et al. (2002). "Pharmacogenetic investigation of the TNF/TNF-receptor system in patients with chronic active Crohn's disease treated with infliximab." *Pharmacogenomics J* 2(2): 127-36.

[64] McNicol, A. and S. J. Israels (1999). "Platelet dense granules: structure, function and implications for haemostasis." *Thromb Res* 95(1): 1-18.

[65] Menasche, G., E. Pastural, et al. (2000). "Mutations in RAB27A cause Griscelli syndrome associated with haemophagocytic syndrome." *Nat Genet* 25(2): 173-6.

[66] Moreland, L. W., M. H. Schiff, et al. (1999). "Etanercept therapy in rheumatoid arthritis. A randomized, controlled trial." *Ann Intern Med* 130(6): 478-86.

[67] Munkholm, P. (2003). "Review article: the incidence and prevalence of colorectal cancer in inflammatory bowel disease." *Aliment Pharmacol Ther* 18 Suppl 2: 1-5.

[68] Nathan, C. (2002). "Points of control in inflammation." *Nature* 420(6917): 846-52.

[69] Negus, R. P., G. W. Stamp, et al. (1997). "Quantitative assessment of the leukocyte infiltrate in ovarian cancer and its relationship to the expression of C-C chemokines." *Am J Pathol* 150(5): 1723-34.

[70] Peetre, C., H. Thysell, et al. (1988). "A tumor necrosis factor binding protein is present in human biological fluids." *Eur J Haematol* 41(5): 414-9.

[71] Pennica, D., G. E. Nedwin, et al. (1984). "Human tumour necrosis factor: precursor structure, expression and homology to lymphotoxin." *Nature* 312(5996): 724-9.

[72] Pikarsky, E., R. M. Porat, et al. (2004). "NF-kappaB functions as a tumour promoter in inflammation-associated cancer." *Nature* 431(7007): 461-6.

[73] Polentarutti, N., M. Introna, et al. (1997). "Expression of monocyte chemotactic protein-3 in human monocytes and endothelial cells." *Eur Cytokine Netw* 8(3): 271-4.

[74] Raposo, G., H. W. Nijman, et al. (1996). "B lymphocytes secrete antigen-presenting vesicles." *J Exp Med* 183(3): 1161-72.

[75] Rosen, H., Y. Gao, et al. (2003). "Artificially controlled aggregation of proteins and targeting in hematopoietic cells." *J Leukoc Biol* 74(5): 800-9.

[76] Sandborn, W. J., S. B. Hanauer, et al. (2001). "Etanercept for active Crohn's disease: a randomized, double-blind, placebo-controlled trial." *Gastroenterology* 121(5): 1088-94.

[77] Seckinger, P., S. Isaaz, et al. (1988). "A human inhibitor of tumor necrosis factor alpha." *J Exp Med* 167(4): 1511-6.

[78] Seldin, D. C., S. Adelman, et al. (1985). "Homology of the rat basophilic leukemia cell and the rat mucosal mast cell." *Proc Natl Acad Sci U S A* 82(11): 3871-5.

[79] Setaluri, V. (2000). "Sorting and targeting of melanosomal membrane proteins: signals, pathways, and mechanisms." *Pigment Cell Res* 13(3): 128-34.

[80] Sica, A., A. Saccani, et al. (2000). "Autocrine production of IL-10 mediates defective IL-12 production and NF-kappa B activation in tumor-associated macrophages." *J Immunol* 164(2): 762-7.

[81] Siegel, S. A., D. J. Shealy, et al. (1995). "The mouse/human chimeric monoclonal antibody cA2 neutralizes TNF in vitro and protects transgenic mice from cachexia and TNF lethality in vivo." *Cytokine* 7(1): 15-25.

[82] Smyth, M. J., Y. Hayakawa, et al. (2002). "New aspects of natural-killer-cell surveillance and therapy of cancer." *Nat Rev Cancer* 2(11): 850-61.

[83] Stinchcombe, J., G. Bossi, et al. (2004). "Linking albinism and immunity: the secrets of secretory lysosomes." *Science* 305(5680): 55-9.

[84] Stoorvogel, W., M. J. Kleijmeer, et al. (2002). "The biogenesis and functions of exosomes." *Traffic* 3(5): 321-30.

[85] Suryaprasad, A. G. and T. Prindiville (2003). "The biology of TNF blockade." *Autoimmun Rev* 2(6): 346-57.

[86] Tardieux, I., P. Webster, et al. (1992). "Lysosome recruitment and fusion are early events required for trypanosome invasion of mammalian cells." *Cell* 71(7): 1117-30.

[87] van Oosten, B. W., F. Barkhof, et al. (1996). "Increased MRI activity and immune activation in two multiple sclerosis patients treated with the monoclonal anti-tumor necrosis factor antibody cA2." *Neurology* 47(6): 1531-4.

[88] Wang, H., M. Yu, et al. (2003). "Nicotinic acetylcholine receptor alpha7 subunit is an essential regulator of inflammation." *Nature* 421(6921): 384-8.

[89] Ward, D. M., S. L. Shiflett, et al. (2003). "Use of expression constructs to dissect the functional domains of the CHS/beige protein: identification of multiple phenotypes." *Traffic* 4(6): 403-15.

[90] Weinblatt, M. E., J. M. Kremer, et al. (1999). "A trial of etanercept, a recombinant tumor necrosis factor receptor:Fc fusion protein, in patients with rheumatoid arthritis receiving methotrexate." *N Engl J Med* 340(4): 253-9.

[91] Yodoi, J., K. Teshigawara, et al. (1985). "TCGF (IL 2)-receptor inducing factor(s). I. Regulation of IL 2 receptor on a natural killer-like cell line (YT cells)." *J Immunol* 134(3): 1623-30.

In: Progress in Immunology Research
Editor: Barbara A. Veskler, pp. 123-157

ISBN 1-59454-380-1
©2005 Nova Science Publishers, Inc.

Chapter VII

The Clinical Significance of Cytokine Genotype Profiles

Laura Koumas and Paul A. Costeas

Immunogenetics Center, Karaiskakio Foundation, Nicosia, Cyprus

Abstract

Cytokines are potent pleiotropic immunomodulatory molecules with extended roles in the direction of immune regulation pathways. Polymorphisms of cytokine genes have been identified and can significantly influence cytokine production levels, thus affecting the outcome of immune balance. Diverse cytokine genotypes suggest differential transcription and protein expression levels among individuals. In addition, several studies have indicated a correlation between cytokine polymorphism distribution and ethnicity. Furthermore, studies suggest that certain cytokine gene polymorphisms are associated with disease progression, for example, systemic lupus erythematosus, rheumatoid arthritis and in patients presenting with sepsis.

Considering that cytokines are inter-dependable and function within complex networks, it is impossible to evaluate cytokines individually in order to assess their involvement with a certain disease condition. An investigative approach recently employed in our laboratory utilizes the strategy of cross-tabulation of cytokine polymorphisms to determine the significance of one cytokine in the context of another. In this manner the development of genetic patterns of cytokine polymorphism combinations, otherwise referred to as immunogenetic profiles, were established. Our laboratory studied the association of cytokine gene polymorphism combinations with various pathologic conditions, namely recurrent spontaneous abortions, cardiovascular disease and leukemia in the Cypriot population. Emerging from our studies is a genetic pattern of cytokine polymorphism combinations that is distinct among patients from these disease groups and subjects used as population control. Furthermore, the importance of cytokine polymorphisms will also be discussed in association with transplantation, graft versus host disease and autoimmunity.

In conclusion, a better understanding of how cytokines may function in concert with one another is warranted and might elucidate their precise roles in disease pathogenesis.

The study of cytokine immunogenetic profiles may provide a useful tool for understanding the pathogenesis and mechanisms involved in various diseases already proposed to be associated with some cytokine polymorphisms.

Cytokine Function and Genetic Polymorphisms

Cytokines are the immune system's soluble agents of cell-to-cell communication. Most cytokines are secreted glycoproteins that act as humoral regulators in an autocrine, paracrine, and sometimes, endocrine manner by modulating the functional activities of individual cells and tissues. These proteins also mediate interactions between cells directly and regulate processes taking place in the extracellular environment. Even though many of the mechanisms of cytokine action have not yet been elucidated in full detail, cytokine function can be used to define cytokines into different groups. In fact, a number of criteria have been utilized for cytokine classification. Cytokines are grouped in those that regulate hematopoiesis, those that affect inflammation and in a group that directs the immune response. Furthermore, cytokines include the interleukins, the interferons, growth factors, colony stimulating factors, the tumor necrosis factor family, the tumor growth factor family and a collection of chemokines that engage G-protein-coupled receptors. The interleukin (IL) group initially included IL-1 though IL-18, thus named as interleukins, originally thought to be produced only by leukocytes. However, other cell types have since been found responsible for their production and the IL number has now reached IL-29. Interleukins are involved in directing immune cell function, for example, proliferation and differentiation of immune cells. Interferons (IFN) are cytokines involved with the inhibition of viral replication in infected cells, and include IFN-α, IFN-β and INF-γ. Interferons share a similar structure and affect the expression of major histocompatibility complex (MHC) Class I and MHC Class II antigens on different cell types. Colony stimulating factors include granulocyte monocyte-colony stimulating factor (GM-CSF), G-CSF and M-CSF. These factors are important in directing hematopoiesis and in the mobilization of precursor immune cells in the bone marrow. Chemokines comprise a group of chemoattractant molecules with potent effects on immune cell adhesion, activation and trafficking, produced by a variety of cell types. Chemokines are classified according to the position of two conserved cysteines, and collectively include monocyte chemoattractant protein (MCP)-1, macrophage inflammatory protein (MIP)-1α, MIP-1β, IL-8 and RANTES.

It would be frivolous to simply consider cytokines within the abovementioned classifications. Cytokine nature is in addition characterized by pleiotropy, redundancy, synergy and antagonism, thus creating a complex and intricate cytokine network. Pleiotropy is the quality of a single cytokine to act on different types of cells and alternatively, the ability of several cell types to produce the same cytokine. Many cytokines are redundant in their activity, that is, they exert similar functions in an immunological response. There is considerable interaction between different cytokines, in both activating and inhibiting directions. Cytokines may synergize to provide a greater response than the additive effect of each cytokine used alone. Alternatively, cytokines may either directly inhibit one another's activity or may act as natural antagonists to a specific cytokine or its receptor, for example IL-1 receptor antagonist (IL-1RA).

A further and most common classification of cytokines is based on their production by different groups of T helper lymphocytes (CD4$^+$), namely Th1 and Th2. Th1 cells differentiate from Th0 cells to secrete pro-inflammatory cytokines such as IFN-γ, TNF-α, IL-1 and IL-12, while Th2 cells are discriminated by their production of anti-inflammatory cytokines, such as IL-10 and IL-4 [137]. However, this classification cannot be considered stringently as some cytokines such as IL-6 can act either as pro- or anti-inflammatory depending on the cell system involved in mediating its production. Th3 cells are those that secrete TGF-β, thereby thought of as immunosuppressor cells. Cytokines can also be referred to as Type-1, -2 or -3, since their production is not restricted to T lymphocytes but extended to a variety of cell types, including structural cells. It is worth mentioning that commitment to Th1 or Th2 lineages is irreversible, and that some Th1/Type-1 cytokines may inhibit the production of Th2/Type-2 cytokines, and vice versa [137]. For example, IL-10 (Type-2) is capable of inhibiting the production of IL-12 (Type-1) to ensure the direction of the immune response is skewed to that of Th2.

Cytokines exert their biological activities via specific membrane receptors, which are expressed on a diverse group of cell types. Cytokine receptor expression is subject to several regulatory mechanisms although some receptors are expressed constitutively. Cytokine receptor proteins have been shown to share a number of characteristics and are grouped into cytokine receptor families that indicate genetic, structural and functional similarities. The receptors are grouped into the immunoglobulin superfamily, the hematopoietic receptor family, the tumor necrosis factor (TNF) family and the chemokine receptor family. Many receptors are multi-subunit structures that are activated by ligand binding and transduce the signal via their intrinsic tyrosine kinase activity (Janus Kinase/STAT signal transduction pathway) [120]. Many receptors within the same family often share common signal transducing receptor components, explaining in part, the functional redundancy of cytokines. This cross-communication between different signaling systems allows the integration of a great diversity of stimuli to reach a cell under varying physiological situations. Considering the ubiquitous cellular distribution of certain cytokine receptors and furthermore, the vast complexity of the cytokine signaling network, the precise physiological function of cytokines *in vivo* is still not fully elucidated.

Given the complexity by which cytokines inter-relate, further complication arises by the fact that cytokines and cytokine receptors are genetically polymorphic [98]. Thereby, different forms of cytokine genes may yield molecules with slightly different but biologically significant bioactivities. Cytokine polymorphisms take the forms of single nucleotide polymorphisms (SNPs), variable number of tandem repeats (VNTRs) and microsatellites. A vast array of cytokines has been shown to be polymorphic at various sites. Some cytokine genes have been well characterized and have a number of SNPs as well as microsatellites described, for example TNF-α and IL-10 [98]. Most of the polymorphisms found in cytokines and their receptors are located in the promoter region, in intronic and in 3' untranslated regions. Even though some cytokine polymorphic loci have been found to alter cytokine production, the remaining described polymorphisms require further investigation. It is often difficult to assess the importance of cytokine polymorphisms in various studies, as the frequency of the polymorphisms may be different according to the ethnic background of the study cohort. Even though the precise role of cytokine polymorphisms remains unclear, some cytokine polymorphisms have been associated with disease, for example IL-1α −889 with Alzheimer's disease [56]. An intriguing and challenging paradigm would consider that cytokine polymorphisms are responsible for the possible outcome of an immune response and could thus render susceptibility to a disease that required an alternative immune reaction. An interesting proposal could suggest that the optimal functioning of the immune system is dependent on inherited immunologic factors, which include among others, the cytokines.

Cytokine Involvement in Normal Physiology

Cytokine Role in Lymphocyte Hematopoiesis

Hematopoiesis is the process by which pluripotent hematopoietic stem cells (HSC) self-renew in the bone marrow and differentiate to give rise to a diversity of cell lineages that constitute the immune system. The proliferative and self-renewal capacity of HSCs diminishes progressively as cells differentiate. The expansion of immature HSCs is believed to be driven by a complex of growth factors, including among others, IL-6, stem cell factor (SCF) and Flt-3 [57, 84, 198, 211]. TGF-β is another growth factor that confers stem cell survival by promoting stem cell quiescence [17, 72, 170, 211]. However, the effects of TGF-β on hematopoiesis are different depending on which cells are involved. For example, TGF-β induces erythroid progenitor cell differentiation, but promotes HSCs to remain in a quiescent undifferentiated state [71]. While the mode of action in both cases is by inhibition of cell cycle activation, cell-specific effects might be due to differences in downstream regulation pathways. A role of TNF-α in HSC proliferation is possible, but its precise effects on HSCs are conflicting, perhaps reflecting a dosage-dependent restriction.

As HSCs divide and differentiate, a decision of myeloid versus lymphoid lineage commitment must be achieved. The regulation of this decision has been under debate, even though the involvement of certain transcription factors has been suggested, such as PU.1 and GATA-1, in early events of lineage commitment decisions [211]. However, for the purposes of this review, we will focus on the impact of cytokines in the hematopoiesis process. HSCs first give rise to a common progenitor cell of either lymphoid or myeloid lineage. The receptor for IL-7 (IL-7R) has been observed on common lymphoid progenitors, but not on myeloid progenitors [153]. Committed IL-7R[+] common lymphoid progenitors generate T lymphocytes, B lymphocytes and NK cells [105], while IL-7R[-] common myeloid precursors further differentiate to monocytes, granulocytes, erythrocytes and megakaryocytes [2]. Therefore, IL-7 appears to be important in lineage commitment decisions through the down-regulation of its receptor. The receptor for GM- CSF (GM-CSFR) also seems important in the initial decision of lymphoid versus myeloid precursor lineage, as its down-regulation is suggested to be among the initial events of the commitment process to common lymphoid progenitors [104]. A possible role for TGF-β has been suggested in the inhibition of the myeloid lineage decision, but further studies need to be performed *in vivo*. Furthermore, an involvement of TGF-β has been speculated in the bias toward erythroid commitment at the expense of granulocytic and monocytic development from common myeloid precursors [55, 72]. An emerging role of IL-11 in early hematopoiesis has also been suggested, mostly involved in various stages of myeloid differentiation, megakaryocytopoiesis and multiple stages of erythropoiesis [159, 195].

Once the decision has been made toward lymphoid commitment, the differentiation process begins for T or B lymphocyte development. Keeping strictly in mind the cytokines implicated in the process and not discussing key transcription factors involved such as Notch-1 and Pax5 [211], steps in the differentiation into T lymphocyte subsets and B lymphocytes involve a variety of cytokines. CD4$^+$ T lymphocytes differentiate into Th1 and Th2 subsets and are subject to cytokine control [137]. Th1 commitment from Th0 bi-potential precursor T cells is mediated by IFN-γ and subsequent upregulation of the IL-12 receptor-β, resulting in the induction of relevant transcription factors [139]. Alternatively, cytokines such as IL-4 and IL-10 are believed to skew early differentiation away from the specifications for the Th1 reaction through inhibition of the transcription factors implicated.

In brief, cytokines are capable of regulatory control of hematopoiesis by modulating the expression of HSC transcription factors. Mesenchymal cells within the bone marrow, constituting the stroma, secrete cytokines involved in the regulation of hematopoiesis and provide the microenvironment that will dictate specific stem cell triggers. The earliest events involved in HSC renewal and differentiation are still under debate. It remains to be elucidated whether the cytokine bone marrow microenvironment is affected by genetic factors such as cytokine polymorphisms or other external factors, leading to changes in hematopoietic control.

Cytokine Role in Lymphocyte Homeostasis

The lymphocyte pool in a healthy individual is of relatively constant size, continuously replenished by entry of new cells and replication of existing cells. Lymphocyte homeostasis is not controlled by production from stem cells, but rather involves regulation of lymphocyte numbers by replication and apoptosis [102]. Lymphocytes acquire extracellular signals instructing them to survive or die. Lymphocyte survival or trophic signals that have been identified thus far include cytokines IL-7 and IL-15 [102]. In T cell development, precursor cells from the bone marrow enter the thymus where they undergo T cell receptor (TCR) rearrangement and positive selection. IL-7 is required for the growth and development of thymocytes through the transition to the pre-T cell stage [142, 153]. Furthermore, IL-7 is involved in the homeostasis of naive (antigen-inexperienced) CD4$^+$ and CD8$^+$ T cells as a second signal of homeostatic survival and proliferation, in addition to TCR signaling [28, 199].

After antigen encounter, T cells are activated to proliferate rapidly and produce cytokines, for example IL-2, which functions as a temporary survival signal until their elimination through the apoptosis pathway. The generation of memory T cells from the pool of activated cells is still under scrutiny, believed to either arise from a subpopulation of cells trained to avoid apoptosis or from a distinctive subset of effector cells [183]. Memory CD4$^+$ or CD8$^+$ T cells do not require antigen stimulation for survival like naïve T cells. However, IL-7 and IL-15 have been implicated as homeostatic regulators of memory CD8$^+$ T cells [175, 184, 200]. A homeostatic factor for memory CD4$^+$ T cells has yet to be identified, creating a gap in our understanding of the regulation of their survival [102].

IL-7 is required for the development of γδ T cells [127] and IL-15 seems to be involved in natural killer (NK) cell maintenance [38], but other homeostatic regulators are possible and currently under study. B cell homeostatic factors are still unclear and not as well understood as those of T cells, but IL-6 is known to play a role in the survival and activation of differentiated antibody-secreting plasma cells [87]. In general, the precise microenvironment for delivery of cytokine signals and the intracellular pathways that regulate lymphocyte homeostasis merit further investigation.

Inflammatory Regulation by Cytokines

Inflammation is a physiologic adaptive response of living tissue to a variety of stimuli and is necessary for survival. The ultimate goal of this response is to repair damaged tissue, limit tissue invasion and restore normal tissue integrity. The early stages of a local inflammatory response involve vasodilation, microvascular structural changes and plasma protein leakage from the bloodstream, and lastly, leukocyte transmigration to the site of injury or infection [161]. Leukocyte recruitment occurs in response to changes in cellular adhesion molecules, the expression of which can be affected by cytokines [161]. The timely resolution of the inflammatory response occurs when the recruited effector leukocytes are cleared and the inflammatory mediator profiles, responsible for propagating the response, return to normal levels. A chronic inflammatory condition resulting from disorder of the resolution phase is emerging as the leading cause of many disease pathologies. The negative regulation of cytokine signaling is important for the resolution of inflammation [210]. Th1 cytokines, such as IL-1, IL-8, MCP-1, TNF-α and IFN-γ, are generally considered pro-inflammatory and associated with cellular immunity. However, Th2 cytokines associated with humoral immunity, such as IL-6, are also involved with inflammation. Induction and suppression of the signal transduction pathways of these cytokines are also involved in inflammatory response regulation. The JAK/STAT pathway, associated with cytokine signaling, is negatively regulated by a family of endogenous JAK inhibitor proteins referred to as suppressors of cytokine signaling (SOCS), each member being responsible for the inhibition of a particular STAT pathway [185]. As different JAKs and STATs are involved in cytokine signal transduction, downstream regulatory pathways of Th1 and Th2 cytokines may work in association or in opposition to each other in order to reach a particular response. For example, since IL-12 activates the STAT4 pathway and IL-4 the STAT6 pathway, STAT4 and STAT6 are essential for Th1 and Th2 lymphocyte development respectively [210]. It is interesting to note, however, that the pro-inflammatory cytokine IL-6 and the anti-inflammatory cytokine IL-10 both activate STAT3 [210]. The level of inflammatory regulation lies in part on the interaction of these cytokines and their pathways and the suppressive control they have on each other.

Inflammatory cytokines act by enhancing the expression of selectins and integrins, adhesion molecules necessary for capture, rolling, adhesion and eventually recruitment of leukocytes to the site of inflammation or injury [161]. For example, IL-1 and TNF-α upregulate the transcription of E-selectin and P-selectin [65, 161], thus facilitating leukocyte capture and rolling along the vessel margin. Chemokines, such as IL-8 and MCP-1, cause rolling of monocytes and firm adherence to E-selectin [77, 116, 121]. Cytokines are also involved in the induction of integrins and intercellular adhesion molecules, responsible for the transmigration of leukocytes across endothelial vessel walls [161]. The involvement of cytokines in the inflammatory pathway has led to the association of specific cytokines to inflammatory disorders, some of which will be discussed in the sections below.

Regulation of Normal Reproductive Function

Evidence suggests that several reproductive tract processes, such as menstruation, ovulation, implantation and parturition, are associated with inflammation [36, 76, 97, 178, 204]. Cytokines, therefore, play a key role in the normal functioning of the reproductive system. The menstrual cycle may be considered as a series of cyclical hormone-driven changes resembling an inflammatory process. Even though Th1 cytokines are mainly considered as pro-inflammatory, Th2 cytokines, such as IL-6, are associated with reproductive tract processes [97]. Elevated levels of IL-8 and consequently increased neutrophil accumulation have been observed pre-menstrually [7]. MCP-1, a potent chemoattractant for monocytes is also found at increased levels in the endometrial perivascular region before menstruation [6]. A role for pro-inflammatory cytokines IL-1α and IL-1β has also been suggested in the pre-ovulatory stage of the menstrual cycle, probably due to induction of IL-8 [76, 169]. Other investigators reported that IL-6, IL-8 and IL-1 receptor antagonist (IL-1ra) are involved in peri-ovulatory cellular interactions [36], suggesting a further association of inflammation to ovulation. This suggests that ovulation may be an inflammatory process regulated by cytokines IL-6 and IL-8, followed by an anti-inflammatory phase (resolution), which is mediated by IL-1ra [36].

Another level of cytokine control is through their effect on prostaglandin (PG) production. Prostaglandins are lipid mediators, which are essential for the normal functioning of the reproductive tract, the actions of which have been extensively studied and detailed by other investigators [14, 97, 100, 101]. Cytokine-prostaglandin interaction is crucial in the reproductive system. For example, IL-1β is a known inducer of PGE_2 and PGF_{2a}, a result mediated by upregulation of COX-2 and phospholipase A2, thus activating the PG production cascade [79, 90]. Since PGs are also involved in most normal reproductive processes including the ovulatory process, implantation, successful pregnancy and parturition, cytokines may guide these reproductive functions via PG interaction.

One important step in achieving a successful pregnancy is embryonic implantation. The IL-1 system has been associated with implantation, taking part in the appropriate regulation of integrins and other adhesion molecules at the implantation site [110]. In addition to regulation by PGs, the embryonic invasiveness is believed to result in part from extracellular matrix remodeling by matrix metalloproteinases (MMPs) [25, 144], influenced by IL-1β. As a counterbalance to the IL-1β response, TGF-β is believed to reverse MMP upregulation and to further induce the expression of tissue inhibitors to MMPs (TIMPs), limiting the potential for endometrial invasion [91]. Even though human implantation is a highly complex subject and cannot be covered in detail herein, it is clear that cytokine balance is involved in this process.

Interactions between cytokines and prostaglandins have emerged as critical components of successful pregnancy [97, 100]. The existing paradigm supports that the balance of Th1/Type-1 and Th2/Type-2 cytokines is crucial for normal reproductive functions, including implantation and pregnancy [154, 167]. It is widely believed that Th2/Th3 cytokines and thus the humoral arm of the immune response provide an environment necessary for successful pregnancy. Alternatively, the actions of Th1 cytokines are thought of as detrimental to the fetus. In particular, IL-10 is a prototypic Th2 cytokine and a key regulator of Th2 immune responses, since it downregulates the expression of Th1 cytokines, such as IL-12 and IFN-γ. Inappropriate Th1 activation either due to infection or injury, may thus lead to adverse conditions such as preterm labor. Proinflammatory cytokines such as IL-1β, IL-6, TNF-α and GM-CSF have been found to increase with preterm labor [97] and are responsible for inducing the cellular arm (Th1) of the immune response, hindering the maintenance of pregnancy. In fact, the process of parturition is considered as inflammatory and involves pro-inflammatory cytokines that together with PGs and MMPs, provide an environment in the uterus to uphold parturition. The chemokine IL-8 is a strong activator and chemoattractant for neutrophils, found at high levels at the onset of parturition [61]. When activated, neutrophils release their specific collagenase-containing granules, taking part in the degradation of collagen fibers and cervical ripening. MCP-1 is also upregulated in the cervix [64] and may cause accumulation of leukocytes contributing to cervical ripening. In addition, activated macrophages and neutrophils infiltrating the myometrium are a rich source of proinflammatory mediators, including IL-1.

Cytokine involvement in reproductive functions is highly complex and involves many levels of regulation and interactions with other key molecules. However, it is obvious that cytokine balance affects the microenvironment that directs reproductive system functions and is thus important for human existence.

Cytokine Levels in Disease Pathology

Autoimmunity

Severe skewing of the Th1/Th2 cytokine balance is characteristic of several common human diseases, including acute and chronic infections, autoimmunity and allergy [62, 63, 137]. The induction of autoimmunity is a complex process, commonly associated with the breaching of several regulatory checkpoints. In general, autoimmunity may result from the activation of a significant number of functional auto-reactive lymphocytes or by presentation of self-antigens on activated antigen-presenting cells (APCs) [172]. Several autoimmune diseases result from the balance being weighed toward Th1 immune reactions and away from Th2 responses. Therapeutic intervention may thus target to suppress the autoimmune effector cells or to deviate the immune response to that of Th2.

Particularly, in rheumatoid arthritis (RA) proinflammatory cytokines IL-1 and TNF-α are strongly associated with synovial hypertrophy, degradation of cartilage and resorption of bone [73]. TNF-α has been found at elevated levels in patients with severe RA and blocking its activity has been suggested as possible therapeutic intervention of autoimmune diseases [5]. Increased TNF-α and IL-12 and the resulting induction of pro-inflammatory transcription factor NFκB (c-Rel and p50) have been described as major mediators of inflammation in other Th1-associated autoimmune diseases, including multiple sclerosis (MS), type I diabetes mellitus and systemic lupus erythematosus (SLE) [62, 207]. The distinction between Th1 versus Th2 cytokine characterization in autoimmunity is dependent on various factors, including disease progression. For example, the multisystem inflammation and production of autoantibodies that characterizes SLE, may take initially a Th1 nature as deduced by elevated IL-12 and IL-18 levels in patient sera, but reverts in a percentage of disease-active patients to a Th2 expression as suggested by persistent IL-10 levels [81, 134, 136, 166].

In addition to its inflammatory properties, TNF-α may also control leukocyte trafficking via the regulation of chemokine receptors [96]. Chemokine receptor expression has been associated with the immune activity of type I diabetes [119]. Chemokine receptors CCR5 and CXCR3 are associated with Th1 and CCR3 together with CCR4 are Th2-related. The effects of TNF-α have been somewhat controversial in the progress of autoimmunity and it appears that the timing of TNF-α expression during the course of the disease is a critical component for the occurrence of autoimmunity [207]. Even though the abovementioned cytokines have been extensively studied in the context of autoimmune diseases, a recently identified cytokine, IL-23, has been shown to share its p40 subunit with that of IL-12 [149]. It is suggested that redundancy in the functions of IL-12 and IL-23 may be the cause of some autoimmune diseases, thus far attributed solely to IL-12 [45]. Constant emergence of new findings indicates how limited our understanding is of these complicated cytokine interactions resulting in autoimmune diseases.

Atopy/Allergy

Dominant Th2 responses have been characterized in several allergic diseases, including asthma, eczema and IgE-mediated food allergy. Overproduction of Th2/Type-2 cytokines, including IL-4, IL-5, IL-9 and IL-13, histamine and a shift to IgE production have been associated with allergy [62, 122]. Asthma is thought to be a complex chronic inflammatory disorder of the airways resulting in symptoms of coughing, wheezing and shortness of breath associated with evidence of bronchial hyper-reactivity [27]. The factors resulting in disease persistence or resolution remain poorly understood, but the skewing of immune responses away from Th1 and toward Th2 is a promising paradigm [146, 180, 182]. Increasing levels of IL-13 coupled with impaired IL-12 production appear to be closely associated with asthma and may underlie the Th2-biased response [205]. In addition, investigators have implicated defective IFN-γ production as a predisposing factor to the development of atopy [190, 202]. These studies have shown that children with asthma had reduced IFN-γ and increased IL-4 reactions, while persistence of asthma in adults involved a reduced Th1 and increased Th2 cytokine environment in response to allergen. In addition, the ratio of IFNγ to IL-4 producing T cells was determined to be lower in atopic asthma patients compared to healthy controls [180]. Furthermore, a positive correlation between total IgE levels and cells producing IL-4 was observed in asthma patients [180]. As these observations suggest that amplified Th2 and diminished Th1 responses are associated with ongoing disease, it is tempting to suggest that the suppression of Th2 and induction of Th1 reactions may be beneficial for disease remission.

Atherosclerosis

Atherosclerosis, the hallmark of coronary artery disease (CAD), is considered to be a slow, complex and multifactorial disease that is determined by the interaction of environmental and genetic risk factors [42]. Increasing evidence suggests that atherosclerosis represents a low-grade chronic inflammatory disorder with immunologic nature that affects the walls of medium to large-sized arteries [40, 117, 147]. As inflammation results from the balance between pro- and anti-inflammatory mediators, the involvement of immune inflammatory mediators may be associated with disease pathogenesis. The emerging paradigm suggests that Th1 and Th2 responses are involved at different stages of the inflammatory processes that lead to atheroma formation [51, 52, 143, 208]. In particular, the pro-inflammatory cytokine IL-6 is involved in a diverse range of biological activities and specifically the synthesis of acute-phase reactants in the liver [22, 26]. Whether IL-6 is an independent predictor for atheroma formation or it has indirect effects through acute-phase protein production, remains to be elucidated. Elevated IL-6 levels have been associated with unstable angina [208]. In addition to C-reactive protein (CRP) production, actions performed by IL-6 include increases in fibrinogen levels and changes in adhesion properties of neutrophils and monocytes, thus contributing to an atherosclerotic-prone environment [3, 26, 209]. The pro-inflammatory actions of TNF-α justify its suggested involvement with CAD and its possible role in disease pathogenesis. The conventional cardiovascular disease risk factors, including hypertension, smoking and diabetes are capable of instigating the release of IL-6 and TNF-α, thereby promoting the initiation of low-grade inflammation [69]. In turn, these pro-inflammatory cytokines may cause the increase of CRP serum levels, currently recognized to have an inflammatory role in acute coronary syndromes [69, 117]. Elevated levels of pro-inflammatory cytokine IL-18 have been implicated in unstable angina and acute coronary syndromes [128, 208]. IL-18 stimulates IFN-γ production in synergy with IL-12 and also promotes the release of other pro-inflammatory mediators, such as TNF-α, IL-1β, IL-6, IL-8 and MMPs [3]. During the inflammatory response accompanying atherogenesis, anti-inflammatory cytokines are also produced for the regulation of the inflammatory process [69]. IL-10, a Th2 anti-inflammatory mediator and suppressor of the immune response, has been suggested as a therapeutic target for atherosclerosis due to its capability to suppress the Th1 immune response [143], the latter being associated with the initiation of atherogenesis. IL-10 also inhibits the NFκB pathway and consequently a critical pro-inflammatory transcription factor cascade [193]. A study by Heeschen and colleagues described reduced levels of IL-10 in patients with acute coronary syndrome and further indicated an inverse correlation between IL-10 and CRP levels in these patients [83]. In addition, other investigators observed a negative correlation between IL-10 and levels of pro-inflammatory mediators IL-6, IL-12 and CRP, suggesting Th1 dominance in unstable angina patients [3, 208]. Despite the fact that the aforementioned cytokines are recognized to play a role in the regulation of inflammation accompanying atherosclerosis, the initiation of inflammation is still elusive and the underlying regulation and molecular mechanisms of cytokine involvement remain unclear.

Reproductive Dysfunction

Since Th1/Th2 cytokine balance within the reproductive tract microenvironment appears to be crucial in several normal reproductive tract processes, it is tempting to assume that skewing of this balance may lead to adverse effects in reproductive function. Inappropriate Th1 activation either due to infection or injury has been associated with pregnancy loss in the first trimester [48, 160, 164]. However, the reports on recurrent pregnancy loss and cytokine balance have thus far been inconsistent. The consensus supports that normal pregnancy is Th2-biased and Th1 cytokines, such as IL-1β, IL-6, TNF-α and IFN-γ, may be detrimental for the maintenance of pregnancy and may be the subsequent cause for miscarriage [48, 126, 164]. Studies involving recurrent spontaneous abortion (RSA) have been conflicting [18, 82], partly due to differences in patient sampling and perhaps in the timing of sample collection. Some reports indicate that women with RSA have significantly elevated levels of IL-12 compared to normal pregnancy controls [160]. Other investigators observed increased levels of IFN-γ and TNF-α in women with RSA [48, 164], further supporting the hypothesis of Th1 cytokine involvement in RSA pathogenesis. In light of recent evidence suggesting that implantation requires the involvement of inflammatory mediators [36], it appears that the cytokine balance may shift during pregnancy according to the different stages concerned. TGF-β and IL-10 levels have been observed to vary throughout pregnancy, reflecting a greater need of immune suppression towards the last trimester of pregnancy than during the first 18 weeks of gestation [157]. This may in part explain the findings by Bates and colleagues, indicating that increased IL-4 and IL-10, but not IFN-γ were observed in RSA women compared to normal pregnancy controls [18]. Excessive Th2 activation has also been suggested to be harmful for pregnancy, as it may trigger alloantibody and autoantibody production [82]. It is apparent that the fine balance of interacting cytokines and the timing of cytokine manifestation is critical for RSA, but the involvement of cytokines with the pathogenesis of RSA merits further investigation.

Cytokine balance has also been studied in women suffering from endometriosis. Endometriosis is characterized by the growth of endometrial tissue in ectopic sites and has recently been attributed an inflammatory character. Increased inflammation, increased leukocytes and activated macrophages have been observed in the peritoneal fluid of women with endometriosis [15, 112]. The findings of aberrant expression of IL-1, IL-6, IL-8 and TNF-α in the peritoneal fluid of women with endometriosis suggest that endometriosis may be facilitated by a peritoneal microenvironment of specific cytokine equilibrium [15, 112, 165]. Indeed, IL-8 and TNF-α were shown to directly stimulate endometrial cell proliferation, endometrial adhesion and angiogenesis and may thus actively support the endometriosis process [34, 112]. IL-8 concentrations also correlated with the severity of disease [8]. TNF-α reflects enhanced peritoneal macrophage activity and has also been associated with disease progression [34]. Elevated levels of TGF-β have been involved with endometriosis and in particular with induction of angiogenesis and the recruitment of monocytes [15, 148]. High peritoneal fluid levels of IL-12 have recently been associated with moderate or severe endometriosis [74]. Alternatively, the same study showed significantly lower peritoneal fluid IL-13 levels in women with endometriosis than in healthy controls, suggesting the existence of reciprocal modulation between Th1 and Th2 peritoneal fluid cytokines in endometriosis patients. Whether the changes in inflammatory and immunologic mediators are the cause or consequence of endometriosis remains to be elucidated, but the importance of cytokines is highlighted in new therapeutic treatment options considered for future management of endometriosis.

Hematological Malignancies

Leukemias are defined by neoplastic proliferations of immature cells of the hematopoietic system, which are characterized by arrested or abnormal differentiation. The precise role of the cytokine network in the development of hematological malignancies is not clear and in some cases controversial [135]. Cytokines act as microenvironmental immunomodulators that may affect leukemia development and outcome [176]. The involvement of cytokines in leukemia has been suggested after observations were made of certain cytokines being detected at elevated levels in leukemia patients, for example IL-6, IL-10 and TNF-α in chronic lymphocytic leukemia (CLL) [66, 68] and IL-6 in multiple myeloma [114]. Elevated IL-6 serum levels were observed in patients with B-CLL, but IL-1β and IL-1ra serum levels were suggested by the same study to be lower in B-CLL patients [92]. Elevated IL-10 serum levels have been reported in other hematological malignancies, including Hodgkin's lymphoma, non-Hodgkin's lymphoma, and acute myelocytic leukemia (AML) [54]. In addition, IL-4 was found at higher levels in sera of patients with T-CLL and IFN-γ was observed at increased levels in patients with T-cell CLL and T-cell acute lymphocytic leukemia (T-ALL) [162]. Studies have also shown that TGF-β serum levels were altered in patients with various types of leukemia, including AML [43] and CLL [78] and further suggested that serum TGF-β may be a valuable parameter in monitoring the prognosis of leukemia [43]. These observations suggest that further studies investigating cytokine involvement in leukemogenesis are warranted, as they could provide invaluable insight to the complicated mechanisms surrounding leukemia development and may eventually lead to new therapeutic intervention.

Research Focus on Cytokine Genetic Polymorphisms and Disease Association

Inflammation-Associated Diseases

After the association of altered cytokine production levels with the expression of common genetic variants, a genetic regulatory role has been suggested in the production of cytokines. Cytokine polymorphisms have been suggested as predisposing factors in disease manifestation, especially in diseases that are facilitated by inflammatory or skewed Th1/Th2 immune responses [98, 197]. Studies associating genetic polymorphisms to disease must be performed in a cautious manner, as most diseases are heterogeneous in their manifestation and pinpointing the cause to a single component can be perhaps unconvincing. Several studies investigating inflammation-associated diseases, such as atherosclerosis and periodontal disease, have demonstrated associations of pro-inflammatory cytokine polymorphisms with these diseases [50, 93, 113, 151]. Care should be taken, however, in the interpretation of these studies with respect to study design, patient and control population sampling, number of subjects used and the disease stage (onset, progression, remission, exacerbation, amelioration) associated with the polymorphism under investigation.

Pathogenesis of atherosclerosis and coronary artery disease involve inflammatory and genetic components. Studies have associated elevated CRP levels with the IL-1β +3954 genotype [23, 60]. In addition, the combination of IL-1β −511*1 and IL-1β +3954*2 polymorphisms was more frequent in subjects with low CRP levels, thus associating IL-1β genetic control with regulation of CRP. The involvement of IL-6 in the inflammation accompanying atherosclerosis led to association studies of IL-6 genotypes in cardiovascular disease patients. Distribution of the IL-6 −174 C genotype was demonstrated to differ among patients with peripheral artery occlusive disease and control subjects [70]. Alternatively, allele 2 of anti-inflammatory IL-1ra was associated with susceptibility to the development of carotid atherosclerosis after patients with atherosclerosis were found to express a significantly higher frequency of IL-1ra allele 2 than did healthy controls [206]. An investigative approach employed by Koumas and colleagues utilizes the strategy of cross-tabulation of cytokine polymorphisms to determine the significance of one cytokine in the context of another. The association of cytokine gene polymorphism combinations was studied in relation to various pathologic conditions including cardiovascular disease in the Cypriot population [107]. In this study, polymorphism frequencies of pro-inflammatory cytokines IL-6 and TNF-α in combination with genotype frequencies of anti-inflammatory cytokine IL-10 were significantly altered in the group of coronary artery disease (CAD) patients compared to healthy controls. The distribution of the IL-10 ATA haplotype was also significantly different in CAD patients compared to the control population. Less CAD patients had the TNFα -308 A(-)/IL-10 ATA(-) genotype combination than controls, but no differences were observed between patients and controls with the TNFα -308 A(+) genotype. Even though no differences were observed among CAD patients and healthy controls with the IL-6 -174 G(-) genotype, significantly less patients had the genotype combination IL-6 -174 G(+)/IL-10 ATA(-). The importance of cytokine interaction in immune-regulated diseases is reflected in this study by the association of IL-10 haplotypes with IL-6 and TNFα functional polymorphisms, as the significance of IL-10 ATA haplotype frequencies was retained only with the presence of certain IL-6 and TNFα polymorphisms. In conclusion, differential cytokine genetic polymorphism profiles may be associated with the development of atherosclerosis, but further studies are required to elucidate the precise interactions of cytokine polymorphisms in cardiovascular disease.

A similar association of pro- and anti-inflammatory cytokine gene polymorphisms was observed in patients with periodontal disease. The pathogenesis of periodontal disease is inflammatory in character, resulting in gingivitis and destruction of the supporting tissues of the teeth [173]. Polymorphisms within the IL-1 gene cluster have been associated with chronic periodontitis. Evidence suggested a significant association of allele 1 of IL-1β with risk of aggressive periodontitis [151, 191]. The IL-1RN (IL-1ra gene) intron2 VNTR genotype was also associated with periodontal disease pathogenesis [188]. Alternatively, polymorphisms in the promoter region of IL-10 were found to be associated with periodontal disease, where the ATA low-expressing haplotype was more prevalent in patients with chronic periodontitis than in healthy controls [173].

A role for inflammation-related genes has been associated with the pathogenesis of Alzheimer's disease. In particular, the IL-1α -889 TT genotype has been associated with sporadic early onset of Alzheimer's disease [177]. IL-1β +3953 genotype distribution varied significantly between patients with early and late onset Alzheimer's disease, further suggesting a role for inflammatory genes in Alzheimer's [177]. Other investigators, however, did not reach the same conclusions with respect to the polymorphism associations described above for cardiovascular disease [10, 133], periodontitis [151] or Alzheimer's disease [111, 123], indicating limitations in study design among different centers. The existing studies, however, support the hypothesis that the balance between pro- and anti- inflammatory cytokines might be reflected in the individual's genetic code and might predispose to certain conditions that are associated with inflammation.

Autoimmunity and Allergy/Atopy

Classification of chronic inflammatory disorders according to the predominance of Th1 or Th2 immune responses has provided insight to possible genetic associations involved in disease pathogenesis. Common genetic linkages associated with some chronic inflammatory diseases, such as diabetes mellitus [59], rheumatoid arthritis (RA) [80], SLE [152] and bronchial asthma [85], have led to the hypothesis that there might be common underlying genes for different inflammatory disorders. An interesting study by Heinzmann and colleagues, compared a single polymorphism of IL-13 (Arg110Gln), associated with heightened IL-13 serum levels and increased Th2 response, in Th1 and Th2 chronic inflammatory diseases. The study demonstrated that the variant Arg110Gln of IL-13 was associated with increased levels of total serum IgE in asthmatic patients and was found at higher frequency among patients with bronchial asthma compared to patients with RA [86]. Considering that autoimmune Th1 diseases include RA and diabetes mellitus and that Th2-associated diseases comprise allergy and asthma, it is tempting to speculate that an inverse association between these disease groups exists within the same individual.

Considering that asthma involves a complex polymorphic clinical picture, it appears that gene polymorphisms are important in representing immunological profiles of an individual that might confer disease susceptibility rather than simple genetic predisposition. Studies investigated the role of polymorphisms of IL-4, a prototypic Th2 cytokine essential for development of airway inflammation and IgE synthesis, in conferring susceptibility to the development of atopy or asthma [20]. Haplotype analysis revealed a strong association between asthma and the IL-4 -34T/IL-4 -589T haplotype as well as the IL-4Rα I50/IL-4Rα Q567 haplotype of the IL-4 receptor [20]. TGF-β genetic control was also associated with asthma diagnosis, since a positive correlation was found with the disease and the TGF-β -509T polymorphism, an association independent of total IgE serum levels [181]. Collectively, it appears that Th2 and Th3 cytokine genes, such as those for IL-4 and TGF-β, might play a role in conferring susceptibility to atopy and asthma.

Studies on genetic influences on RA have focused primarily on polymorphisms in the TNF-α and IL-1 genes, with conflicting results. Even though these pro-inflammatory cytokines are found at elevated levels in patients with RA [174], several investigators have failed to attribute a role for TNF-α genotypes to the disease [138, 150]. Some studies, however, suggested that the TNF-α -308 A1 allele as well as presence of the IL-1β A2A2 genotype were associated with RA and disease severity [47]. It is apparent that direct comparison of the different studies is difficult, since the different ethnic populations investigated and the different RA subgroups sampled impose various limitations [150, 168]. Inconsistent findings have also been reported for IL-1β and IL-1RN promoter polymorphisms and their association to severe forms of RA. The IL-1β +3954 polymorphism was associated with severity of joint destruction in RA [32], but other studies claim that there were no significant differences [47]. Genetic polymorphisms in the Th1-opposing cytokines IL-1RN*2 and IL-4 VNTR in intron 3 were suggested to be protective factors for severe joint destruction and RA pathogenesis [33, 115]. Gene polymorphism frequencies of IL-10 -1087 were reported to be altered in female RA patients, where the IL-10 −1087 A/A low-expressing genotype was found at higher frequency in patients than controls [150]. In addition, the functional promoter polymorphism of MMP-1, 2G/2G, was suggested to be related to clinical phenotypes of RA, but not to disease susceptibility [115]. It is clear that further studies need to be performed in order to reach a consensus on genetic determinants associated with RA pathogenesis.

Association studies of cytokine gene polymorphisms and diabetes are conflicting and vary among different ethnic populations. Studies suggested that the frequency of the TNF-R1 −383C allele, conferring high expression, was significantly increased in diabetes patients compared to healthy controls [145]. As this polymorphism confers high TNF-R1 expression, the TNF-R1 gene region might be associated with increased TNF-α activity and susceptibility to diabetes. The IL-10 low-expressing ATA haplotype was associated with the age-at-onset in patients with type-1 diabetes as it was found at higher frequency in adult-onset than early-onset patients [94]. In addition, the frequency of the IL-6 -174 GG genotype was increased in patients with type-1 diabetes compared to healthy controls [95]. As these findings suggest genetic control of low-expressing Th2 and normal/high-expressing Th1 reactions in the predisposition to type-1 diabetes, other studies are in disagreement [41, 179], creating inconsistencies in the literature.

Cytokine Polymorphisms in Reproductive Disorders

When the determination of a disorder phenotype is dependent on net cytokine balance, it is difficult to pinpoint a single factor as the culprit in disease pathogenesis. Investigations concerning women suffering from recurrent spontaneous abortion (RSA) are difficult to compare since patient sampling greatly differs among studies. Several reports suggested association of IFN-γ +874 T/T, TNF-α −308A and IL-10 −1082G/G high-expressing genotypes with RSA, but associations with the IL-6 −174 G to C polymorphism were not observed [49]. Involvement of the IFN-γ +874 A/T genotype was also found at higher frequency in RSA patients compared to controls, indicative of a Th1-favored environment [158]. Other studies also agreed that certain cytokine gene polymorphisms, namely the rare forms of Th1 cytokines IL-1β and TNF-α, were more pronounced in women with two or more miscarriages compared to women with normal pregnancies [163, 201]. The findings by other investigators, however, indicating no significant association in the frequencies of IFN-γ +874 A/T, TNF-α -308 G/A and IL-10 –1082 G/A polymorphisms in RSA patients [13, 155], indicates inconsistency in patient and control population sampling and perhaps in different perception of the definition of RSA.

Since cytokines function in a complex, interacting microenvironment, Costeas and colleagues studied cytokine polymorphisms in association to one another and referred to these combinations as immunogenetic profiles. Even though no statistical difference was observed in individual cytokine polymorphism frequencies among RSA patients and multiparous controls, the combinations of cytokine polymorphisms studied were distinct between women with successful pregnancies and those prone to miscarriages. Women suffering from RSA were characterized by cytokine genotypes that suggested either inadequate or excessive Th2/Th3 immunity [44]. In particular, the IL-10 ATA(+)/TGFβc25 C(+) genotype, reflecting insufficient Th2/Th3 immunity, was more frequent in RSA patients than in the control population. Excessive Th2/Th3 reactions were represented in the IL-10 ATA(-)/TGFβc10 C(-) immunogenetic profile, which was found to be more frequent in women suffering from RSA [44]. Furthermore, more RSA patients had the IL-6 genotype conferring low IL-6 expression (IL-6 –174 C/C) together with IL-10 ATA(-) haplotypes, indicating that balance between these Th2 cytokines is crucial in fertility outcome. Within the context of IL-10 haplotypes, IFN-γ +874 A/T and TNF-α –308 G/A polymorphism frequencies were significantly different in women suffering from RSA, with the polymorphisms conferring the highest cytokine expression being more prevalent [44]. In particular, a combination of low expression of Th2 cytokines with high expression of Th1 cytokines was more prevalent in RSA patients, suggesting intrinsic ability to mount excessive Th1 immune responses. Other reports are in agreement with these observations as gene polymorphisms of Th1 cytokines IL-1β and TNF-α were associated in women with recurrent pregnancy loss, which may predispose them to vigorous Th1 responses [88, 163]. It is thus possible that RSA patients have a propensity for miscarriages, reflected in their immunogenetic makeup.

Endometriosis is another reproductive tract disorder with possible genetic predisposition, in addition to environmental factors, for its pathogenesis. Studies involving the implication of cytokine gene polymorphisms in endometriosis focused again on the effect of the Th1/Th2 cytokine balance [103]. The frequencies of the IFN-γ intron 1 microsatellite $(CA)_n$ repeat polymorphism, but not IL-4 –590C/T genotype frequencies, were recently associated with susceptibility to endometriosis in a Japanese population [103]. Another current study suggested another Th1 cytokine, TNF-α, may have genetic influence on endometriosis susceptibility [194]. In particular, polymorphisms at positions –1031, -863 and –857 of the TNF-α gene promoter region were investigated and their frequencies were statistically significant in patients with endometriosis compared to healthy controls [194]. It seems likely that the balance between cytokine polymorphisms is vital in reproductive functions and it is possible that different immunogenetic profiles may lead to different types of reproductive problems.

Cytokine Genetic Variability and Transplantation

Even though organ transplantation increases life expectancy and life style advantages, acute allograft rejection remains a major complication and cause of morbidity. As immune response regulators, cytokines may be implicated in acute or chronic allograft rejection by providing a pro-inflammatory environment [130]. Alternatively, cytokines may allow allograft tolerance through an anti-inflammatory immune balance. Several studies have indicated that genetic control of cytokines may be involved with the outcome of transplantation. Studies have shown that the high-producing TNF-α –308A allele was associated with rejection of kidney and liver grafts [9, 130, 156], but other investigators have failed to demonstrate this effect [75, 131]. More specifically, the presence of the TNF-α –308 A/A genotype was significantly associated with acute cellular rejection in liver transplantation [19]. Significant increases were also observed in IFN-γ microsatellite CA-repeat allele #2 in association with fibrosis complications after lung transplantation [11]. IL-10 has been studied as an anti-inflammatory cytokine and associations have been suggested between the IL-10 low-producer –1082 AA genotype and acute rejection episodes in renal transplantation [75]. Some investigators have also focused on the integrated effects of cytokine polymorphisms and the outcome of transplantation. In renal transplantation, recipients with the TNF-α –308G allele and high-producing IL-10 –1082G allele were shown to experience a significantly higher number of acute rejection episodes after HLA-DR-mismatched transplants [171]. Conversely, recipients with the low-producing TNF-α –308G/G genotype and high or intermediate-producing IL-10 GCC haplotypes were associated with the lowest risk of acute rejection episodes after pediatric heart transplantation [12]. The conflicting results may be partly explained by tissue-specific effects determined by the transplanted organ. The association of certain recipient cytokine genotypes and acute rejections after organ transplantation may allow future predictions of the course of the allograft and greatly improve transplantation success.

Graft-versus-host disease (GVHD) is the principal serious complication of allogeneic bone marrow transplantation (BMT) [39]. Pro-inflammatory cytokines are thought to be important mediators of the regulation and induction of GVHD [109]. On the other hand, anti-inflammatory cytokines may be associated with transplantation tolerance and decreased GVHD [89, 106]. Studies involving sibling BMT donor/recipient pairs indicated that the recipient response was critical in the outcome of BMT, especially with respect to acute GVHD. At the genetic level, polymorphisms in IFN-γ intron 1 microsatellite (CA)ₙ repeats and IL-6 –174 were associated with acute GVHD severity [39]. Recipients with the IFN-γ intron 1-3/3 homozygous genotype were more likely to develop severe acute GVHD, whereas recipients with the IL-6 –174G/G genotype had a greater incidence of chronic GVHD [39]. Studies involving the IL-1 family, suggested that donor possession of the IL-1α –889 allele*2 and IL-1α VNTR polymorphisms in sibling allogeneic BMT was associated with the occurrence of chronic, but not acute GVHD [46]. In unrelated BMT, IL-1 polymorphisms were suggested to influence survival, but not to alter risk for GVHD [124]. IL-10 has been associated with sibling BMT success, since the IL-10 –592A/A genotype has been suggested to decrease the risk of grade III or IV acute GVHD and thus to confer a favorable outcome after transplantation [118]. The inflammatory destruction of host tissue in GVHD is also affected by the pro-inflammatory cytokine IL-2. The high-producing genotype of IL-2 -330 has been suggested to confer increased risk for acute GVHD after unrelated BMT [124]. TNF-α polymorphisms together with IL-10 genetic variants have also been shown to correlate with unrelated BMT outcome. Investigators suggested that the TNFd4 and TNF-α – 1031C alleles were associated with high risk for transplant-related mortality [99]. In addition, an association was made between presence of donor IL-10 GCC haplotype (high-expressing) and decreased risk for transplant-related mortality, supporting an anti-inflammatory protective role within the graft [99]. Alternatively, in another recent study, recipient IL-10 ATA haplotype (low-expressing) was found to be an independent risk factor for chronic GVHD [140]. Collectively, these studies should ultimately lead to a compilation of genetic risk factor assignment in BMT and in the development of a predictive model of possible BMT success.

Hematological Malignancies

The involvement of certain cytokines in leukemia has been suggested after observations were made of altered serum levels in patients suffering from hematological malignancies. Several studies addressed the possible genetic influence on elevated cytokines found in leukemia patient sera. One report suggested the genetic association of TNF-α –308G with CLL [53], but subsequent studies have failed to support such an association in patients with CLL, ALL, non-Hodgkin's lymphoma (NHL) and Hodgkin's lymphoma (HL) [125, 189, 203]. The frequency of the high-producing TNF-α –857T allele was, however, associated with adult T-cell leukemia/lymphoma in individuals infected with human T-lymphotropic virus type-1 (HTLV-1) as compared to healthy carriers [196]. The observed alterations in serum levels of IL-6, IL-1β and IL-1ra in patients with B-CLL were suggested to be independent of genetic cytokine control [92]. Studies investigating IL-10 –1082G/A, IL-10 – 592C/A and IFN-γ intron 1 CA-repeat polymorphisms as predisposing factors in the development of HL, also failed to attribute statistical significance in genotype association to the disease [141].

Even though the majority of current research does not appear to support the involvement of cytokine genetic polymorphisms in the development of hematological malignancies, an investigative approach of integrated cytokine polymorphisms appears promising in understanding the net effects of genetic cytokine association in leukemia. Preliminary evidence by Costeas and colleagues suggests that the genotype distribution of IL-10 haplotypes in combination with TGF-β and IL-6 polymorphisms was statistically significant when leukemia patients and healthy control subjects were compared (Costeas *et al.*, unpublished observations). These observations suggest that when the combination of low-producing TGF-β and low-producing IL-10 polymorphisms were involved, leukemia development may be facilitated. In addition, combinations of genotypes conferring high IL-6 and high IL-10 production appeared to be reduced in leukemia patients, indicating excessive Th2 immunity (Costeas *et al.*, unpublished observations). Conversely, insufficient Th2/Th3 immunity, reflected by low-producing TGF-β and low-producing IL-10 genotypes, was observed in the genetic code of more leukemia patients than control population subjects. This preliminary data supports observations made by other investigators suggesting that abnormalities in cytokines and the cytokine network may be involved in hematological malignancies [108, 135, 211]. However, cytokine network complexity arising from cytokine pleiotropy, redundancy, synergy and antagonism should not be ignored and further studies are warranted involving other cytokine polymorphisms in order to create a complete perception of possible cytokine genetic involvement in leukemia.

Cytokines in Therapeutic Intervention

Considering the plethora of diseases with immunologic character, strategies for intervening with cytokine pathways have been developed and many are currently approved as therapeutic treatments. Several possible approaches for cytokine inhibition exist, including drugs inhibiting cytokine synthesis (glucocorticoids, cyclosporin A, rapamycin, Th1/Th2 selective inhibitors), humanized blocking antibodies to cytokines or cytokine receptors, soluble receptors to adsorb specific secreted cytokines and receptor antagonists or drugs that block cytokine signal transduction pathways [16].

In diseases with a dominant Th character, the mere skewing of cytokine balance is seriously considered as therapeutic treatment [16, 132, 187]. Inhibition of Th2 cytokines has been suggested in the treatment of atopic disease, and more specifically, humanized IL-13Rα2 is now in clinical development as a therapeutic approach for asthma [16]. Employment of IL-4 appears promising as therapeutic treatment of the autoimmunity caused by psoriasis and is currently investigated in clinical trials [132]. Th1 cytokine inhibition can be applied for the treatment of inflammation-related diseases. A new era in inflammatory and autoimmune disease management has emerged after the validation of TNF inhibitors for clinical use, namely infliximab (Remicade^{TM}), a blocking humanized monoclonal antibody to TNF-α and etanercept (Enbrel^{TM}), a genetically engineered soluble TNF receptor dimer [5, 192]. Both agents act by competitive inhibition with TNF for its receptors. TNF inhibitors are currently approved for clinical use in the treatment of rheumatoid arthritis (RA) and anti-TNF therapy was recently confirmed to protect joints from structural damage [5, 192]. Inhibiting TNF-α activity has also been a target in the prevention and management of endometriosis and etanercept is currently under consideration as therapeutic treatment [112].

The pro-inflammatory tissue-destructive role of IL-1 has been associated with several human diseases, thus justifying the targeting of the IL-1 pathway and its receptors for clinical therapeutic intervention. Anakinra (Kineret^{TM}) is a recombinant IL-1 receptor antagonist with approved use for the treatment of RA, currently in Phase II/III clinical trials for the management of other inflammatory diseases [30]. Anakinra was shown to have an effect on CRP levels and to improve clinical signs of RA [31, 37]. However, combination therapy of RA with IL-1 inhibitors and TNF-α antagonists appears to confer enhanced therapeutic benefits [21, 67]. IL-1 trap, a recombinant fusion protein with IL-1 neutralizing action, is currently in Phase II clinical trials for the treatment of RA [30].

IFNs were among the first proteins shown to be effective in cancer treatment. In particular, one type of IFN-α, IFN-α2, appears to have clinical activity in hematological malignancies, in particular, CML and hairy-cell leukemia [29]. Studies have shown that IFN-α2b treatment in CML patients correlated with higher rates of hematologic and cytogenetic responses [4, 29]. IFN-α may also restore defective adhesion of CML progenitors to bone marrow stroma, thus reinstating normal growth inhibitory signals to myeloid precursor proliferation [24]. IFN-α2α has therapeutic activity in a percentage of patients with advanced cutaneous T-cell lymphoma [35] and IFN-α was also demonstrated to have potent anti-proliferative effects in patients with multiple myeloma [129]. Anti-IFN antibodies have been implicated in the treatment of multiple sclerosis, but clinical trials await further validation [1, 58].

Recent advances in therapeutics have also given a role for downstream cytokine pathway regulation factors in the treatment of disease. The inhibition of the JAK/STAT pathway at different levels has suggested that tyrosine kinase inhibitors may act as anti-cancer, anti-inflammation and anti-allograft rejection agents [120]. Contrary to other tyrosine kinase inhibitors, JAK inhibitors have yet to enter clinical trials, but the JAK3 inhibitor is currently being developed and appears promising in the prevention of organ transplant rejection [186].

Although the development of biologic response modifiers described in this section merits further research, their current clinical use in debilitating diseases opens new avenues in therapeutic intervention. The use of cytokine genetic polymorphism typing may be feasible as a clinical application in the future, thus indicating groups of patients who will probably benefit most from cytokine therapy; customized therapy treatments may prove more efficient in combating disease. The current therapeutic agents are only the beginning in the possible evolution of intricate cytokine therapy in diseases with an immunologic component. The current review only scratches the surface of a highly complex field, as new discoveries will lead to new clinical applications for cytokines and identify new and more efficient therapeutic targets.

References

[1]   Adorini, L. (2004). Immunotherapeutic approaches in multiple sclerosis. *J Neurol Sci 223*, 13-24.

[2]   Akashi, K., Traver, D., Miyamoto, T. & Weissman, I. L. (2000). A clonogenic common myeloid progenitor that gives rise to all myeloid lineages. *Nature 404*, 193-197.

[3]   Alam, S. E., Nasser, S. S., Fernainy, K. E., Habib, A. A. & Badr, K. F. (2004). Cytokine imbalance in acute coronary syndrome. *Curr Opin Pharmacol 4*, 166-170.

[4]   Alimena, G., Morra, E., Lazzarino, M., Liberati, A. M., Montefusco, E., Inverardi, D., Bernasconi, P., Mancini, M., Donti, E., Grignani, F. & et al. (1988). Interferon alpha-2b as therapy for Ph'-positive chronic myelogenous leukemia: a study of 82 patients treated with intermittent or daily administration. *Blood 72*, 642-647.

[5]   Anderson, D. L. (2004). TNF inhibitors: a new age in rheumatoid arthritis treatment. *Am J Nurs 104*, 60-68; quiz 68-69.

[6]   Arici, A., MacDonald, P. C. & Casey, M. L. (1995). Regulation of monocyte chemotactic protein-1 gene expression in human endometrial cells in cultures. *Mol Cell Endocrinol 107*, 189-197.

[7]   Arici, A., Seli, E., Senturk, L. M., Gutierrez, L. S., Oral, E. & Taylor, H. S. (1998). Interleukin-8 in the human endometrium. *J Clin Endocrinol Metab 83*, 1783-1787.

[8] Arici, A., Seli, E., Zeyneloglu, H. B., Senturk, L. M., Oral, E. & Olive, D. L. (1998). Interleukin-8 induces proliferation of endometrial stromal cells: a potential autocrine growth factor. *J Clin Endocrinol Metab 83*, 1201-1205.

[9] Asano, H., Kobayashi, T., Uchida, K., Hayashi, S., Yokoyama, I., Inoko, H. & Takagi, H. (1997). Significance of tumor necrosis factor microsatellite polymorphism in renal transplantation. *Tissue Antigens 50*, 484-488.

[10] Auer, J., Weber, T., Berent, R., Lassnig, E., Lamm, G. & Eber, B. (2003). Genetic polymorphisms in cytokine and adhesion molecule genes in coronary artery disease. *Am J Pharmacogenomics 3*, 317-328.

[11] Awad, M., Pravica, V., Perrey, C., El Gamel, A., Yonan, N., Sinnott, P. J. & Hutchinson, I. V. (1999). CA repeat allele polymorphism in the first intron of the human interferon-gamma gene is associated with lung allograft fibrosis. *Hum Immunol 60*, 343-346.

[12] Awad, M. R., Webber, S., Boyle, G., Sturchioc, C., Ahmed, M., Martell, J., Law, Y., Miller, S. A., Bowman, P., Gribar, S., Pigula, F., Mazariegos, G., Griffith, B. P. & Zeevi, A. (2001). The effect of cytokine gene polymorphisms on pediatric heart allograft outcome. *J Heart Lung Transplant 20*, 625-630.

[13] Babbage, S. J., Arkwright, P. D., Vince, G. S., Perrey, C., Pravica, V., Quenby, S., Bates, M. & Hutchinson, I. V. (2001). Cytokine promoter gene polymorphisms and idiopathic recurrent pregnancy loss. *J Reprod Immunol 51*, 21-27.

[14] Baird, D. T., Cameron, S. T., Critchley, H. O., Drudy, T. A., Howe, A., Jones, R. L., Lea, R. G. & Kelly, R. W. (1996). Prostaglandins and menstruation. *Eur J Obstet Gynecol Reprod Biol 70*, 15-17.

[15] Barcz, E., Kaminski, P. & Marianowski, L. (2000). Role of cytokines in pathogenesis of endometriosis. *Med Sci Monit 6*, 1042-1046.

[16] Barnes, P. J. (2003). Cytokine-directed therapies for the treatment of chronic airway diseases. *Cytokine Growth Factor Rev 14*, 511-522.

[17] Batard, P., Monier, M. N., Fortunel, N., Ducos, K., Sansilvestri-Morel, P., Phan, T., Hatzfeld, A. & Hatzfeld, J. A. (2000). TGF-(beta)1 maintains hematopoietic immaturity by a reversible negative control of cell cycle and induces CD34 antigen up-modulation. *J Cell Sci 113*, 383-390.

[18] Bates, M. D., Quenby, S., Takakuwa, K., Johnson, P. M. & Vince, G. S. (2002). Aberrant cytokine production by peripheral blood mononuclear cells in recurrent pregnancy loss? *Hum Reprod 17*, 2439-2444.

[19] Bathgate, A. J., Pravica, V., Perrey, C., Therapondos, G., Plevris, J. N., Hayes, P. C. & Hutchinson, I. V. (2000). The effect of polymorphisms in tumor necrosis factor-alpha, interleukin-10, and transforming growth factor-beta1 genes in acute hepatic allograft rejection. *Transplantation 69*, 1514-1517.

[20] Beghe, B., Barton, S., Rorke, S., Peng, Q., Sayers, I., Gaunt, T., Keith, T. P., Clough, J. B., Holgate, S. T. & Holloway, J. W. (2003). Polymorphisms in the interleukin-4 and interleukin-4 receptor alpha chain genes confer susceptibility to asthma and atopy in a Caucasian population. *Clin Exp Allergy 33*, 1111-1117.

[21] Bendele, A. M., Chlipala, E. S., Scherrer, J., Frazier, J., Sennello, G., Rich, W. J. & Edwards, C. K., 3rd. (2000). Combination benefit of treatment with the cytokine

inhibitors interleukin-1 receptor antagonist and PEGylated soluble tumor necrosis factor receptor type I in animal models of rheumatoid arthritis. *Arthritis Rheum 43*, 2648-2659.

[22] Bennet, A. M., Prince, J. A., Fei, G. Z., Lyrenas, L., Huang, Y., Wiman, B., Frostegard, J. & Faire, U. (2003). Interleukin-6 serum levels and genotypes influence the risk for myocardial infarction. *Atherosclerosis 171*, 359-367.

[23] Berger, P., McConnell, J. P., Nunn, M., Kornman, K. S., Sorrell, J., Stephenson, K. & Duff, G. W. (2002). C-reactive protein levels are influenced by common IL-1 gene variations. *Cytokine 17*, 171-174.

[24] Bhatia, R., McCarthy, J. B. & Verfaillie, C. M. (1996). Interferon-alpha restores normal beta 1 integrin-mediated inhibition of hematopoietic progenitor proliferation by the marrow microenvironment in chronic myelogenous leukemia. *Blood 87*, 3883-3891.

[25] Bischof, P. & Campana, A. (2000). Molecular mediators of implantation. *Baillieres Best Pract Res Clin Obstet Gynaecol 14*, 801-814.

[26] Blake, G. J. & Ridker, P. M. (2001). High sensitivity C-reactive protein for predicting cardiovascular disease: an inflammatory hypothesis. *Eur Heart J 22*, 349-352.

[27] Blumenthal, M. (2004). The immunopathology and genetics of asthma. *Minn Med 87*, 53-56.

[28] Boise, L. H., Minn, A. J., June, C. H., Lindsten, T. & Thompson, C. B. (1995). Growth factors can enhance lymphocyte survival without committing the cell to undergo cell division. *Proc Natl Acad Sci U S A 92*, 5491-5495.

[29] Borden, E. C., Lindner, D., Dreicer, R., Hussein, M. & Peereboom, D. (2000). Second-generation interferons for cancer: clinical targets. *Semin Cancer Biol 10*, 125-144.

[30] Braddock, M. & Quinn, A. (2004). Targeting IL-1 in inflammatory disease: new opportunities for therapeutic intervention. *Nat Rev Drug Discov 3*, 330-339.

[31] Bresnihan, B., Alvaro-Gracia, J. M., Cobby, M., Doherty, M., Domljan, Z., Emery, P., Nuki, G., Pavelka, K., Rau, R., Rozman, B., Watt, I., Williams, B., Aitchison, R., McCabe, D. & Musikic, P. (1998). Treatment of rheumatoid arthritis with recombinant human interleukin-1 receptor antagonist. *Arthritis Rheum 41*, 2196-2204.

[32] Buchs, N., di Giovine, F. S., Silvestri, T., Vannier, E., Duff, G. W. & Miossec, P. (2001). IL-1B and IL-1Ra gene polymorphisms and disease severity in rheumatoid arthritis: interaction with their plasma levels. *Genes Immun 2*, 222-228.

[33] Buchs, N., Silvestri, T., di Giovine, F. S., Chabaud, M., Vannier, E., Duff, G. W. & Miossec, P. (2000). IL-4 VNTR gene polymorphism in chronic polyarthritis. The rare allele is associated with protection against destruction. *Rheumatology (Oxford) 39*, 1126-1131.

[34] Bullimore, D. W. (2003). Endometriosis is sustained by tumour necrosis factor-alpha. *Med Hypotheses 60*, 84-88.

[35] Bunn, P. A., Jr., Foon, K. A., Ihde, D. C., Longo, D. L., Eddy, J., Winkler, C. F., Veach, S. R., Zeffren, J., Sherwin, S. & Oldham, R. (1984). Recombinant leukocyte A interferon: an active agent in advanced cutaneous T-cell lymphomas. *Ann Intern Med 101*, 484-487.

[36] Buscher, U., Chen, F. C., Kentenich, H. & Schmiady, H. (1999). Cytokines in the follicular fluid of stimulated and non-stimulated human ovaries; is ovulation a suppressed inflammatory reaction? *Hum Reprod 14*, 162-166.

[37] Campion, G. V., Lebsack, M. E., Lookabaugh, J., Gordon, G. & Catalano, M. (1996). Dose-range and dose-frequency study of recombinant human interleukin-1 receptor antagonist in patients with rheumatoid arthritis. The IL-1Ra Arthritis Study Group. *Arthritis Rheum 39*, 1092-1101.

[38] Carson, W. E., Fehniger, T. A., Haldar, S., Eckhert, K., Lindemann, M. J., Lai, C. F., Croce, C. M., Baumann, H. & Caligiuri, M. A. (1997). A potential role for interleukin-15 in the regulation of human natural killer cell survival. *J Clin Invest 99*, 937-943.

[39] Cavet, J., Dickinson, A. M., Norden, J., Taylor, P. R., Jackson, G. H. & Middleton, P. G. (2001). Interferon-gamma and interleukin-6 gene polymorphisms associate with graft-versus-host disease in HLA-matched sibling bone marrow transplantation. *Blood 98*, 1594-1600.

[40] Cesari, M., Penninx, B. W., Newman, A. B., Kritchevsky, S. B., Nicklas, B. J., Sutton-Tyrrell, K., Rubin, S. M., Ding, J., Simonsick, E. M., Harris, T. B. & Pahor, M. (2003). Inflammatory markers and onset of cardiovascular events: results from the Health ABC study. *Circulation 108*, 2317-2322. Epub 2003 Oct 2320.

[41] Chang, Y. H., Huang, C. N. & Shiau, M. Y. (2004). The C-174G promoter polymorphism of the interleukin-6 (IL-6) gene that affects insulin sensitivity in Caucasians is not involved in the pathogenesis of Taiwanese type 2 diabetes mellitus. *Eur Cytokine Netw 15*, 117-119.

[42] Chapman, C. M., Beilby, J. P., Humphries, S. E., Palmer, L. J., Thompson, P. L. & Hung, J. (2003). Association of an allelic variant of interleukin-6 with subclinical carotid atherosclerosis in an Australian community population. *Eur Heart J 24*, 1494-1499.

[43] Chen, Y., Lu, L. & Wang, L. (1998). [Study on gene expression of TGF beta 1 and its receptor in leukemia cells and the serum TGF beta 1 level in the patients with acute leukemia]. *Zhonghua Xue Ye Xue Za Zhi 19*, 576-580.

[44] Costeas, P. A., Koumouli, A., Giantsiou-Kyriakou, A., Papaloizou, A. & Koumas, L. (2004). Th2/Th3 cytokine genotypes are associated with pregnancy loss. *Hum Immunol 65*, 135-141.

[45] Cua, D. J., Sherlock, J., Chen, Y., Murphy, C. A., Joyce, B., Seymour, B., Lucian, L., To, W., Kwan, S., Churakova, T., Zurawski, S., Wiekowski, M., Lira, S. A., Gorman, D., Kastelein, R. A. & Sedgwick, J. D. (2003). Interleukin-23 rather than interleukin-12 is the critical cytokine for autoimmune inflammation of the brain. *Nature 421*, 744-748.

[46] Cullup, H., Dickinson, A. M., Cavet, J., Jackson, G. H. & Middleton, P. G. (2003). Polymorphisms of interleukin-1alpha constitute independent risk factors for chronic graft-versus-host disease after allogeneic bone marrow transplantation. *Br J Haematol 122*, 778-787.

[47] Cvetkovic, J. T., Wallberg-Jonsson, S., Stegmayr, B., Rantapaa-Dahlqvist, S. & Lefvert, A. K. (2002). Susceptibility for and clinical manifestations of rheumatoid arthritis are associated with polymorphisms of the TNF-alpha, IL-1beta, and IL-1Ra genes. *J Rheumatol 29*, 212-219.

[48] Daher, S., de Arruda Geraldes Denardi, K., Blotta, M. H., Mamoni, R. L., Reck, A. P., Camano, L. & Mattar, R. (2004). Cytokines in recurrent pregnancy loss. *J Reprod Immunol 62*, 151-157.

[49] Daher, S., Shulzhenko, N., Morgun, A., Mattar, R., Rampim, G. F., Camano, L. & DeLima, M. G. (2003). Associations between cytokine gene polymorphisms and recurrent pregnancy loss. *J Reprod Immunol 58*, 69-77.

[50] D'Aiuto, F., Parkar, M., Andreou, G., Brett, P. M., Ready, D. & Tonetti, M. S. (2004). Periodontitis and atherogenesis: causal association or simple coincidence? *J Clin Periodontol 31*, 402-411.

[51] Daugherty, A. & Rateri, D. L. (2002). T lymphocytes in atherosclerosis: the yin-yang of Th1 and Th2 influence on lesion formation. *Circ Res 90*, 1039-1040.

[52] Davenport, P. & Tipping, P. G. (2003). The role of interleukin-4 and interleukin-12 in the progression of atherosclerosis in apolipoprotein E-deficient mice. *Am J Pathol 163*, 1117-1125.

[53] Demeter, J., Porzsolt, F., Ramisch, S., Schmidt, D., Schmid, M. & Messer, G. (1997). Polymorphism of the tumour necrosis factor-alpha and lymphotoxin-alpha genes in chronic lymphocytic leukaemia. *Br J Haematol 97*, 107-112.

[54] Denizot, Y., Turlure, P., Bordessoule, D., Trimoreau, F. & Praloran, V. (1999). Serum IL-10 and IL-13 concentrations in patients with haematological malignancies. *Cytokine 11*, 634-635.

[55] Drexler, H. G., Meyer, C., Zaborski, M., Uphoff, C. C. & Quentmeier, H. (1998). Growth-inhibitory effects of transforming growth factor-beta 1 on myeloid leukemia cell lines. *Leuk Res 22*, 927-938.

[56] Du, Y., Dodel, R. C., Eastwood, B. J., Bales, K. R., Gao, F., Lohmuller, F., Muller, U., Kurz, A., Zimmer, R., Evans, R. M., Hake, A., Gasser, T., Oertel, W. H., Griffin, W. S., Paul, S. M. & Farlow, M. R. (2000). Association of an interleukin 1 alpha polymorphism with Alzheimer's disease. *Neurology 55*, 480-483.

[57] Duarte, R. F. & Franf, D. A. (2002). The synergy between stem cell factor (SCF) and granulocyte colony-stimulating factor (G-CSF): molecular basis and clinical relevance. *Leuk Lymphoma 43*, 1179-1187.

[58] Durelli, L. & Ricci, A. (2004). Anti-interferon antibodies in multiple sclerosis. molecular basis and their impact on clinical efficacy. *Front Biosci 9*, 2192-2204.

[59] Eerligh, P., Koeleman, B. P., Dudbridge, F., Jan Bruining, G., Roep, B. O. & Giphart, M. J. (2004). Functional genetic polymorphisms in cytokines and metabolic genes as additional genetic markers for susceptibility to develop type 1 diabetes. *Genes Immun 5*, 36-40.

[60] Eklund, C., Jahan, F., Pessi, T., Lehtimaki, T. & Hurme, M. (2003). Interleukin 1B gene polymorphism is associated with baseline C-reactive protein levels in healthy individuals. *Eur Cytokine Netw 14*, 168-171.

[61] el Maradny, E., Kanayama, N., Maehara, K., Kobayashi, T. & Terao, T. (1996). Expression of interleukin-8 receptors in the gestational tissues before and after initiation of labor: immunohistochemical study. *Acta Obstet Gynecol Scand 75*, 790-796.

[62] Elenkov, I. J. (2004). Glucocorticoids and the Th1/Th2 balance. *Ann N Y Acad Sci 1024*, 138-146.

[63] Elenkov, I. J. & Chrousos, G. P. (1999). Stress Hormones, Th1/Th2 patterns, Pro/Anti-inflammatory Cytokines and Susceptibility to Disease. *Trends Endocrinol Metab 10*, 359-368.

[64] Esplin, M. S., Romero, R., Chaiworapongsa, T., Kim, Y. M., Edwin, S., Gomez, R., Gonzalez, R. & Adashi, E. Y. (2003). Amniotic fluid levels of immunoreactive monocyte chemotactic protein-1 increase during term parturition. *J Matern Fetal Neonatal Med 14*, 51-56.

[65] Etter, H., Althaus, R., Eugster, H. P., Santamaria-Babi, L. F., Weber, L. & Moser, R. (1998). IL-4 and IL-13 downregulate rolling adhesion of leukocytes to IL-1 or TNF-alpha-activated endothelial cells by limiting the interval of E-selectin expression. *Cytokine 10*, 395-403.

[66] Fayad, L., Keating, M. J., Reuben, J. M., O'Brien, S., Lee, B. N., Lerner, S. & Kurzrock, R. (2001). Interleukin-6 and interleukin-10 levels in chronic lymphocytic leukemia: correlation with phenotypic characteristics and outcome. *Blood 97*, 256-263.

[67] Feige, U., Hu, Y. L., Gasser, J., Campagnuolo, G., Munyakazi, L. & Bolon, B. (2000). Anti-interleukin-1 and anti-tumor necrosis factor-alpha synergistically inhibit adjuvant arthritis in Lewis rats. *Cell Mol Life Sci 57*, 1457-1470.

[68] Ferrajoli, A., Keating, M. J., Manshouri, T., Giles, F. J., Dey, A., Estrov, Z., Koller, C. A., Kurzrock, R., Thomas, D. A., Faderl, S., Lerner, S., O'Brien, S. & Albitar, M. (2002). The clinical significance of tumor necrosis factor-alpha plasma level in patients having chronic lymphocytic leukemia. *Blood 100*, 1215-1219.

[69] Fichtlscherer, S., Heeschen, C. & Zeiher, A. M. (2004). Inflammatory markers and coronary artery disease. *Curr Opin Pharmacol 4*, 124-131.

[70] Flex, A., Gaetani, E., Pola, R., Santoliquido, A., Aloi, F., Papaleo, P., Dal Lago, A., Pola, E., Serricchio, M., Tondi, P. & Pola, P. (2002). The -174 G/C polymorphism of the interleukin-6 gene promoter is associated with peripheral artery occlusive disease. *Eur J Vasc Endovasc Surg 24*, 264-268.

[71] Fortunel, N. O., Hatzfeld, A. & Hatzfeld, J. A. (2000). Transforming growth factor-beta: pleiotropic role in the regulation of hematopoiesis. *Blood 96*, 2022-2036.

[72] Fortunel, N. O., Hatzfeld, J. A., Monier, M. N. & Hatzfeld, A. (2003). Control of hematopoietic stem/progenitor cell fate by transforming growth factor-beta. *Oncol Res 13*, 445-453.

[73] Gabay, C. (2002). Cytokine inhibitors in the treatment of rheumatoid arthritis. *Expert Opin Biol Ther 2*, 135-149.

[74] Gallinelli, A., Chiossi, G., Giannella, L., Marsella, T., Genazzani, A. D. & Volpe, A. (2004). Different concentrations of interleukins in the peritoneal fluid of women with endometriosis: relationships with lymphocyte subsets. *Gynecol Endocrinol 18*, 144-151.

[75] George, S., Turner, D., Reynard, M., Navarrete, C., Rizvi, I., Fernando, O. N., Powis, S. H., Moorhead, J. F. & Varghese, Z. (2001). Significance of cytokine gene polymorphism in renal transplantation. *Transplant Proc 33*, 483-484.

[76] Gerard, N., Caillaud, M., Martoriati, A., Goudet, G. & Lalmanach, A. C. (2004). The interleukin-1 system and female reproduction. *J Endocrinol 180*, 203-212.

[77] Gerszten, R. E., Garcia-Zepeda, E. A., Lim, Y. C., Yoshida, M., Ding, H. A., Gimbrone, M. A., Jr., Luster, A. D., Luscinskas, F. W. & Rosenzweig, A. (1999). MCP-1 and IL-8 trigger firm adhesion of monocytes to vascular endothelium under flow conditions. *Nature 398*, 718-723.

[78] Gora-Tybor, J., Blonski, J. Z. & Robak, T. (2003). Circulating proangiogenic cytokines and angiogenesis inhibitor endostatin in untreated patients with chronic lymphocytic leukemia. *Mediators Inflamm 12*, 167-171.

[79] Gray, T., Nettesheim, P., Loftin, C., Koo, J. S., Bonner, J., Peddada, S. & Langenbach, R. (2004). Interleukin-1beta-induced mucin production in human airway epithelium is mediated by cyclooxygenase-2, prostaglandin E2 receptors, and cyclic AMP-protein kinase A signaling. *Mol Pharmacol 66*, 337-346.

[80] Harney, S. M., Newton, J. L. & Wordsworth, B. P. (2003). Molecular genetics of rheumatoid arthritis. *Curr Opin Pharmacol 3*, 280-285.

[81] Hase, K., Tani, K., Shimizu, T., Ohmoto, Y., Matsushima, K. & Sone, S. (2001). Increased CCR4 expression in active systemic lupus erythematosus. *J Leukoc Biol 70*, 749-755.

[82] Hayakawa, S., Fujikawa, T., Fukuoka, H., Chisima, F., Karasaki-Suzuki, M., Ohkoshi, E., Ohi, H., Kiyoshi Fujii, T., Tochigi, M., Satoh, K., Shimizu, T., Nishinarita, S., Nemoto, N. & Sakurai, I. (2000). Murine fetal resorption and experimental pre-eclampsia are induced by both excessive Th1 and Th2 activation. *J Reprod Immunol 47*, 121-138.

[83] Heeschen, C., Dimmeler, S., Hamm, C. W., Fichtlscherer, S., Boersma, E., Simoons, M. L. & Zeiher, A. M. (2003). Serum level of the antiinflammatory cytokine interleukin-10 is an important prognostic determinant in patients with acute coronary syndromes. *Circulation 107*, 2109-2114. Epub 2003 Mar 2131.

[84] Heike, T. & Nakahata, T. (2002). Ex vivo expansion of hematopoietic stem cells by cytokines. *Biochim Biophys Acta 1592*, 313-321.

[85] Heinzmann, A. & Deichmann, K. A. (2001). Genes for atopy and asthma. *Curr Opin Allergy Clin Immunol 1*, 387-392.

[86] Heinzmann, A., Jerkic, S. P., Ganter, K., Kurz, T., Blattmann, S., Schuchmann, L., Gerhold, K., Berner, R. & Deichmann, K. A. (2003). Association study of the IL13 variant Arg110Gln in atopic diseases and juvenile idiopathic arthritis. *J Allergy Clin Immunol 112*, 735-739.

[87] Hilbert, D. M., Kopf, M., Mock, B. A., Kohler, G. & Rudikoff, S. (1995). Interleukin 6 is essential for in vivo development of B lineage neoplasms. *J Exp Med 182*, 243-248.

[88] Hill, J. A., 3rd & Choi, B. C. (2000). Immunodystrophism: evidence for a novel alloimmune hypothesis for recurrent pregnancy loss involving Th1-type immunity to trophoblast. *Semin Reprod Med 18*, 401-405.

[89] Holler, E., Roncarolo, M. G., Hintermeier-Knabe, R., Eissner, G., Ertl, B., Schulz, U., Knabe, H., Kolb, H. J., Andreesen, R. & Wilmanns, W. (2000). Prognostic significance of increased IL-10 production in patients prior to allogeneic bone marrow transplantation. *Bone Marrow Transplant 25*, 237-241.

[90] Homaidan, F. R., Chakroun, Dbaibo, G. S., El-Assaad, W. & El-Sabban, M. E. (2001). IL-1 activates two phospholipid signaling pathways in intestinal epithelial cells. *Inflamm Res 50*, 375-381.

[91] Huang, H. Y., Wen, Y., Irwin, J. C., Kruessel, J. S., Soong, Y. K. & Polan, M. L. (1998). Cytokine-mediated regulation of 92-kilodalton type IV collagenase, tissue inhibitor or metalloproteinase-1 (TIMP-1), and TIMP-3 messenger ribonucleic acid expression in human endometrial stromal cells. *J Clin Endocrinol Metab 83*, 1721-1729.

[92] Hulkkonen, J., Vilpo, J., Vilpo, L., Koski, T. & Hurme, M. (2000). Interleukin-1 beta, interleukin-1 receptor antagonist and interleukin-6 plasma levels and cytokine gene polymorphisms in chronic lymphocytic leukemia: correlation with prognostic parameters. *Haematologica 85*, 600-606.

[93] Humphries, S. E., Luong, L. A., Ogg, M. S., Hawe, E. & Miller, G. J. (2001). The interleukin-6 -174 G/C promoter polymorphism is associated with risk of coronary heart disease and systolic blood pressure in healthy men. *Eur Heart J 22*, 2243-2252.

[94] Ide, A., Kawasaki, E., Abiru, N., Sun, F., Takahashi, R., Kuwahara, H., Fujita, N., Kita, A., Oshima, K., Sakamaki, H., Uotani, S., Yamasaki, H., Yamaguchi, Y. & Eguchi, K. (2002). Genetic association between interleukin-10 gene promoter region polymorphisms and type 1 diabetes age-at-onset. *Hum Immunol 63*, 690-695.

[95] Jahromi, M. M., Millward, B. A. & Demaine, A. G. (2000). A polymorphism in the promoter region of the gene for interleukin-6 is associated with susceptibility to type 1 diabetes mellitus. *J Interferon Cytokine Res 20*, 885-888.

[96] Jawa, R. S., Quaid, G. A., Williams, M. A., Cave, C. M., Robinson, C. T., Babcock, G. F., Lieberman, M. A., Witt, D. & Solomkin, J. S. (1999). Tumor necrosis factor alpha regulates CXC chemokine receptor expression and function. *Shock 11*, 385-390.

[97] Keelan, J. A., Blumenstein, M., Helliwell, R. J., Sato, T. A., Marvin, K. W. & Mitchell, M. D. (2003). Cytokines, prostaglandins and parturition--a review. *Placenta 24*, S33-46.

[98] Keen, L. J. (2002). The extent and analysis of cytokine and cytokine receptor gene polymorphism. *Transpl Immunol 10*, 143-146.

[99] Keen, L. J., DeFor, T. E., Bidwell, J. L., Davies, S. M., Bradley, B. A. & Hows, J. M. (2004). Interleukin-10 and tumor necrosis factor alpha region haplotypes predict transplant-related mortality after unrelated donor stem cell transplantation. *Blood 103*, 3599-3602. Epub 2003 Dec 3530.

[100] Kelly, R. W. (2002). Inflammatory mediators and cervical ripening. *J Reprod Immunol 57*, 217-224.

[101] Kelly, R. W., King, A. E. & Critchley, H. O. (2001). Cytokine control in human endometrium. *Reproduction 121*, 3-19.

[102] Khaled, A. R. & Durum, S. K. (2002). The role of cytokines in lymphocyte homeostasis. *Biotechniques Suppl*, 40-45.

[103] Kitawaki, J., Koshiba, H., Kitaoka, Y., Teramoto, M., Hasegawa, G., Nakamura, N., Yoshikawa, T., Ohta, M., Obayashi, H. & Honjo, H. (2004). Interferon-gamma gene dinucleotide (CA) repeat and interleukin-4 promoter region (-590C/T) polymorphisms

in Japanese patients with endometriosis. *Hum Reprod 19*, 1765-1769. Epub 2004 May 1727.

[104] Kondo, M., Scherer, D. C., Miyamoto, T., King, A. G., Akashi, K., Sugamura, K. & Weissman, I. L. (2000). Cell-fate conversion of lymphoid-committed progenitors by instructive actions of cytokines. *Nature 407*, 383-386.

[105] Kondo, M., Weissman, I. L. & Akashi, K. (1997). Identification of clonogenic common lymphoid progenitors in mouse bone marrow. *Cell 91*, 661-672.

[106] Korholz, D., Kunst, D., Hempel, L., Sohngen, D., Heyll, A., Bonig, H., Gobel, U., Zintl, F. & Burdach, S. (1997). Decreased interleukin 10 and increased interferon-gamma production in patients with chronic graft-versus-host disease after allogeneic bone marrow transplantation. *Bone Marrow Transplant 19*, 691-695.

[107] Koumas, L., Economides, C., Papaloizou, A., Giantsiou Kyriakou, A., Koumouli, A. & Costeas, P. A. (2004). Association of cytokine gene polymorphisms and cardiovascular disease. *Immunology 2004* 39-44.

[108] Krause, D. S. (2002). Regulation of hematopoietic stem cell fate. *Oncogene 21*, 3262-3269.

[109] Krenger, W., Hill, G. R. & Ferrara, J. L. (1997). Cytokine cascades in acute graft-versus-host disease. *Transplantation 64*, 553-558.

[110] Krussel, J. S., Bielfeld, P., Polan, M. L. & Simon, C. (2003). Regulation of embryonic implantation. *Eur J Obstet Gynecol Reprod Biol 110*, S2-9.

[111] Kuo, Y. M., Liao, P. C., Lin, C., Wu, C. W., Huang, H. M., Lin, C. C. & Chuo, L. J. (2003). Lack of association between interleukin-1alpha polymorphism and Alzheimer disease or vascular dementia. *Alzheimer Dis Assoc Disord 17*, 94-97.

[112] Kyama, C. M., Debrock, S., Mwenda, J. M. & D'Hooghe, T. M. (2003). Potential involvement of the immune system in the development of endometriosis. *Reprod Biol Endocrinol 1*, 123.

[113] Lamblin, N., Bauters, C. & Helbecque, N. (2001). Gene polymorphisms of pro- (or anti-) inflammatory cytokines and vascular disease. *Eur Heart J 22*, 2219-2220.

[114] Lauta, V. M. (2003). A review of the cytokine network in multiple myeloma: diagnostic, prognostic, and therapeutic implications. *Cancer 97*, 2440-2452.

[115] Lee, Y. H., Kim, H. J., Rho, Y. H., Choi, S. J., Ji, J. D. & Song, G. G. (2004). Interleukin-1 receptor antagonist gene polymorphism and rheumatoid arthritis. *Rheumatol Int 24*, 133-136. Epub 2004 Feb 2011.

[116] Ley, K., Allietta, M., Bullard, D. C. & Morgan, S. (1998). Importance of E-selectin for firm leukocyte adhesion in vivo. *Circ Res 83*, 287-294.

[117] Libby, P., Ridker, P. M. & Maseri, A. (2002). Inflammation and atherosclerosis. *Circulation 105*, 1135-1143.

[118] Lin, M. T., Storer, B., Martin, P. J., Tseng, L. H., Gooley, T., Chen, P. J. & Hansen, J. A. (2003). Relation of an interleukin-10 promoter polymorphism to graft-versus-host disease and survival after hematopoietic-cell transplantation. *N Engl J Med 349*, 2201-2210.

[119] Lohmann, T., Laue, S., Nietzschmann, U., Kapellen, T. M., Lehmann, I., Schroeder, S., Paschke, R. & Kiess, W. (2002). Reduced expression of Th1-associated chemokine

receptors on peripheral blood lymphocytes at diagnosis of type 1 diabetes. *Diabetes 51*, 2474-2480.

[120] Luo, C. & Laaja, P. (2004). Inhibitors of JAKs/STATs and the kinases: a possible new cluster of drugs. *Drug Discov Today 9*, 268-275.

[121] Luscinskas, F. W., Gerszten, R. E., Garcia-Zepeda, E. A., Lim, Y. C., Yoshida, M., Ding, H. A., Gimbrone, M. A., Jr., Luster, A. D. & Rosenzweig, A. (2000). C-C and C-X-C chemokines trigger firm adhesion of monocytes to vascular endothelium under flow conditions. *Ann N Y Acad Sci 902*, 288-293.

[122] Lynch, E. L., Little, F. F., Wilson, K. C., Center, D. M. & Cruikshank, W. W. (2003). Immunomodulatory cytokines in asthmatic inflammation. *Cytokine Growth Factor Rev 14*, 489-502.

[123] Ma, S. L., Tang, N. L., Lam, L. C. & Chiu, H. F. (2003). Lack of association of the interleukin-1beta gene polymorphism with Alzheimer's disease in a Chinese population. *Dement Geriatr Cogn Disord 16*, 265-268.

[124] MacMillan, M. L., Radloff, G. A., Kiffmeyer, W. R., DeFor, T. E., Weisdorf, D. J. & Davies, S. M. (2003). High-producer interleukin-2 genotype increases risk for acute graft-versus-host disease after unrelated donor bone marrow transplantation. *Transplantation 76*, 1758-1762.

[125] Mainou-Fowler, T., Dickinson, A. M., Taylor, P. R., Mounter, P., Jack, F., Proctor, S. J., Nordon, J. & Middleton, P. G. (2000). Tumour necrosis factor gene polymorphisms in lymphoproliferative disease. *Leuk Lymphoma 38*, 547-552.

[126] Makhseed, M., Raghupathy, R., Azizieh, F., Farhat, R., Hassan, N. & Bandar, A. (2000). Circulating cytokines and CD30 in normal human pregnancy and recurrent spontaneous abortions. *Hum Reprod 15*, 2011-2017.

[127] Maki, K., Sunaga, S., Komagata, Y., Kodaira, Y., Mabuchi, A., Karasuyama, H., Yokomuro, K., Miyazaki, J. I. & Ikuta, K. (1996). Interleukin 7 receptor-deficient mice lack gammadelta T cells. *Proc Natl Acad Sci U S A 93*, 7172-7177.

[128] Mallat, Z., Henry, P., Fressonnet, R., Alouani, S., Scoazec, A., Beaufils, P., Chvatchko, Y. & Tedgui, A. (2002). Increased plasma concentrations of interleukin-18 in acute coronary syndromes. *Heart 88*, 467-469.

[129] Mandelli, F., Avvisati, G., Amadori, S., Boccadoro, M., Gernone, A., Lauta, V. M., Marmont, F., Petrucci, M. T., Tribalto, M., Vegna, M. L. & et al. (1990). Maintenance treatment with recombinant interferon alfa-2b in patients with multiple myeloma responding to conventional induction chemotherapy. *N Engl J Med 322*, 1430-1434.

[130] Marder, B., Schroppel, B. & Murphy, B. (2003). Genetic variability and transplantation. *Curr Opin Urol 13*, 81-89.

[131] Marshall, S. E., McLaren, A. J., Haldar, N. A., Bunce, M., Morris, P. J. & Welsh, K. I. (2000). The impact of recipient cytokine genotype on acute rejection after renal transplantation. *Transplantation 70*, 1485-1491.

[132] Martin, R. (2003). Interleukin 4 treatment of psoriasis: are pleiotropic cytokines suitable therapies for autoimmune diseases? *Trends Pharmacol Sci 24*, 613-616.

[133] McGlinchey, P. G., Spence, M. S., Patterson, C. C., Allen, A. R., Murphy, G., Savage, D. A., Maxwell, A. P. & McKeown, P. P. (2004). Cytokine gene polymorphisms in

ischaemic heart disease: investigation using family-based tests of association. *J Mol Med 18*, 18.

[134] Mok, C. C. & Lau, C. S. (2003). Pathogenesis of systemic lupus erythematosus. *J Clin Pathol 56*, 481-490.

[135] Moqattash, S. & Lutton, J. D. (2004). Leukemia cells and the cytokine network: therapeutic prospects. *Exp Biol Med (Maywood) 229*, 121-137.

[136] Morimoto, S., Tokano, Y., Kaneko, H., Nozawa, K., Amano, H. & Hashimoto, H. (2001). The increased interleukin-13 in patients with systemic lupus erythematosus: relations to other Th1-, Th2-related cytokines and clinical findings. *Autoimmunity 34*, 19-25.

[137] Mosmann, T. R. & Sad, S. (1996). The expanding universe of T-cell subsets: Th1, Th2 and more. *Immunol Today 17*, 138-146.

[138] Mugnier, B., Balandraud, N., Darque, A., Roudier, C., Roudier, J. & Reviron, D. (2003). Polymorphism at position -308 of the tumor necrosis factor alpha gene influences outcome of infliximab therapy in rheumatoid arthritis. *Arthritis Rheum 48*, 1849-1852.

[139] Mullen, A. C., High, F. A., Hutchins, A. S., Lee, H. W., Villarino, A. V., Livingston, D. M., Kung, A. L., Cereb, N., Yao, T. P., Yang, S. Y. & Reiner, S. L. (2001). Role of T-bet in commitment of TH1 cells before IL-12-dependent selection. *Science 292*, 1907-1910.

[140] Mullighan, C., Heatley, S., Doherty, K., Szabo, F., Grigg, A., Hughes, T., Schwarer, A., Szer, J., Tait, B., To, B. & Bardy, P. (2004). Non-HLA immunogenetic polymorphisms and the risk of complications after allogeneic hemopoietic stem-cell transplantation. *Transplantation 77*, 587-596.

[141] Munro, L. R., Johnston, P. W., Marshall, N. A., Canning, S. J., Hewitt, S. G., Tveita, K. & Vickers, M. A. (2003). Polymorphisms in the interleukin-10 and interferon gamma genes in Hodgkin lymphoma. *Leuk Lymphoma 44*, 2083-2088.

[142] Murray, R., Suda, T., Wrighton, N., Lee, F. & Zlotnik, A. (1989). IL-7 is a growth and maintenance factor for mature and immature thymocyte subsets. *Int Immunol 1*, 526-531.

[143] Namiki, M., Kawashima, S., Yamashita, T., Ozaki, M., Sakoda, T., Inoue, N., Hirata, K., Morishita, R., Kaneda, Y. & Yokoyama, M. (2004). Intramuscular gene transfer of interleukin-10 cDNA reduces atherosclerosis in apolipoprotein E-knockout mice. *Atherosclerosis 172*, 21-29.

[144] Nardo, L. G., Nikas, G. & Makrigiannakis, A. (2003). Molecules in blastocyst implantation. Role of matrix metalloproteinases, cytokines and growth factors. *J Reprod Med 48*, 137-147.

[145] Nishimura, M., Obayashi, H., Mizuta, I., Hara, H., Adachi, T., Ohta, M., Tegoshi, H., Fukui, M., Hasegawa, G., Shigeta, H., Kitagawa, Y., Nakano, K., Kaji, R. & Nakamura, N. (2003). TNF, TNF receptor type 1, and allograft inflammatory factor-1 gene polymorphisms in Japanese patients with type 1 diabetes. *Hum Immunol 64*, 302-309.

[146] Nurse, B., Haus, M., Puterman, A. S., Weinberg, E. G. & Potter, P. C. (1997). Reduced interferon-gamma but normal IL-4 and IL-5 release by peripheral blood mononuclear cells from Xhosa children with atopic asthma. *J Allergy Clin Immunol 100*, 662-668.

[147] Okopien, B., Hyper, M., Kowalski, J., Belowski, D., Madej, A., Zielinski, M., Tokarz, D., Kalina, Z. & Herman, Z. S. (2001). A new immunological marker of atherosclerotic injury of arterial wall. *Res Commun Mol Pathol Pharmacol 109*, 241-248.

[148] Oosterlynck, D. J., Meuleman, C., Waer, M. & Koninckx, P. R. (1994). Transforming growth factor-beta activity is increased in peritoneal fluid from women with endometriosis. *Obstet Gynecol 83*, 287-292.

[149] Oppmann, B., Lesley, R., Blom, B., Timans, J. C., Xu, Y., Hunte, B., Vega, F., Yu, N., Wang, J., Singh, K., Zonin, F., Vaisberg, E., Churakova, T., Liu, M., Gorman, D., Wagner, J., Zurawski, S., Liu, Y., Abrams, J. S., Moore, K. W., Rennick, D., de Waal-Malefyt, R., Hannum, C., Bazan, J. F. & Kastelein, R. A. (2000). Novel p19 protein engages IL-12p40 to form a cytokine, IL-23, with biological activities similar as well as distinct from IL-12. *Immunity 13*, 715-725.

[150] Padyukov, L., Hytonen, A. M., Smolnikova, M., Hahn-Zoric, M., Nilsson, N., Hanson, L. A., Tarkowski, A. & Klareskog, L. (2004). Polymorphism in promoter region of IL10 gene is associated with rheumatoid arthritis in women. *J Rheumatol 31*, 422-425.

[151] Parkhill, J. M., Hennig, B. J., Chapple, I. L., Heasman, P. A. & Taylor, J. J. (2000). Association of interleukin-1 gene polymorphisms with early-onset periodontitis. *J Clin Periodontol 27*, 682-689.

[152] Parks, C. G., Pandey, J. P., Dooley, M. A., Treadwell, E. L., St Clair, E. W., Gilkeson, G. S., Feghali-Botswick, C. L. & Cooper, G. S. (2004). Genetic polymorphisms in tumor necrosis factor (TNF)-alpha and TNF-beta in a population-based study of systemic lupus erythematosus: associations and interaction with the interleukin-1alpha-889 C/T polymorphism. *Hum Immunol 65*, 622-631.

[153] Peschon, J. J., Morrissey, P. J., Grabstein, K. H., Ramsdell, F. J., Maraskovsky, E., Gliniak, B. C., Park, L. S., Ziegler, S. F., Williams, D. E., Ware, C. B. & et al. (1994). Early lymphocyte expansion is severely impaired in interleukin 7 receptor-deficient mice. *J Exp Med 180*, 1955-1960.

[154] Piccinni, M. P., Scaletti, C., Maggi, E. & Romagnani, S. (2000). Role of hormone-controlled Th1- and Th2-type cytokines in successful pregnancy. *J Neuroimmunol 109*, 30-33.

[155] Pietrowski, D., Bettendorf, H., Keck, C., Burkle, B., Unfried, G., Riener, E. K., Hefler, L. A. & Tempfer, C. (2004). Lack of association of TNFalpha gene polymorphisms and recurrent pregnancy loss in Caucasian women. *J Reprod Immunol 61*, 51-58.

[156] Poli, F., Boschiero, L., Giannoni, F., Tonini, M., Scalamogna, M., Ancona, G. & Sirchia, G. (2000). Tumour necrosis factor-alpha gene polymorphism: implications in kidney transplantation. *Cytokine 12*, 1778-1783.

[157] Power, L. L., Popplewell, E. J., Holloway, J. A., Diaper, N. D., Warner, J. O. & Jones, C. A. (2002). Immunoregulatory molecules during pregnancy and at birth. *J Reprod Immunol 56*, 19-28.

[158] Prigoshin, N., Tambutti, M., Larriba, J., Gogorza, S. & Testa, R. (2004). Cytokine gene polymorphisms in recurrent pregnancy loss of unknown cause. *Am J Reprod Immunol 52*, 36-41.

[159] Quesniaux, V. F., Clark, S. C., Turner, K. & Fagg, B. (1992). Interleukin-11 stimulates multiple phases of erythropoiesis in vitro. *Blood 80*, 1218-1223.

[160] Raghupathy, R. (2001). Pregnancy: success and failure within the Th1/Th2/Th3 paradigm. *Semin Immunol 13*, 219-227.

[161] Rankin, J. A. (2004). Biological mediators of acute inflammation. *AACN Clin Issues 15*, 3-17.

[162] Raziuddin, S., Sheikha, A., Abu-Eshy, S. & al-Janadi, M. (1994). Circulating levels of cytokines and soluble cytokine receptors in various T-cell malignancies. *Cancer 73*, 2426-2431.

[163] Reid, J. G., Simpson, N. A., Walker, R. G., Economidou, O., Shillito, J., Gooi, H. C., Duffy, S. R. & Walker, J. J. (2001). The carriage of pro-inflammatory cytokine gene polymorphisms in recurrent pregnancy loss. *Am J Reprod Immunol 45*, 35-40.

[164] Rezaei, A. & Dabbagh, A. (2002). T-helper (1) cytokines increase during early pregnancy in women with a history of recurrent spontaneous abortion. *Med Sci Monit 8*, CR607-610.

[165] Richter, O. N., Dorn, C., Rosing, B., Flaskamp, C. & Ulrich, U. (2004). Tumor necrosis factor alpha secretion by peritoneal macrophages in patients with endometriosis. *Arch Gynecol Obstet 27*, 27.

[166] Robak, E., Robak, T., Wozniacka, A., Zak-Prelich, M., Sysa-Jedrzejowska, A. & Stepien, H. (2002). Proinflammatory interferon-gamma--inducing monokines (interleukin-12, interleukin-18, interleukin-15)--serum profile in patients with systemic lupus erythematosus. *Eur Cytokine Netw 13*, 364-368.

[167] Romagnani, P., Annunziato, F., Piccinni, M. P., Maggi, E. & Romagnani, S. (2000). Th1/Th2 cells, their associated molecules and role in pathophysiology. *Eur Cytokine Netw 11*, 510-511.

[168] Rosen, P., Thompson, S. & Glass, D. (2003). Non-HLA gene polymorphisms in juvenile rheumatoid arthritis. *Clin Exp Rheumatol 21*, 650-656.

[169] Runesson, E., Ivarsson, K., Janson, P. O. & Brannstrom, M. (2000). Gonadotropin- and cytokine-regulated expression of the chemokine interleukin 8 in the human preovulatory follicle of the menstrual cycle. *J Clin Endocrinol Metab 85*, 4387-4395.

[170] Ruscetti, F. W. & Bartelmez, S. H. (2001). Transforming growth factor beta, pleiotropic regulator of hematopoietic stem cells: potential physiological and clinical relevance. *Int J Hematol 74*, 18-25.

[171] Sankaran, D., Asderakis, A., Ashraf, S., Roberts, I. S., Short, C. D., Dyer, P. A., Sinnott, P. J. & Hutchinson, I. V. (1999). Cytokine gene polymorphisms predict acute graft rejection following renal transplantation. *Kidney Int 56*, 281-288.

[172] Sarvetnick, N. & Ohashi, P. S. (2003). Autoimmunity. *Curr Opin Immunol 15*, 647-650.

[173] Scarel-Caminaga, R. M., Trevilatto, P. C., Souza, A. P., Brito, R. B., Camargo, L. E. & Line, S. R. (2004). Interleukin 10 gene promoter polymorphisms are associated with chronic periodontitis. *J Clin Periodontol 31*, 443-448.

[174] Schlaak, J. F., Pfers, I., Meyer Zum Buschenfelde, K. H. & Marker-Hermann, E. (1996). Different cytokine profiles in the synovial fluid of patients with osteoarthritis, rheumatoid arthritis and seronegative spondylarthropathies. *Clin Exp Rheumatol 14*, 155-162.

[175] Schluns, K. S., Kieper, W. C., Jameson, S. C. & Lefrancois, L. (2000). Interleukin-7 mediates the homeostasis of naive and memory CD8 T cells in vivo. *Nat Immunol 1*, 426-432.

[176] Schwarzmeier, J. D. (1996). The role of cytokines in haematopoiesis. *Eur J Haematol Suppl 60*, 69-74.

[177] Sciacca, F. L., Ferri, C., Licastro, F., Veglia, F., Biunno, I., Gavazzi, A., Calabrese, E., Martinelli Boneschi, F., Sorbi, S., Mariani, C., Franceschi, M. & Grimaldi, L. M. (2003). Interleukin-1B polymorphism is associated with age at onset of Alzheimer's disease. *Neurobiol Aging 24*, 927-931.

[178] Sherer, D. M. & Abulafia, O. (2001). Angiogenesis during implantation, and placental and early embryonic development. *Placenta 22*, 1-13.

[179] Shiau, M. Y., Wu, C. Y., Huang, C. N., Hu, S. W., Lin, S. J. & Chang, Y. H. (2003). TNF-alpha polymorphisms and type 2 diabetes mellitus in Taiwanese patients. *Tissue Antigens 61*, 393-397.

[180] Shirai, T., Suzuki, K., Inui, N., Suda, T., Chida, K. & Nakamura, H. (2003). Th1/Th2 profile in peripheral blood in atopic cough and atopic asthma. *Clin Exp Allergy 33*, 84-89.

[181] Silverman, E. S., Palmer, L. J., Subramaniam, V., Hallock, A., Mathew, S., Vallone, J., Faffe, D. S., Shikanai, T., Raby, B. A., Weiss, S. T. & Shore, S. A. (2004). Transforming growth factor-beta1 promoter polymorphism C-509T is associated with asthma. *Am J Respir Crit Care Med 169*, 214-219. Epub 2003 Nov 2003.

[182] Smart, J. M., Horak, E., Kemp, A. S., Robertson, C. F. & Tang, M. L. (2002). Polyclonal and allergen-induced cytokine responses in adults with asthma: resolution of asthma is associated with normalization of IFN-gamma responses. *J Allergy Clin Immunol 110*, 450-456.

[183] Sprent, J. & Surh, C. D. (2001). Generation and maintenance of memory T cells. *Curr Opin Immunol 13*, 248-254.

[184] Sprent, J. & Surh, C. D. (2003). Cytokines and T cell homeostasis. *Immunol Lett 85*, 145-149.

[185] Starr, R., Willson, T. A., Viney, E. M., Murray, L. J., Rayner, J. R., Jenkins, B. J., Gonda, T. J., Alexander, W. S., Metcalf, D., Nicola, N. A. & Hilton, D. J. (1997). A family of cytokine-inducible inhibitors of signalling. *Nature 387*, 917-921.

[186] Stepkowski, S. M., Erwin-Cohen, R. A., Behbod, F., Wang, M. E., Qu, X., Tejpal, N., Nagy, Z. S., Kahan, B. D. & Kirken, R. A. (2002). Selective inhibitor of Janus tyrosine kinase 3, PNU156804, prolongs allograft survival and acts synergistically with cyclosporine but additively with rapamycin. *Blood 99*, 680-689.

[187] Suarez-Pinzon, W. L. & Rabinovitch, A. (2001). Approaches to type 1 diabetes prevention by intervention in cytokine immunoregulatory circuits. *Int J Exp Diabetes Res 2*, 3-17.

[188] Tai, H., Endo, M., Shimada, Y., Gou, E., Orima, K., Kobayashi, T., Yamazaki, K. & Yoshie, H. (2002). Association of interleukin-1 receptor antagonist gene polymorphisms with early onset periodontitis in Japanese. *J Clin Periodontol 29*, 882-888.

[189] Takeuchi, S., Takeuchi, N., Tsukasaki, K., Bartram, C. R., Zimmermann, M., Schrappe, M., Taguchi, H. & Koeffler, H. P. (2002). Genetic polymorphisms in the tumour necrosis factor locus in childhood acute lymphoblastic leukaemia. *Br J Haematol 119*, 985-987.

[190] Tang, M. L., Kemp, A. S., Thorburn, J. & Hill, D. J. (1994). Reduced interferon-gamma secretion in neonates and subsequent atopy. *Lancet 344*, 983-985.

[191] Taylor, J. J., Preshaw, P. M. & Donaldson, P. T. (2000). Cytokine gene polymorphism and immunoregulation in periodontal disease. *Periodontol 35*, 158-182.

[192] Taylor, P. C. (2003). Antibody therapy for rheumatoid arthritis. *Curr Opin Pharmacol 3*, 323-328.

[193] Tedgui, A. & Mallat, Z. (2001). Anti-inflammatory mechanisms in the vascular wall. *Circ Res 88*, 877-887.

[194] Teramoto, M., Kitawaki, J., Koshiba, H., Kitaoka, Y., Obayashi, H., Hasegawa, G., Nakamura, N., Yoshikawa, T., Matsushita, M., Maruya, E., Saji, H., Ohta, M. & Honjo, H. (2004). Genetic contribution of tumor necrosis factor (TNF)-alpha gene promoter (-1031, -863 and -857) and TNF receptor 2 gene polymorphisms in endometriosis susceptibility. *Am J Reprod Immunol 51*, 352-357.

[195] Teramura, M., Kobayashi, S., Hoshino, S., Oshimi, K. & Mizoguchi, H. (1992). Interleukin-11 enhances human megakaryocytopoiesis in vitro. *Blood 79*, 327-331.

[196] Tsukasaki, K., Miller, C. W., Kubota, T., Takeuchi, S., Fujimoto, T., Ikeda, S., Tomonaga, M. & Koeffler, H. P. (2001). Tumor necrosis factor alpha polymorphism associated with increased susceptibility to development of adult T-cell leukemia/lymphoma in human T-lymphotropic virus type 1 carriers. *Cancer Res 61*, 3770-3774.

[197] Vandenbroeck, K. & Goris, A. (2003). Cytokine gene polymorphisms in multifactorial diseases: gateways to novel targets for immunotherapy? *Trends Pharmacol Sci 24*, 284-289.

[198] Veiby, O. P., Jacobsen, F. W., Cui, L., Lyman, S. D. & Jacobsen, S. E. (1996). The flt3 ligand promotes the survival of primitive hemopoietic progenitor cells with myeloid as well as B lymphoid potential. Suppression of apoptosis and counteraction by TNF-alpha and TGF-beta. *J Immunol 157*, 2953-2960.

[199] Vella, A., Teague, T. K., Ihle, J., Kappler, J. & Marrack, P. (1997). Interleukin 4 (IL-4) or IL-7 prevents the death of resting T cells: stat6 is probably not required for the effect of IL-4. *J Exp Med 186*, 325-330.

[200] Waldmann, T. A., Dubois, S. & Tagaya, Y. (2001). Contrasting roles of IL-2 and IL-15 in the life and death of lymphocytes: implications for immunotherapy. *Immunity 14*, 105-110.

[201] Wang, Z. C., Yunis, E. J., De los Santos, M. J., Xiao, L., Anderson, D. J. & Hill, J. A. (2002). T helper 1-type immunity to trophoblast antigens in women with a history of

recurrent pregnancy loss is associated with polymorphism of the IL1B promoter region. *Genes Immun 3*, 38-42.

[202] Warner, J. A., Miles, E. A., Jones, A. C., Quint, D. J., Colwell, B. M. & Warner, J. O. (1994). Is deficiency of interferon gamma production by allergen triggered cord blood cells a predictor of atopic eczema? *Clin Exp Allergy 24*, 423-430.

[203] Wihlborg, C., Sjoberg, J., Intaglietta, M., Axdorph, U., Pisa, E. K. & Pisa, P. (1999). Tumour necrosis factor-alpha cytokine promoter gene polymorphism in Hodgkin's disease and chronic lymphocytic leukaemia. *Br J Haematol 104*, 346-349.

[204] Willis, C., Morris, J. M., Danis, V. & Gallery, E. D. (2003). Cytokine production by peripheral blood monocytes during the normal human ovulatory menstrual cycle. *Hum Reprod 18*, 1173-1178.

[205] Wills-Karp, M. (2001). IL-12/IL-13 axis in allergic asthma. *J Allergy Clin Immunol 107*, 9-18.

[206] Worrall, B. B., Azhar, S., Nyquist, P. A., Ackerman, R. H., Hamm, T. L. & DeGraba, T. J. (2003). Interleukin-1 receptor antagonist gene polymorphisms in carotid atherosclerosis. *Stroke 34*, 790-793. Epub 2003 Feb 2020.

[207] Yadav, D. & Sarvetnick, N. (2003). Cytokines and autoimmunity: redundancy defines their complex nature. *Curr Opin Immunol 15*, 697-703.

[208] Yamashita, H., Shimada, K., Seki, E., Mokuno, H. & Daida, H. (2003). Concentrations of interleukins, interferon, and C-reactive protein in stable and unstable angina pectoris. *Am J Cardiol 91*, 133-136.

[209] Yeh, E. T., Anderson, H. V., Pasceri, V. & Willerson, J. T. (2001). C-reactive protein: linking inflammation to cardiovascular complications. *Circulation 104*, 974-975.

[210] Yoshimura, A., Mori, H., Ohishi, M., Aki, D. & Hanada, T. (2003). Negative regulation of cytokine signaling influences inflammation. *Curr Opin Immunol 15*, 704-708.

[211] Zhu, J. & Emerson, S. G. (2002). Hematopoietic cytokines, transcription factors and lineage commitment. *Oncogene 21*, 3295-3313.

In: Progress in Immunology Research
Editor: Barbara A. Veskler, pp. 159-206

ISBN 1-59454-380-1
©2005 Nova Science Publishers, Inc.

Chapter VIII

Schistosomiasis: The Immunology of Exposure and Infection

Takafira Mduluza, *Francisca Mutapi and Patricia D. Ndhlovu*
Department of Biochemistry, University of Zimbabwe
Mount Pleasant, Harare, Zimbabwe

Abstract

Parasitic infections are prevalent in both tropical and sub-tropical areas. Most of the affected areas are in the developing countries of the world where control measures are lacking or are intermittently applied. Vaccinations against parasitic diseases have been the major talk over the past decades yet there has been very little advancement in developing effective vaccines for diseases such as malaria and schistosomiasis. Although, much research was done to understand the patterns of infections in the human host, little advancement in vaccine development has been witnessed. The main outcome has been the immunological implications of exposure and infection to malaria and schistosomiasis whose correlates of immunity have been reported to be dependent on age and exposure. The phenomena of age and exposure dependent immunity was investigated in details and development of partial immunity has been identified in older children compared to young children living in endemic areas. Evidence is accumulating that the pathology observed in most parasitic infection cases is not caused directly by parasite products but by normal components of the immune response, especially anti-parasitic IgE antibody, IgG4 the blocking antibody and cytokines like TNF-α/β, IL-10, IFN-γ and IL-6. The virulence of parasitic infection appears to be inversely related to the capacity of the parasite to induce production of protective host's cytokines and IL-4, IL-5, TNF-α and IFN-γ are central to this process. Transmission patterns are core to the severity of infection and sometimes this is related to host immune development. Immunological

* Takafira Mduluza, Department of Biochemistry, University of Zimbabwe, P.O. Box MP167, Mount Pleasant, Harare, Zimbabwe. Tel: 263-4-303211 ext.1344; Fax: 263-4-308046. Email: mduluza@medic.uz.ac.zw or tmduluza@yahoo.com

markers of susceptibility and resistance to infection are often useful in predicting progression or severity and development of protection. There is therefore a need to understand immune mechanism during infection and the immune response in individuals in the same locality who resist development of clinical illness. The immunological implication of infection and exposure is discussed relative to humoral and cellular immunity and the possible boosting of protective responses by chemotherapy and the possible negative impacts. The discussion gives insight into the understanding of infection and exposure to schistosome infections necessary in the development and current search for vaccines to schistosome infections.

Introduction

Schistosomiasis remains a global programme and affects more than 200 million people (WHO, 1993). In Africa, there are three species that cause schistosomiasis in humans, namely *Schistosoma mansoni, S. haematobium,* and *S. intercalatum. S. mansoni* and *S. haematobium* are widely distributed, while *S. intercalatum* has a rather limited distribution, being found only in Zaire, Cameroon and Gabon. *S.mansoni* and *S.haematobium* are both endemic in Zimbabwe, the latter being the most prevalent (Taylor and Makura, 1985; Ndhlovu *et al* 1992). Mortality due to schistosomiasis was estimated at 11,000 deaths per year and the burden of disease DALY's was 1,713,000 (WHO report 1995). In Zimbabwe, schistosomiasis is one of the top ten causes for hospital admissions (Ministry of Health, 1992), which is an indication of its public health importance. From an immunological view point, important stages of the life cycle are those that occur in the human host i.e. the schistosomula, the adult worms and the eggs excreted by the female adult worm.

The immune system of the definitive host is exposed to a range of antigenic stimulations, including somatic and surface antigens of the different developmental stages of the schistosome (schistosomulum, adult worm, egg), and also to excretory antigens from each of these stages (Dunne *et al.*, 1988). This results in a wide range of complex immunological responses and interactions being initiated (Phillips, 1992). The situation becomes even more complicated as the parasite appears to have some capacity to protect itself against immune attack (McLaren, 1985; McLaren and Terry, 1982; Smithers *et al.*, 1969; McLaren *et al.*, 1975). This continuous confrontation leads to a variety of immuno-regulatory interactions (Colley, 1981a,b). Nevertheless, ample evidence has been provided for the existence of an immunologically mediated and acquired resistance to infection in mammalian schistosomiasis. This resistance may be of the "concomitant" type with immune effector responses directed against the invading schistosomula, whilst adult worms continue to survive (Smithers and Terry, 1969a). At the same time, there is a modulation of the granulomatous reaction towards the tissue deposited *Schistosoma* eggs (Phillips *et al.*, 1978).

Antigens

Interests in immunodiagnostic, sero-immunological and vaccine aspects of human schistoso-miasis infections have led to numerous studies being conducted on antibody responses to schistosome antigens in man. As compared with *S.mansoni*, much less is known about antigens of

S. haematobium, primarily due to the fact that *S. haematobium* is difficult to keep, particularly in laboratories outside the endemic areas. The difficulty in maintaining the life cycle of *S. haematobium* in the laboratory, primarily due to lack of an appropriate rodent definitive experimental host, is one of the explanations for the limited work of this nature done on *S. haematobium* as compared to *S. mansoni*. For example, while *S. mansoni* egg antigens have been prepared and characterized in numerous studies (Dunne and Doenhoff, 1983; Dunne and Bickle, 1987; Dunne *et al.*, 1981; McLaren *et al.*, 1981), few studies have dealt with *S. haematobium* egg antigens (Hillyer and Pacheco,1986). The study identified a 55 kDa antigen specific for *S. haematobium*. Similarly, studies carried out to characterize *S. haematobium* schistosomulum and adult worm antigens are rather few in number (Simpson *et al.*, 1985; Kelly *et al.*, 1987). Hillyer and Pacheco (1986) have isolated and characterized a *S. haematobium* egg antigen which shares epitopes with an adult worm antigen, and which is specific for *S. haematobium*. Such an antigen could prove useful for immunodiagnostic purposes. *S. haematobium* schistosomula antigens have been characterized by Simpson *et al.* (1985) and Kelly *et al.* (1987). Some of these were shown to cross-react with schistosomula antigens of *S. mansoni* and *S. bovis*, although one appeared to be specific for *S. haematobium*. Hayunga *et al.* (1981) and Trottein *et al.* (1992) studied adult *S. haematobium* antigens. Trottein *et al.* (1992) also cloned a glutathione S-transferase suggested to be a potential vaccine candidate

Realizing the paucity of information regarding *S. haematobium* antigens, Ndhlovu *et al* (1996) prepared and partially characterized *S. haematobium* antigens. Since studies on *S. mansoni* had shown that antigens from all stages in the life cycle could be recognized by human antibodies (Butterworth *et al.*,1985; Dunne *et al.*, 1988), the antigens produced and partially characterized comprised soluble worm antigens (SWA), soluble egg antigens (SEA) and soluble schistosomulum antigens (SSA). A Zimbabwean isolate of *S. haematobium* was used to limit possible negative influences of geographical strain variations in antigenicity (Hackett *et al.*, 1987), and issues related to the species and stage specificity of the different antigens were addressed. The molecular weights of the soluble *S. haematobium* egg antigens (SEA) identified ranged from 23.4 kDa to just over 100 kDa. At least 16 different antigens were identified in the SDS-PAGE. Immunoblotting using serum from *S. mansoni* and *S. haematobium* infected baboons showed that all the antigens were recognized with the exception of the 50.1 kDa antigen. This antigen was weakly recognized by the *S. haematobium* serum but not recognized by the *S. mansoni* serum. Immunoblotting using *S. haematobium* adult worm, egg and schistosomulum anti-sera raised in rabbits showed that the 100.0, 90.0, 69.2, 66.2, 60.2, 49.0 and 36.3 kDa antigens were egg specific in not being recognized by anti-worm and anti-schistosomulum serum. The 74.1, 40.7, 28.8 and 23.4 kDa egg antigens were not recognized by the anti-egg serum. Cross-reacting antigens, i.e. antigens recognized by anti-egg, anti-worm and anti-schistosomulum serum, comprised the 35.5 and 77.6 kDa antigens. Anti-worm serum also recognized a 40.7 kDa antigen that was not recognized by any of the other anti-sera..

Studies on cross-reacting and common antigens between different stages of schistosomes (Dunne and Bickle, 1987) have provided important information regarding the age-related, acquired resistance to schistosome infection, and recent studies have led to the identification of some candidate antigens for vaccine development. Emphasis here has been put on the 28 kDa glutathione S-transferase of *S. mansoni* which has been cloned and which is now ready for phase 1 human trials (Auriault et al., 1990; Trottein et al., 1992). It is well recognized that some antigens are shared by different parasite species and by different stages in the life cycle of a given species (Pammenter *et al.*, 1992). That most *S. haematobium* egg antigens in our studies were recognized by *S. mansoni* baboon serum was in agreement with the earlier demonstration by Doenhoff *et al.* (1993), and this also applies for soluble *S. mansoni* egg antigens being recognized by human *S. haematobium* serum. The 50.1 kDa antigen recognized faintly in the study by *S. haematobium* baboon serum and by *S. haematobium* anti-egg serum, but not by *S. mansoni* human serum, provided some evidence for this antigen being specific for *S. haematobium* and therefore of potential use in species specific immuno-diagnosis of *S. haematobium* infection. This antigen was presumably related to the 55 kDa antigen identified by Hillyer and Pacheco (1986) as being *S. haematobium* species specific. If species specificity is not needed, the whole range of egg antigens could be a potential candidate for immunodiagnosis of *Schistosoma* infection.

The blocking antibody concept has been paid considerable interest in explaining the lack of resistance to *Schistosoma* infection in young children (Butterworth *et al.*, 1987). This concept was based on shared/common antigens between eggs and schistosomula, and the demonstration that the 89.1, 77.6 and 35.5 kDa *S. haematobium* egg antigens are recognized by *S. haematobium* schistosomulum anti-serum raised in rabbits was of particular interest. Further studies will show whether these particular cross-reacting antigens are potential targets for the blocking antibodies.

The finding of our studies that many of the soluble *S. haematobium* adult worm antigens were recognized by baboon *S. mansoni* serum, reflecting that the two *Schistosoma* species share common antigens, confirms numerous previous studies. Hayunga *et al.* (1981) showed that the majority of *S. haematobium* worm antigens were recognized by *S. mansoni* sera and that only the 25.0 kDa antigen was *S. haematobium* specific. Shah *et al.* (1985) showed that *S. mansoni* adult worm antigens of molecular weights of 85.0, 77.6 and 44.1 kDa were recognized by *S. haematobium* serum, and Aronstein and Strand (1983) reported that the majority of soluble worm antigens recognized by sera from *S. haematobium* infected patients were also recognized by sera from *S. mansoni* infected patients. This has in fact led some researchers to use the readily available *S. mansoni* antigens in studies on humoral responses in *S. haematobium* infections (Hagi *et al.*, 1990; Sathe *et al.*, 1991). As shown in the Zimbabwean studies, however, not all *S. haematobium* adult worm antigens, namely the 18.0, 18.6 and 26.0 kDa antigens, were recognized by *S. mansoni* baboon serum. This shows that *S. mansoni* may not be the best choice in obtaining target antigens for assaying immune reactivity in *S. haematobium* infection, and that the possibility might still exist for achieving species specificity in immuno-diagnosis of *Schistosoma* infection on the basis of adult worm antigens. Kelly *et al.* (1987) has previously identified a 28 kDa and a 19 kDa adult worm antigen being specific for *S. haematobium*. The 18.6 kDa antigen identified in the study as being specific for *S. haematobium* is presumably identical with the 19 kDa antigen identified by Kelly *et al.* (1987).

The recognition by *S. haematobium* anti-schistosomulum sera of *S. haematobium* adult worm antigens of molecular weights of >90.0, 65.5, 57.0 and 32.0 kDa points to some sharing of antigens between adult worms and schistosomula. Such sharing may nourish the concept of concomitant immunity in *Schistosoma* infections (Smithers and Terry, 1969; Simpson and Smithers, 1985), but further studies are obviously needed.

Simpson *et al.* (1985) used iodogen-catalyzed labelling followed by immunoprecipitation to identify schistosomulum surface antigens. The major antigens identified were a 17 kDa antigen and a complex of 24-30 kDa antigens. In our study, based on SDS-PAGE and Immunoblotting,

14 antigens were identified. The demonstration that human *S. mansoni* serum may recognize soluble *S. haematobium* schistosomulum antigen was not supported by the study carried out by Simpson *et al.* (1985). Although the studies on schistosomulum antigen were not that comprehensive, the general impression of comprehensive cross- reactivity and antigen sharing in human *Schistosoma* infections was confirmed.

The results of the Zimbabwean studies clearly revealed a low species and stage specificity in the antigen characteristics. This supports the general feeling of a close antigenic familiarity between *S. mansoni* and *S. haematobium* (Kelly, 1987), and expresses the close immunological interaction between different stages in the *Schistosoma* life cycle in the definitive host. The similarity of *S.haematobium* and *S.mansoni* of the Zimbabwean species may be helpful in a country where resources are scarce as far as having a schistosomiasis vaccine. However care must be taken in using humans in the trials. Robertson (1985), suggested that trials could be done on monkeys and baboons, which are many in Zimbabwe. She further alluded to the fact that we have plenty helminth and their intermediate snail host and thus could benefit from collaboration with laboratories working on developing a schistosomiasis vaccine.

Immunodiagnosis

The circulating antigen detection approach was introduced as a very promising tool in sero- and immuno-epidemiological studies and in diagnosis of intestinal and urinary human schistosomiasis (Deelder *et al.,* 1989a; De Jonge *et al.,*1989a,d). The CAA detection techniques, including both the original ELISA technique (Deelder *et al.,* 1989a) and the Magnetic Bead Antigen Capture Enzyme-Linked Immuno Assay technique (Gundersen et al., 1992a) have, however, been developed for use under strictly controlled conditions in well-equipped laboratories. Such conditions are seldom found in field laboratories in schistoso-miasis endemic areas, and from the point of view of the practical applicability of these techniques in control programme implementation, some simplification and adaptation of the techniques to such conditions is a necessity. It is in the light of these considerations that we optimised the MBAC-EIA (Ndhlovu et al., 1995).

As a result of the study, simplification and adaptation of the Magnetic Bead Antigen Capture Enzyme-Linked Immuno Assay (MBAC-EIA) technique to field conditions was achieved and has been applied to other studies that we conducted (Ndhlovu et al., 2002). The assay may be performed successfully within the broad temperature range of 18-37^{0}C in that the slightly lower sensitivity at low temperatures may be adjusted for by prolonging the incubation period. At higher temperatures, a possible need for shortening the incubation time was indicated, but adjusting incubation time to temperature will without doubt allow maximum assay sensitivity over a rather broad temperature range. The demonstration that beads can be re-used, that hand-shaking is as good as automatic titerteck shaking, that aspiration of the supernatant before the addition of conjugate is not necessary, and that the use of whole blood and serum offers similar assay sensitivity, all add to the conclusion that the modified MBAC-EIA technique, as developed by Ndhlovu et al., (1995), is highly suitable for use under field conditions in Zimbabwe and elsewhere in tropical, in schistosomiasis endemic environments.

Urinary and intestinal schistosomiasis commonly overlaps in distribution, and future community-based schistosomiasis morbidity control programmes may need to address both types of schistosome infections concurrently, with praziquantel being effective against both species. To that end, a diagnostic technique, which at minimal costs could diagnose both *S. mansoni* and *S. haematobium* infections would be very useful. CAA and CCA are excreted by all known *Schistosoma* species (De Jonge *et al.*, 1988, 1989a-d; Deelder *et al.*, 1989a; Van't Wout *et al.*, 1992) so their detection, seen from a diagnostic point of view, could be very useful in areas endemic for both types of schistosomiasis. Under such conditions, some people will harbour double infections while others would harbour single infections with either of the species.

The sensitivity of the MBAC-EIA technique in diagnosis of schistosomiasis in an area endemic for both urinary and intestinal schistosomiasis was very high with an overall sensitivity of 95%. This level of sensitivity was comparable to that of earlier studies reported in pure *S. mansoni* and *S. haematobium* infections (De Jonge *et al.*, 1989d; Gundersen *et al.*, 1992b). That 12 out of 14 non-egg excreters were positive for CAA presumably reflects a low sensitivity of the parasitological diagnostic technique used, although the possibility exists that the infection consists of either immature worms or worms of only one sex.

It is well realized that the simple addition of *S. mansoni* and *S. haematobium* egg counts in double infected people is a simplification, especially from the point of view of using egg outputs in estimating morbidity. However, a positive, although weak association between CAA and *Schistosoma* egg counts was demonstrated by Ndhlovu *et al.,* (1995). This might be taken to support the suggestion that CAA measurements in the future might be useful also in morbidity assessment in double infected people.

The results from the application of the MBAC-EIA technique in *S. haematobium* infection revealed a very high assay sensitivity. The sensitivity is similar to, or even higher than that obtained using the traditional ELISA technique for assaying CAA levels in *S. haematobium* infections (De Jonge *et al.*, 1989d, 1991a). Only three egg positive people were negative for CAA while 76.7% of egg negatives were scored positive for CAA. This presumably mainly reflects the low sensitivity of the parasitological diagnostic technique used although immature or single sex infections could also be considered responsible. The study on *S. haematobium* thus confirms previous results from studies on *S.mansoni* (Gundersen *et al.*, 1992b) that the sensitivity of the MBAC-EIA exceeds that of the traditional parasitological diagnostic technique based on detection of eggs in excreta.

Further evaluation of the detection of the circulating antigen levels in serum and urine samples was done using the following techniques.

(a) Magnetic Bead Antigen Capture Enzyme Immuno-Assay (MBAC - EIA), for the detection of Circulating Anodic Antigens (CAA) and

(b) The traditional ELISA on detection of CAA and Circulating Cathodic Antigen (CCA).

(c) The ATO Rapid reagent strips for the detection of (CCA)

(d) Traditional ELISA for detection of Soluble Egg Antigens (SEA).

The study was in 2 parts in Mashonaland Central area of Zimbabwe. One study (Mupfure) evaluated the circulating antigens prior to treatment and at specific time points up to day 64 after treatment. The second study (Kaziro) evaluated 3 of the five techniques in a cohort of school children in a re-infection study for 2 years. Details of these results are reported in Ndhlovu et al.(2002a,b). Results of these studies showed that using serum samples, the sensitivity of the MBAC- EIA for the detection of CAA was 100%. While that of the Traditional ELISA technique (TCAA) was 90%. Urine samples were used to detect CCA and SEA by the traditional ELISA. The CCA was 93.8%. While that for detection of SEA, was 81% and that for the ATO RAPID TEST 84 %. The specificity of each of the tests was above 50%.

Prior to treatment, urine egg counts correlated positively with all the tests with the SEA and TCAA having similar correlation coefficients (r = 0.475, p<0.001; r=0.474, p=0.005).

Levels of TCAA in serum and levels of CCA in urine also correlated positively at r=0.665, p<0.0001, n=56. SEA did not correlate with either CCA or TCAA. After treatment, results for the different tests were varied. All the tests performed well in a longitudinal re-infection study and showed good correlations with urine egg counts.

The overall positive correlation between S. haematobium egg excretion and CAA and CCA levels and also morbidity indicators of proteinuria, micro-haematuria is especially important from the morbidity assessment point of view. Thus, CAA, CCA and for that matter SEA may, in addition to demonstrating the presence of the S. haematobium infection, provide information concerning the morbidity situation at least at community level. This is a central finding because the present schistosomiasis haematobium control strategy is on morbidity control.

The field applicability, cost efficacy and high specificity and sensitivity of the MBAC-EIA technique for CAA detection and the ATO rapid test for CCA provide a sound basis for their general and widespread use in control programme implementation. The ATO Rapid test for CCA has an added advantage of being non-invasive as it uses urine samples. Without using sophisticated equipment, large populations may be screened in a short time using only limited manpower, and screening for both S. haematobium and S. mansoni and treatment may be provided in a single, integrated procedure.

Immuno-Epidemiology of Infection and Exposure.

Epidemiological evidence of the development of resistance to infection in humans was provided by some studies in Zimbabwe, which showed a decline in prevalence and intensity of infection in older age groups suggesting the development of an acquired, age-related resistance to re-infection (Clarke, 1966; Ndhlovu et al., 1996; Mutapi et al., 1997).

In an effort to bridge the gap between S. mansoni and S. haematobium, Ndhlovu et al., 1996a, b; reported on a study done at Dzwete village situated in Chikwaka communal lands in Mashonaland East Province of Zimbabwe. In this population, the overall prevalence of S. haematobium infection in the population was 43%. Prevalence figures ranged from 9.2% in the 44+ years age group to 65% in the 10-14 years age group with males and females having similar prevalences of S. haematobium infection of 44.3% and 42.0%, respectively. The intensity of infection, as expressed by the geometric mean egg output among positives only, reached the peak of 30.9 eggs per 10 ml of urine in the 10-14 years age group. This peak in the intensity of infection was followed by a gradual decrease with increasing age to reach the minimum value of 3.6 eggs per 10 ml of urine in the 44+ years age group (Figure 1).

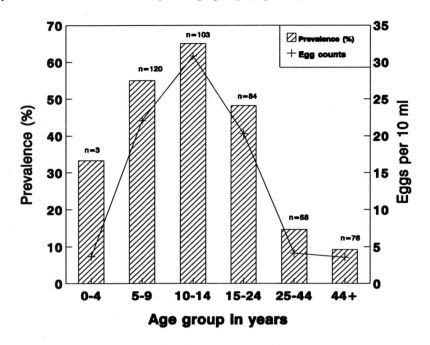

Figure 1. Age-related prevalence and intensity of S.haematobium infection curves.

The only logical step was then to assess how immunity fitted into the pattern. As urine egg counts measurements have their own problems it was decided to measure circulating anodic antigens in the same study population. Thus humoral immunity and circulating anodic antigens (CAA) were measured. The work demonstrated that there was a good correlation between CAA and egg counts. IgE to SEA and SWA levels were lower in *S. haematobium* egg positives than in egg negatives, there was a negative correlation with egg output and CAA but a positive correlation with age, indicating an association between IgE and resistance to infection. In contrast, IgG4 and IgM levels were higher in *S. haematobium* egg positives than in egg negatives. There was also a negative correlation with age indicating an association between IgM and IgG4, and susceptibility to infection.

In the study, a negative correlation between CAA levels and age, being independent of the effect of age-related patterns of intensity of infection, was evident. This finding presumably reflected an inverse relationship between CAA levels and duration of exposure to schistosome infection. The individuals had lived in the endemic area all their lives and the older aged individuals had thus been exposed to schistosome infection for a longer time and hence had lower worm burdens and lower levels of CAA. In previous studies, it has been shown that duration (in years) of exposure to schistosome infection rather than age as such plays a central role in the age-dependent acquisition of schistosome infection in endemic areas (Kloetzel and da Silva, 1967; Hagan *et al.*, 1987; Sturrock *et al.*, 1987; Wilkins *et al.*,

1987). However, Butterworth *et al.* (1992) pointed to a possible need of re-assessing the potential importance of changes in the immune status and infection patterns taking place around the time of puberty.

The *S. haematobium* age-prevalence and age-intensity of infection patterns in the Dzvete village study population were typically convex shaped, with peak prevalence and intensity of infection, as revealed by *S. haematobium* egg counts in urine, in the second decade of life. That the reduction in *S. haematobium* egg excretion is due to a reduction in worm burdens with increasing age was indicated by the fact that levels of CAA showed a positive significant correlation with intensity of infection as measured by egg counts (r=0.6, P<0.0005). These findings pointed towards the fact that immunity reactions are directed primarily towards the worms and not towards their fecundity.

The negative correlation between IgE anti-egg and anti-worm antibodies and CAA levels may suggest IgE antibodies being involved with the immune effector mechanisms resulting in reduced worm burdens. Similarly, the patterns and relationships as regard the anti-SWA IgA responses were suggestive of an antibody associated with resistance to schistosome infection. These results affirmed those of Dunne *et al.* (1993) and Capron *et al.* (1993). However, more studies are needed to further elucidate the possible role of IgA in the regulatory response to schistosome infections and the cytokine types response for the stimulation.

On the other hand, the positive correlations demonstrated between IgM and IgG4 antibody responses and CAA levels pointed to the association of these antibodies with susceptibility to schistosome infection through their suggested abilities to block immune effector mechanisms directed against schistosomula (Khalife *et al.*, 1986; Butterworth *et al.*, 1987, 1992; Jassim *et al.*, 1987; Dunne *et al.*, 1988; Hagan *et al.*, 1991). The shift from susceptibility to resistance with increasing age was thus linked to a shift in the relative occurrence of blocking and protective antibodies. The role of IgE as a protective antibody has been demonstrated in studies on both *S. haematobium* and *S. mansoni* (Hagan *et al.*, 1991; Rihet *et al.*, 1991; Dunne *et al.*, 1992). IgE is believed to play a central role in the antibody-dependent, cellular cytotoxicity (ADCC) towards schistosomula involving eosinophils, platelets and macrophages (Capron *et al.*, 1977a,b, 1982). IgA levels also increase in the second decade of life and there is evidence that IgA antibodies may be involved with immune effector mechanisms directed towards reduction of worm fecundity and egg viability (Capron *et al.*, 1993). Dunne *et al.* (1993) furthermore presented results that demonstrated that IgA antibodies could mediate killing of schistosomula in the presence of activated human eosinophils.

In an attempt to ascertain the role of cell-mediated immunity, the granuloma index was measured in the same population. Results did not demonstrate significant correlations. The Granuloma Index, being an indicator of cellular mediated immunity, and any of the antibodies addressed. However, the increase with age among *S. haematobium* egg negatives in both the Granuloma Index and the IgE levels may have indicated that the capacity to produce an increased granuloma response coincides with the capacity to mount a high IgE antibody response. This may lead to the tentative suggestion that the stimulation of production of some of the cells taking part in granuloma formation (eosinophils) and the production of IgE may be triggered by the same lymphokines. The production of eosinophils and IgE, both being associated with resistance to *S. haematobium* infection, is triggered by

Th2 derived lymphokines (IL-5 and IL-4) (Sanderson *et al.*, 1985; Pene *et al.*, 1988a,b), which, at least in the mouse model, are present at peak granuloma formation (Pearce *et al.*, 1991). These Th2 derived lymphokines could play a role in resistance to *S. haematobium* infection in the older segment of the endemic human populations. The similarity in the increase with age of production of eosinophils and IgE also support the multifactorial nature of immune reactions in schistosome infections (Butterworth and Hagan, 1987).

The presence of an age-related, immunologically mediated acquired resistance, the progress made in antigen identification, the recent developments in recombinant DNA technology, and the accumulating knowledge regarding the immunological reactivity in human and animal infections give hope for development of a schistosomiasis vaccine. However, the crucial factor will be to stimulate protective and not irrelevant and inappropriate responses (Butterworth *et al.*, 1992). This might be achieved by pre-selection of antigens stimulating IgE production through screening with IgE-enriched serum from immune individuals, as suggested by Hagan (1993). Anaphylactic reactions would, in theory, not occur as IgG4 production would probably also be stimulated by these antigens. The efficacy of the vaccine would, however, depend on the IgE production exceeding that of IgG4. Thus, the vaccine should aim to induce the production of Th2 derived lymphokines, especially IL-4, which triggers production of both IgE and IgG4, in a way that favours switching the B cell clones to IgE production. However, much research is still needed as a background for the development of an effective vaccine against schistosomiasis.

IgE is involved in many of the ADCC reactions. Several studies have shown that various sub-populations of monocytes, basophils, mast cells, macrophages, eosinophils and platelets from both rodents and humans carry specific IgE Fc receptors, which facilitate the binding of IgE to these cells. By cross-linking to schistosomula surface antigens, cell-parasite contact is established. The subsequent cellular degranulation leads to release of various subcellular components and mediators, which can damage and kill the schistosomula. This IgE antibody dependent cell mediated cytotoxicity was first shown by Capron *et al.* (1977) and was confirmed in humans by *in vitro* studies (Joseph *et al.*, 1983). Immuno-epidemiological studies have also confirmed the role of IgE in resistance to human schistosome infections (Hagan *et al.*, 1991; Rihet *et al.*, 1991; Dunne *et al.*, 1992). The important role of IgE in the regulatory response to schistosome infection in various non-human models has been demonstrated, for example by Joseph *et al.* (1978), Kigoni *et al.* (1986), Capron *et al.* (1977), and McLaren and Smithers (1987). The effector cell in these situations is most commonly the eosinophil.

Indirect evidence of the central role of the eosinophils in resistance to reinfection in human schistosomiasis has been provided by several other field studies in which correlations between the level of blood eosinophilia and susceptibility to re-infection have been reported. Such a relationship has been shown in the case of *S. mansoni* in studies in Kenya by Sturrock *et al.* (1983) and in the case of *S. haematobium* in studies in the Gambia by Hagan *et al.* (1985a,b). Further evidence of this has been obtained by Butterworth *et al.* (1979) and Capron *et al.* (1984).

The failure of young people to mount an effective response to *Schistosoma* infection, in spite of activation of both the humoral and cellular arms of the immune system, was rather puzzling until the existence of blocking antibodies was demonstrated and their function recognized. It is thought that antibodies with specificity for polysaccharide antigens in the schistosome egg cross-react with antigens on the surface of the schistosomulum. The binding of these "blocking" antibodies interferes with the immune effector activities of potentially protective antibodies. The existence of such blocking antibodies is at least one factor responsible for the fact that resistance to reinfection does not exist in the younger segment of the human population (Butterworth *et al.*, 1987, 1988, 1992). Antibodies, which have been shown to play the role as blocking antibodies comprise human IgG2 (Butterworth *et al.*, 1984, 1987, 1988), human IgM (Khalife *et al.*, 1986; Jassim *et al.*, 1987) and human IgG4 (Dunne *et al.*, 1988; Hagan *et al.*, 1991). Resistance developing in older children/young adults may be due to a relative shift in the quantitative occurrence of the different antibodies with IgE now dominating over IgG2, IgG4 and IgM (Magnusson and Johansson, 1989; Hagan *et al.*, 1991; Dunne *et al.*, 1992). All these findings, except those of Hagan *et al.* (1991), relate to *S. mansoni*, and very limited attention has in fact been paid to *S. haematobium*.

Chemotherapy Accelerates the Development of Immunity

Schistosome infections are a public health concern in tropical and sub-tropical countries (Webbe, 1981). Various field and experimental studies have suggested that hosts develop some acquired immunity to the parasite (Terry, 1994; Hagan, 1992; Hagan, 1991; Dunne, 1992; Butterworth, 1985). In humans acquired immunity develops slowly over several years (Woolhouse, 1999), and treatment with praziquantel has been suggest speeding up the development of schistosome-specific acquired immunity (Mutapi, 1998). Praziquantel or oxamniquine treatment of schistosome-infected people alters parasite specific cellular and humoral responses particularly in children (Webster, 1997; Grogan, 1996; Mutapi, 1998). These treatment-induced changes have been associated with resistance to re-infection (Correa-Oliveira, 2000). Furthermore, immune responses following treatment in children have been shown to mimic those of exposed but untreated adults who have lower levels of infection. For example, levels of anti-egg IgE and IgG4 in *S. haematobium*-infected children in the Gabon were shown to increase and decrease respectively following treatment to reach levels similar to those in untreated, exposed adults (Grogan, 1996). In Zimbabwe we showed that levels of anti-egg IgG1 increased and IgA decreased in *S. haematobium*-infected children matching levels in untreated, exposed adults (Mutapi, 1998). The study also showed a significant negative correlation between levels of IgG1 and IgA, both before and after treatment. This suggests that treatment-induced changes are more complicated than independent increases or decreases in each antibody.

Previous studies have examined factors, which affect the changes in antibody levels following treatment, and how they affect re-infection rates (Dunne, 1992; Butterworth, 1985; Correa-Oliveira, 2000; Feldmeier, 1983; Hagan, 1987; Hagan, 1991; Mutapi, 1998; Satti, 1996). However, there have been no studies to date of the nature of treatment-induced changes. Characterization of these changes is important for a better understanding of how schistosome-specific acquired immunity develops and is essential for vaccine development and schistosomiasis control. The consensus is to use treatment synergistically with vaccination as is being done with the Clinical trials of Bilvax© in Senegal. In addition, many of the vaccine target populations will have been previously treated as part of a national control or research programme.

Older children have the highest infection intensities, which are attributable to cumulative infection in this age range. There is a correlation between age and infection intensity. Therefore, it is not surprising that there is a positive correlation between pre-treatment antibody levels and age, as immune responses are mounted in response to infection. As infection intensity increases levels of antibodies will increase until a point is reached when antibody responses begin to limit infection levels. This pattern will be most apparent in quick-evolving, short-lived immune responses with little or no immunological memory such as IgM responses against carbohydrate or polysaccharide antigens, and less apparent in slower evolving immune responses with strong immunological memory such as those involving IgE (Woolhouse, 1996). It is therefore not surprising that pre-treatment levels of both IgE and IgM show a positive association with pre-treatment infection intensity, although this is only significant for IgM. Significant positive correlations between infection intensity and IgM have been reported both for *S. mansoni* and *S. haematobium* (Butterworth, 1992; Ndhlovu, 1996)

Treatment of schistosome-infected individuals is now widely accepted to alter schistosome-specific cellular (Colley, 1986; Feldmeier, 1988; Grogan, 1998; Ottesen, 1978] and humoral immune responses (Feldmeier, 1983; Gryzch, 1993; Grogan, 1996; Mutapi, 1998; Naus, 1998). We reported that treatment accelerated the development of acquired immunity in *S. haematobium*-infected children (Mutapi, 1998). The changes are due to increased antigen presented to the immune system following treatment. Treatment with anti-helminthic drugs kills adult schistosomes. Praziquantel acts quickly and kills worms by inducing worm paralysis (Andrews, 1985) while oxamniquine acts more slowly, binding on to the worm DNA and RNA stopping protein synthesis (Pica-Mattoccia, 1989, Fallon, 1996). Worm death results in introduction of antigens not normally available to the immune system (Fallon, 1995). The increases in levels of IgE and IgM are likely to be responding to this antigen increase. Such increases in antibody levels have been reported from studies of *S. mansoni*, *S. haematobium* and *S. japonicum* infected children of similar age groups (Li, 1999; Grogan, 1996; Mutapi, 1998; Naus, 1998; Satti, 1996; Webster, 1997).

The finding that showed the changes in levels of IgE and IgM decreases significantly with age. However, when the influence of other confounding factors such as sex, age and pre-treatment infection intensity are taken into account, the relationship is not significant, although the negative correlation still exists. The relationship between pre-treatment infection intensity and the changes in levels of IgE and IgM is not significant, showing that pre-treatment infection intensity alone does not sufficiently explain the changes in levels of IgE and IgM. In addition, there was a significant negative correlation between the change in antibody levels and pre-treatment antibody levels. These three observations suggest that the changes in levels of the two antibodies are related to the development of these same responses before treatment and not to current levels of infection before treatment or age. The interaction between the development of the antibody responses before treatment and the magnitude of change following treatment would work through several mechanisms such as activation of memory cells in the case of responses which induce immunological memory or stimulation of new specific B-cells as suggested by Grogan (1996).

The study also showed a significant difference in the age-antibody profiles before and after treatment in the children treated with either of the two drugs. This analysis also showed that the alteration in the age-antibody relationship in not systematic but is related to pre-treatment antibody levels. This result suggests that the development of acquired immunity in children is related more to the antigens they have been exposed to rather than any other age-related process, which would not be altered by treatment. Two other age related factors have been suggested to affect immune responses in children; a) hormonal changes at puberty; and b) age-related maturity of the immune system (Gryseels, 1994). If the former affected immune responses in these children, then a difference should have been observed in the changes between the youngest children (6 years old) and the oldest children (14 years old) independent of pre-treatment infection intensity. Such a difference was not observed. If the latter was the case, and children were unable to mount some of the immune responses (inability to recognize some of the epitopes, for example), then treatment would not have altered their ability to mount immune responses and thus should not have altered the immune responses. The results suggest that before treatment the older children have experienced enough antigens to mount immune responses to some tegumental or sub-tegumental and internal antigens so that treatment only augments the response slightly while younger children have experienced lower levels of such antigens so that treatment greatly increases the levels of these antigens available to the immune system. Field work and work on vaccine candidate antigens has suggested that antigens mostly associated with resistance to infection are located within or beneath the worm tegument, thereby making it difficult for the immune system to access them in live worms (Berquist, 1998; Capron, 1994; Fallon, 1995; Dupre, 1999). Since worms have a long life span, 3-7years (Fulford, 1995), it may take several years for the antigens to be released in sufficient amounts for immune responses to be mounted.

These findings give the first indication of a relationship between the magnitude of treatment-induced antibody changes and pre-treatment antibody levels as well as an alteration between the age-antibody profiles following treatment. Three major conclusions can be drawn; first, the relationships between treatment-induced changes in levels of IgE and IgM are related to the level of acquired immunity before treatment. Second, the relationship between age and antibody levels can be altered by treatment. Third, the level of acquired immunity before treatment is related to experience of parasite antigens. Taken together these results support the two hypotheses that acquired immunity to schistosomiasis develops as a function of experience of parasitic antigens and that treatment speeds up the development of acquired immunity to tegumental/sub-tegumental and internal antigens.

Cellular Immunity of *Schistosoma* Infection and Exposure

Although much work has been done on the role of antibodies in resistance to infection with schistosomes, less work has been done on the role of cellular factors in resistance. Levels of proliferation of PBMC in response to specific and non-specific stimulants have been exploited as a measure of cellular immunity in individuals with schistosomiasis. It has been a repeated finding that only low levels of proliferation of PBMC could be detected from heavily or chronically infected patients following schistosome antigen stimulation (Domingo and Warren, 1968; Ottesen *et al.*, 1977; Colley *et al.*, 1977a,b, 1986). However, similar studies by Gazzinelli *et al.*, (1985), Roberts *et al.*, (1993), Ribeiro de Jesus *et al.*, (1993) on *S.mansoni* infected individuals and Grogan *et al.*, (1996) on *S.haematobium* infected individuals reported that low levels of proliferation to schistosome antigens take place before treatment. The same PBMC responded well to unrelated non-schistosome stimulants before treatment. The PBMC were observed to show elevated levels of proliferation after treatment. This unresponsiveness of PBMC, before treatment, to schistosome-specific antigen stimulation appears to be related to an immunomodulation that takes place during schistosome infection. The immunoregulation was said to involve several other poorly defined agents in sera, reportedly causing anergy in T cell responses during heavy and chronic infections (Colley *et al.*, 1977b, 1983; Fieldmeier, Gastl and Poggensee, 1985a,b; Ellner *et al.*, 1985).

Immunoregulation is believed to be involved in the responses of PBMC from individuals living in high transmission area. Before treatment, PBMC from individuals with heavy infection intensity showed lower levels of proliferation than PBMC from uninfected individuals from high transmission area. However, after treatment proliferation levels were reversed, and were observed to be higher among the re-infected than the uninfected individuals. One explanation for this is that praziquantel treatment had an effect on the immune response. The antigens released by dead and damaged schistosome worms as a result of the treatment boosted cellular responsiveness. As a consequence, reinfection is marked by a higher level of cellular reactivity indicated by increased levels of proliferation to schistosome-specific antigen stimulation. In contrast, individuals of low infection intensity and prevalence from a low transmission area do not experience massive schistosome invasion by the parasite. PBMC from the infected individuals from the low transmission area showed higher levels of proliferation to schistosome antigen stimulation than PBMC from uninfected individuals. The same pattern was observed at the follow-up points after treatment.

Cytokine Profiles and T Helper Cell Subsets in Human Schistosoma Infection and Exposure

T lymphocytes are now well understood as the central regulatory cells of the immune system, with their function being mediated by cytokines whose production is induced as a result of antigen stimulated cellular activation. The T lymphocytes involved in this cross regulation have been designated as T helper type 1 (Th1) and Th2 subsets, and it is recognized that the pattern of cytokines produced may directly determine whether responses are protective in character or may induce susceptibility. The role played by the T cell subsets and cytokines in directing immune responses in schistosomiasis is still unclear. Animal studies have not helped clarify the situation since results of protection studies using mice seem to conflict with those obtained from the human studies. Observing cellular and cytokine responses to *S.haematobium* in humans from areas of different endemicity have added to our understanding of how individuals respond to infection and studies conducted in Zimbabwe helped in the identification of the roles played by particular cytokines in determining susceptibility or resistance to further *S.haematobium* infection. Although members of each of the cytokine classes play important regulatory roles in the immune responses, attention has been focused on IFN-γ as a cytokine, which is a dominant feature of the Th1 response and IL-4, IL-5, and IL-10, which are dominant features of the Th2 response. The possible role of these 'signature' cytokines in the control of *S.haematobium* immunity has been examined, principally in the context of regulation of immune responsiveness by T cells. The Th1-like responses have been found to dominate early during infection while Th2-like responses begin to be elevated during infection and dominate responses post-treatment in uninfected individuals.

Cytokines normally function in concert with one another and may be involved in generating a network or cascade of other cytokines that interact with other cell regulators such as hormones and nueropeptides that are in turn capable of stimulating the generation of many others (as reviewed by Romagnani, 1991, 1997; Abbas, Murphy and Sher, 1996). In all infected individuals examined before treatment, IFN-γ, IL-4, IL-5 and IL-10 were found at high levels. This presumably reflects the 'normal' situation in which the outcome of particular responses is determined by the interaction of a variety of cytokines, which may have cross-regulatory activities, a condition which may be essential to ensure fine modulation of the immune responses as the host immune response deals with newly establishing infection.

Studies on murine schistosomiasis have shown that Th1 responses indicated by IFN-γ production are associated with protection whereas Th2 responses appear to cause immunopathology (Sher and Coffman, 1992). In contrast, studies on reinfection after chemotherapy in humans have shown that resistance is associated with eosinophilia and with production of specific IgE (Hagan *et al.*, 1985a,b, 1991; Rihet *et al.*, 1991). Production of eosinophil is dependent on IL-5 while the production of IgE is dependent on IL-4 both Th2 type responses (Lopez *et al.*, 1988; Sher *et al.*, 1990; Mahanty *et al.*, 1997). However, it may be considered that both Th1 and Th2 may be involved in immunity. Roberts and colleagues, (1993) in a Kenyan study reported high levels of IL-5 and IFN-γ production in response to *S.mansoni* antigens in resistant individuals.

Another approach to such work is to examine T cell lines or clones derived from the PBMC of infected or exposed individuals. This has the advantage that the cells producing the cytokines can be reasonably accurately defined but the disadvantage that cytokines derived from other cell types are no longer represented in the system. Nevertheless despite the efforts involved in generating them, T cell lines and clones are proving to be a valuable source of information on human responses in general and to infections in particular. Th1 clones or cell lines through their production of IFN-γ and TNF-β have been reported to induce enhanced activity in macrophages (enhanced cellular immunity), while the Th2 clones or cell lines make products which are well adapted to promoting B cells to antibody producing cells (Del Prete, Maggi and Romagnani, 1994). The products of Th2 cells IL-4, IL-5, IL-6, IL-10 and IL-13 may each potentially enhance some aspects of the B cell response and thus of antibody production (Coffman *et al.*, 1988). A Th2-like response, by opposing the effects of IFN-γ on macrophages through the action of IL-10 and IL-4, and possibly suppressing production of IFN-γ and other cytokines of Th1 cells, could exert influence on the control of antibody production. Th2 induction may depend on the route of immunization or infection or on the nature and concentration of the antigen (Bradley *et al.*, 1993). IFN-γ secreted by T cells causes induction of MHC class II on macrophages. More APCs expressing MHC class II results in more antigen presentation to T cells and more IFN-γ availability in the environment (Marrack and Kappler, 1994; Sprent, 1994). This may explain the high levels of IFN-γ found in the infected individuals as IFN-γ is suspected to be one of the first cytokines to be produced in abundance during schistosome infection perhaps indicating cellular mobilization to combat newly acquired infection.

The dominance of Th2-like cytokines after treatment allows the speculation that the Th2-like cytokine environment favored immune reactivities, possibly including ADCC mechanisms, which could have contributed to resistance to reinfection. This interpretation would be entirely consistent with the evidence from other studies that indicates that Th2-associated responses appear to be associated with resistance to reinfection (Hagan *et al.*, 1985a,b, 1991; Rihet *et al.*, 1991, 1992, Dunne *et al.*, 1992, Dessein *et al.*, 1992). The Th1- and Th2-like cytokine patterns have been implicated in several other disease states including helminths infections, where the Th2 subset has been linked to pathogenesis. But from the critical analysis of these studies it appears the T helper cell-dependent effector and regulatory mechanisms as well as the critical role played by the cytokines in controlling immune responses during schistosomiasis need further elucidation.

In contrast to intracellular microbes, extracellular pathogens including helminths, trigger Th2-dominated responses (Urban *et al.*, 1994). In some studies these Th2 responses that dominate after infection have been related to immunological unresponsiveness. This may be suspected to be the situation in the individuals from the high transmission area before treatment and in those with heavy infections. The Th2-like responses are involved in driving eosinophil production and activation and in antibody production by B cells. The dominance of the Th2 response results in down regulating the effects of IFN-γ that is involved in DTH. As a result of long-term dominance of this response it is suspected that T cell anergy develops and the T cells lose responsiveness. The PBMC from the high transmission area gave low levels of proliferative responses to stimulation with schistosome-specific antigens and this could have been due to the simultaneously elevated levels of IL-4, IL-5, IL-10 and IFN-γ before treatment.

Cytokine studies carried out over the last decade have demonstrated that T cells produce cytokines that serve as autocrine growth factors for their own lineage (Coffman *et al.*, 1991). This may be the reason that the uninfected people in the study cohort maintained an elevated Th2 type cytokine environment at post-treatment examination points. Prior to treatment, it seems that the T cell responses maintained a balanced state of both Th1- and Th2- like cytokine environment that was observed in both infected and uninfected children having low IL-4:IFN-γ, IL-5:IFN-γ and IL-10:IFN-γ ratios (Figure 2). After treatment it appears that the state of T cell anergy is removed resulting in increased proliferation to specific antigen stimulation, and consequently setting up of a Th2-like cytokine environment (Coffman and von der Weird, 1997).

After early cloning *in vitro* the CD4[+] T cell pool does not always fit easily into the Th1/Th2 classification leading to the proposition that there are Th0 T cells which only differentiate into Th1 and Th2 upon prolonged *in vitro* culture (Romagnani, 1991; Couissinier and Dessein, 1995). This interpretation challenges the importance of the Th1 and Th2 dichotomy and has raised the possibility that it might be an artifact of *in vitro* differentiation. Another view is that, depending on the signals received, the duration of the stimulation and maturity of the immune effector responses, T cells can produce different pattern of cytokines. Various groups have attempted to determine if human T cells fit into the Th1/Th2 dichotomy. To date there is little hard and fast evidence to support this concept, with many groups reporting PBMC cultures producing IL-2, IL-4, IL-5 and IFN-γ. The lack of confirmation of the Th1/Th2 concept in human schistosomiasis may suggest that the observed differences in the cytokine production in the mouse are more likely to be a characteristic unique to the murine models. What is clear and demonstrated in our study is that in human schistosomiasis, no distinct polarized cytokine types for Th1/Th2 subsets exist, due to production by single T cell clones of both 'signature' cytokines which are normally used to identify each T cell subset.

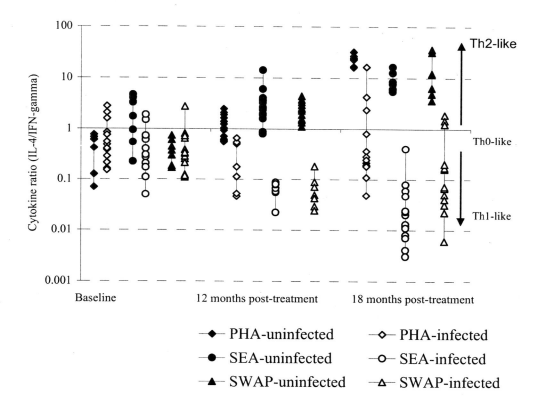

Figure 2. Cytokine ratio at different follow-up time point according to either being infected or uninfected with S.haematobium.

Individuals from areas of different transmission patterns have different infection and re-infection patterns.

Individuals living in *Schistosoma* endemic areas show different patterns of infection prevalence and intensity depending on the transmission pattern in the area. The findings reported by other workers that infection prevalence and intensity peaks in young children. No differences were detected in infection levels for all age groups below 16 years old that were found to be heavily infected. However, the infection prevalence and intensity was found to be low in the adult age groups for both study cohorts living in low and high transmission areas. After treatment the patterns were seen to differ between the transmission areas. The individuals from high transmission area showed a sharp rise in infection prevalence and intensity relative to age, with the infection prevalence peaking earlier than in the age group in the low transmission area. Considering infection intensity, it was found that the lower age groups were re-infected earlier and at higher intensities than the older age groups with no reinfection being recorded in the adult group. For the individuals living in low transmission area the peak reinfection prevalence occurred in an older age group than in high transmission area and in this case the age group was found to be most heavily infected prior to treatment and was most heavily re-infected.

Individuals resident in an area where infection prevalence and intensities are high, develop resistance to infection earlier due to greater 'experience' of parasites.

Cellular immune response levels appear to reflect the different infection patterns and exposure histories of the people in the high and low transmission areas. The individuals from the low transmission area are exposed to low levels of infection and as a result only experience a low antigen load. Consequently their infection status does not progress beyond an 'acute' phase and this is reflected in their continuing susceptibility to infection and immunological reactivity to stimulation with schistosome antigens *in vitro*. Those living in this area that are not infected are probably not infected due to the relatively restricted opportunities for exposure. Therefore they have, only a limited, and possibly no experience of the parasite. Reflecting this 'naive' status PBMC derived from them fail to respond to schistosome antigen (Mduluza et al., 2001).

The failure of cells derived from individuals living in high transmission area to respond to schistosome antigen stimulation probably reflects a state of T cell anergy. This is a characteristic feature of schistosomiasis mansoni and is well documented in the literature. Occurring as a result of intense infection this relatively simple host adjustment to the persistent antigenic stimulation can be regarded as beneficial for the host and thus essential for the long-term maintenance of the adult parasite and of its life cycle. The individuals from the high transmission area who were not infected showed a high response to schistosome antigen stimulation. This may be indicative of some degree of resistance to infection, which is constantly re-stimulated through repeated exposures. Immunological reactivity is maintained while invading parasites are eliminated. In contrast to these infected individuals from high transmission area, infected individual from the low transmission area gave even higher levels of responses presumably reflecting the absence of development of any immunomodulation, whereas the uninfected individuals in low transmission area who gave very low levels of response to stimulation.

Treatment with praziquantel caused some form of immune boosting which was shown to result in increased reactivity to stimulation with all schistosome antigens relative to pre-treatment reactivity levels. Although the overall response to stimulation increased, the infected individuals from the low transmission area remained high responders while the non-infected still showed low responses to schistosome antigen stimulation. The individuals in the high transmission area showed a reverse of their responses possibly indicating the development of protective immunity post-treatment perhaps resulting from the enhanced exposure to parasite antigens induced by treatment. This may also have been reflected in reinfection after treatment with the adults remaining uninfected. In contrast, adults from the low transmission area were re-infected after treatment. This would be consistent with the view that as yet they had failed to develop an effective protective immune response and that the low levels of infection (antigenic load) they had were insufficient to induce an enhanced reactivity following treatment. Taken together these findings suggest that repeated exposure to the parasite is essential for the development of protective immune responses (Mduluza et al., 2003). Although the increased reactivity after treatment of individuals from the high transmission area may in part be explained by the lifting of it of the observed immunoregulation it was noted that following treatment this status was not readily regained as re-infected individuals in the high transmission area showed high responses to specific schistosome antigens post-treatment compared with the pre-treatment infected groups.

The cytokine profiles vary according to infection patterns of each area and the antigens stimulating their production.

Individuals living in different areas of schistosome transmission show different patterns of cytokine production after schistosome-specific antigen stimulation. The individuals from the high transmission area where infection was high produce, overall, high levels of all the cytokines as measured in this study (IL-4, IL-5, IL-10 and IFN-γ), compared to the individuals from low transmission area of schistosomiasis. Even though the PBMC from individuals living in the high transmission area were found to give low levels of proliferation to stimulation, they were capable of producing higher levels of cytokines when compared with the responding PBMC obtained from individuals living in the low transmission area. Within each transmission area, the levels of each cytokine have been found to differ according to infection status. Before treatment, the infected individuals from both areas produced high levels of IFN-γ (a Th1-type cytokine), which may be involved in early stages of the development of immune responses aimed at curbing infection (Mduluza et al., 2003). However, it was clearly shown when examining the PBMC from the high transmission area, that the Th2 cytokines (IL-4, IL-5 and IL-10) were also produced at high levels during infection and before treatment. This raises the possibility of having an immune response geared both to mobilize cell-mediated immunity and drive humoral responses, by involvement of both Th1 and Th2 cytokines. As Th1 and Th2 cytokines are normally considered to be cross-regulatory, further work is needed to confirm if this combined cytokine production is reflected in both cellular and antibody responses to schistosome infection. This revealed that following treatment the individuals who remain uninfected in the high transmission area had a dominating Th2-like response with high levels of IL-4, IL-5 and IL-10 relative to IFN-γ. However, individuals who become re-infected had high levels of IFN-γ compared with those who were uninfected but the ratios of the Th1/Th2 cytokines showed that the cytokines for these classes (IL-4:IFN-γ, IL-5:IFN-γ) were found at a similar level both before and after treatment (Mduluza et al., 2003).

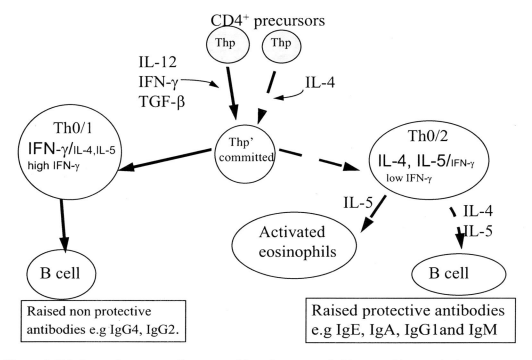

Figure 3: T helper subsets according to cytokine phenotypes in human schistosomiasis susceptibility and protection

The Existence of Th0/1- and Th0/2-Like Cytokine Subsets Evidenced from T Cell Cloning

The work on T cell cloning for the characterization of T cell responses involved in schistosomiasis immune responses produced clearly unpolarised cytokine profiles of the T helper cell subsets. This is in contrast with the findings in other studies cited above, many of them from studies of murine models of *S.mansoni* infections. However, the findings on the T cell cloning mirrored the results reported in human *S.mansoni* infection reported by Couissinier and Dessein, (1995). In their experiments Couissinier and Dessein found that both 'signature' cytokines were co-expressed, with the T cell clones having a dominating Th2-like cytokine profiles being from resistant individuals. The PBMC were obtained from individuals from the high transmission area prior to treatment. Age and duration of infection, (history of exposure to infection) may play a role in influencing the development of unpolarised T helper cells with the combined 'signature' cytokine profiles which have been recorded. In terms of IL-4:IFN-γ and IL-5:IFN-γ the 'signature' cytokines for defining each T helper cell subset, most of the clones derived from all the children showed co-expression of the cytokines, resulting in the subsets being classified as Th0/1 and Th0/2, depending on the ratio (Mduluza et al., 2001). Although the extremes of each subset class, i.e. Th1 and Th2 were recorded, these were relatively rare. Although the sample size is limited, from this work it was concluded that the balance of the cytokines relative to each subset marks the immune state of an individual. Those able to resist parasite invasion and establishment mounted protective immune responses reflected in elevated levels of IL-4, IL-5 and IL-10, cytokines marking Th0/2-like responses. In contrast, the dominance of IFN-γ marks a state of susceptibility to infection of a given individual. However, the balance between these Th1- and Th2-like responses appears to be finely tuned and depending on the timing, quantity and duration of exposure to infection, the cytokine balance may shift, possibly resulting in a change in the immune response from putative protective to susceptibility or *vice versa* (Figure 3). Additional work is required to confirm this hypothesis.

Conclusions

Human immuno-epidemiological studies are always desirable for an improved understanding of the disease conditions in the actual host and ultimately for the final understanding and application of vaccine candidates when they become available. However, such studies have to contend with several hurdles as far as the use of, and working with, the human host is concerned, including the participation and willingness to provide the required biological samples, especially blood. The use of communities in many endemic areas may require some application of visible immediate benefits during implementation and progress of the study, e.g. provision of treatment for the disease under study. In an ideal situation, studies of this type would be fully integrated into the existing health care system of the areas where they are carried out.

The findings on *S.haematobium* seem to be in line with other work reported from other endemic areas on *S.mansoni* and *S.haematobium*. The use of relatively unpurified antigenic extracts of different life cycle stages can be justified on the basis that the organism is so complex, that immunity and immune responsiveness is unlikely to be determined by responses to a single antigen. Instead the net level of immunity may depend on a combination of different responses to different antigens. Although it is still valid to select to study responses to individual purified or recombinant antigens which can provide useful information in the development of vaccines and in eliciting particular immune responses.

The findings from Zimbabwean studies indicate that Th2-like immune effects may be important in mediating human protective immunity to schistosome infection. This does not exclude the possibility that the mechanisms of resistance reported from studies of the mouse model of immunity to *Schistosoma*; with a reported Th1-like bias involving a predominance of IFN-γ are not also important in human schistosomiasis. IFN-γ was produced early and in abundance during infection indicating a useful role during this infection stage.

T helper cell subset determination is still complex and requires more work. Considering the number of patients' samples that can be handled at any one time in the determination of the subsets using single cell assays, the methodology needs to be modified and simplified to make it more practical for routine field use. Although over-modification can sometimes result in a weakening of the specificity of the methods, the use of two or more methods for a determination may help strengthen the significance of the findings, e.g. T cell cloning by cell cultures and the use of mRNA detection of cytokines genes by reverse transcriptase polymerase chain reaction (RT-PCR) at the same time. The use of RT-PCR may help the study of the T helper subsets in a wider age range of a population living in endemic areas. This may be important, as cytokine polarization may be present in a particular age group or group with a particular history of exposure to infection. We looked at an age group within the range where peak infection prevalence and intensity is found and coincidentally this may be the age group where cytokine polarization is lacking due to the effects of infection intensity. Therefore, a cautious interpretation of the data demands that a distinct cytokine polarization of Th1/Th2 subsets should not as yet be excluded, possibly being found in adults living in the high transmission area, who had more experience of infections.

The use of praziquantel as a chemoprophylaxis should now be regarded as the only useful alternative to protect young children living in schistosome endemic areas from infection and subsequent pathology. Since the cost of the drug has now been reduced and significant drug resistance has not been reported, the duration of retreatment can be extended six to 12 month intervals. Several studies have reported reinfection prevalence reaching half the pre-treatment levels from six months post-treatment. Thus a six monthly treatment schedule may make the programmes supportable in terms of the cost of drug purchase. This approach would benefit greatly the young children in these endemic areas by protecting them from further re-infection and the heavy infection levels that result in morbidity and pathology, since it is well documented that the severity of clinical disease is dependent on intensity of infection rather than simply presence or absence of infection.

The findings summarized give some additional information for the development of a vaccine for use against schistosome infections in human. Immunity can be demonstrated in naturally infected individuals after subsequent chemotherapy, and the indications are that it is the nature of the response that is important, and not just the exact identity of a particular antigen that provides protective immunity. The most important and crucial issue appears to be the achievement of the right balance of protective as opposed to inappropriate responses. Considerable work is required on ways to elicit the correct responses in humans already primed to infection, which seem to be different where infections exist naturally and the period post-treatment. This leaves a huge gap that requires filling especially for the possibly developed vaccines, the stages at which they may have to be administered, with or before treatment, and the age groups that may require the vaccines.

References

Abbas, K.A., Murphy, M.K. & Sher, A. (1996). Functional diversity of helper T lymphocytes. *Nature*, 383,787-793.

Alves-Brito, C.F., Simpson, J.G., Bahia-Oliveira, L.M.G., Rabello, A.L.T., Rocha, R.S., Lambertucci, J.R., Gazzinelli, G., Katz, N., Correa-Oliveira, R. (1992). Analysis of anti-keyhole limpet haemocyanin antibody in Brazilians, supports its use for the diagnosis of acute schistosomiasis mansoni. *Transactions of the Royal Society of Tropical Medicine and Hygiene:* 86, 53-56.

Amiri, P., Locksley, R.M., Parslow, T.G., Sadick, M., Rector,E., Ritter, D., McKerrow, J.H. (1992). Tumour necrosis factor alpha restores egg-laying in schistosome-infected SCID mice. *Nature:* 356, 604-607.

Anwar, A.R.E., Smithers, S.R., Kay, A.B. (1979). Killing of schistosomula of *Schistosoma mansoni* coated with antibody and/or complement by human leucocytes *in vitro*: requirement for complement in preferential killing by eosinophils. *Journal of Immunology:* 122, 628-637.

Aronstein, W.S., Strand, M. (1983). Identification of species specific and gender specific proteins and glycoproteins of three human schistosomes. *Journal of Parasitology:* 69, 1006-1017.

Auriault, C., Gras-Masse, H., Pierce, R.J., Butterworth, A.E., Wolowezuk, I., Capron, M., Ouma, J.H., Balloul, J.M., Khalife, J., Neyrinck, J.L., Tartar, A., Koech, D., Capron, A. (1990). Antibody responses of *Schistosoma mansoni* infected human subjects to the recombinant P28 Glutathione-S-Transferase and to synthetic peptides. *Journal of Clinical Microbiology:* 28, 1918-1924.

Agnew, A.M., Murare, H.M., Navarrete Sandival, S., de Jonge, N., Krijer, F.W., Deelder, A.M. & Doenhoff, M.J. (1992). The susceptibility of adult schistosomes to immune attrition. *Mem Inst Oswaldo Cruz, Rio de Janeiro*, 87, 87-93.

Arap Siongok, T.K., Mahmoud, A.A.F., Ouma, J.H., Warren, K.S., Muller, A.S., Hander, A.K. & Houser, H.B. (1976). Morbidity in schistosomiasis mansoni in relation to intensity of infection: study of a community in Machakos, Kenya. *American Journal of Tropical Medicine and Hygiene*, 25, 273-284.

Araujo, M.I., Ribeiro de Jesus, A.M., Bacellar, O., Sabin, E., Pearce, E. & Carvalho, E.M. (1996). Evidence of T helper type 2 activation in human schistosomiasis. *European Journal of Immunology*, 26,1399-1403.

Arthur, R.P. & Mason, D. (1986). T cell that help B cell responses to soluble antigen are distinguishable from those producing interleukin 2 on mitogenic or allogeneic stimulation. *Journal of Experimental Medicine*, 4, 774-786.

Auriault, C., Pancrle, V., Wolowczuk, I., Asseman, C., Ferru, I. & Verwaerde, C. (1996). Cellular immune response and pathology in schistosomiasis. *Parasite Immunology*, 62, 199-208.

Auriault, C., Quaissi, M.A., Torpier, G., Eisen, H. & Capron, A. (1981). Proteolytic cleavage of IgG bound to the Fc receptor of *Schistosoma mansoni* schistosomulae. *Parasite Immunology*, 3, 33-35.

Auriault, C., Wolowczuk, I., Damonneville, M., Velge-Rousel, F., Pancre, V., Grass-Masses, H., Tartar, A. & Capron, A. (1990). T cell antigens and epitopes in schistosomiasis. *Current Topics in Microbiology and Immunology*, 155, 3-8.

Bahia-Oliveira, L.M.G., Gazzinelli, G., Eloi-Santos, S.M., Cunha-Melo, J.R., Alves-Oliveira, L.F., Silveira, A.M.S., Viana, I.R.C., Carmo, J., Souza, A. & Correa-Oliveira, R. (1992). Differential cellular reactivities to adult worm antigens of patients with different clinical forms of Schistosomiasis mansoni. *Transaction of the Royal Society of Tropical Medicine and Hygiene*, 86, 57-61.

Banchereau, J., de Paoli, P., Valle', A., Garcia, E. & Rousset, F. (1991). Long-term human B cell lines dependent on interleukin-4 and antibody to CD40. *Science*, 251, 70-72.

Banchereau, J. & Rousset, F. (1991). Growing human B lymphocytes in the CD40 system. *Nature*, 353, 678-679.

Barbour, A.D. (1985). The importance of age and water contact patterns in relation to *Schistosoma haematobium* infection. *Transaction of the Royal Society of Tropical Medicine and Hygiene*, 79, 151-153.

Barsoum, I.S., Gamil, M.F., Al-Khafif, A.M., Ramzy, M.R., Alamy, A.M. & Colley, D. (1982). Immune responses and immunoregulation in relation to human schistosomiasis in Egypt. I. Effects of treatment on *in vitro* cellular responsiveness. *American Journal of Tropical Medicine and Hygiene*, 31, 1181-1191.

Bashir, M., Bickle, Q., Bushara, H., Cook, L., Shi, F., He, D., Huggins, M., Lin, J., Malik, K. & Moloney, A. (1994). Evaluation of defined antigen vaccines against *Schistosoma bovis* and *S.japonicum* in bovines. *Tropical Geographical Medicine*, 46, 255-258.

Bergquist, N.R. (1995). Controlling schistosomiasis by Vaccination: A realistic Option? *Parasitology Today*, 11, 191-194.

Boulager, D., Warter, A., Trottein, F., Mauny, F., Bremond, P. Audibert, F., Couret, D., Kadri, S. Godin, C., Sellin, E., Pierce, R.J., Lecocq, J.P., Sellin, B. & Capron, A. (1995). Vaccination of patas monkeys experimentally infected with *Schistosoma haematobium* using a recombinant glutathione S-transferase cloned from *S.mansoni*. *Parasite Immunology*, 17, 361-369.

Boulay, J.L. & Paul, W.E. (1992a). The interleukin-4 family of lymphokines. *Current Opinion in Immunology*, 4, 294-298.

Boulay, J.L. & Paul, W.E. (1992b). The interleukin-4-related lymphokines and their binding to hematopiotin receptors. *Journal of Biological Chemistry*, 267, 20525-20528.

Bradley, D.J. & McCullough, F.S. (1973). Egg output stability and the epidemiology of *Schistosoma haematobium*. Part II. An analysis of the epidemiology of endemic *Schistosoma haematobium*. *Transaction of the Royal Society of Tropical Medicine and Hygiene*, 67, 491-500.

Bradley, L.M., Duncan, D.D., Yoshimoto, K. & Swain, S.L. (1993). Memory effectors: a potent IL-4-secreting helper T cell population that develops *in vivo* after restimulation with antigen. *Journal of Immunology*, 150, 3119-3130.

Bushara, H.O., Bashir, M.E., Malik, K.H., Mukhtar, M.M., Trottein, F., Capron, A. & Taylor, M.G. (1994). Suppression of *Schistosoma bovis* egg production in cattle by vaccination with glutathione S-transferase or keyhole limpet haemocyanin. *Parasite Immunology*, 15, 383-390.

Butterworth, A.E. (1990). Immunology of schistosomiasis. In:Wyler J D. ed. *Modern Parasite Biology W H Freeman and CO, New York*, 14, 262-288.

Butterworth, A.E. (1984). Cell-mediated damage to helminths. *Advanced Parasitology*, 23, 143-235.

Butterworth, A.E., Capron, M., Cordingley, J.S., Dalton, P.R., Dunne, D.W., Kariuki, H.C., Kimani, G., Koech, D., Mugambi, M., Ouma, J.H., Prentice, M.A., Richardson, B.A., Arap-Siongok, T.K., Sturrock, R.F. & Taylor, D.W. (1985). Immunity after treatment of human schistosomiasis mansoni. ii. Indentification of resistant individuals, and analysis of their immune responses. *Transactions of the Royal Society of Tropical Medicine and Hygiene*, 79, 393-408.

Butterworth, A.E., Dalton, P.R., Dunne, D.W., Mugambi, M., Ouma, J.H., Richardson, B.A., Arap Siongok, T.K. & Sturrock, R.F. (1984). Immunity after treatment of human schistosomiasis mansoni. I. Study Design, pretreatment observations and the results of treatment. *Transactions of the Royal Society of Tropical Medicine and Hygiene*, 78, 108-123.

Butterworth, A.E. & Hagan, P. (1987). Immunity in human schistosomiasis. *Parasitology Today*, 3, 11-16.

Butterworth, A.E., Sturrock, R.F., Houba, V. & Rees, P.H. (1974). Antibody-dependent cell-mediated damage to schistosomula *in vitro*. *Nature, London*, 352, 503-505.

Butterworth, A.E., Sturrock, R.F., Ouma, J.H., Mbugua, G.G., Fulford, A.J.C., Kariuki, H.C. & Koech, D. (1991). Comparison of different chemotherapy strategies against *Schistosoma mansoni* in Machakos District, Kenya: Effects on human infection and morbidity. *Parasitology*, 103, 339-344.

Butterworth, A.E., Taylor, D.W., Veith, M.C., Vadas, M.A., Dessein, A., Sturrock, R.F. & Wells, E. (1982). Studies on the mechanisms of immunity in human schistosomiasis. *Immunological Reviews*, 61, 5-39.

Butterworth, A.E., Bensted-Smith, R., Capron, A., Capron, M., Dalton, P.R., Dunne, D.W., Gryzch, J., Kariuki, H.C., Khalife, J., Koech, D., Mugambi, M., Ouma, J.H., Arap-Siongok, T.K. & Sturrock, R.F. (1987). Immunity in human schistosomiasis mansoni. Prevention by blocking antibodies of expression of immunity in young children. *Parasitology*, 94, 281-300.

Butterworth, A.E., Dunne, D.W., Fulford, A.J.C., Capron, M., Khalife, J., Capron, A., Koech, D., Ouma, J.H. & Sturrock, R.F. (1988). Immunity in human schistosomiasis mansoni: Cross-reactive IgM and IgG2 anti-carbohydrate antibodies block the expression of immunity. *Biochemie*, 70, 1053-1063.

Butterworth, A.E., Dunne, D.W., Fulford, A.J.C., Thorne, K.J.I., Gachuhi, K., Ouma, J.H. & Sturrock, R.F. (1992). Human immunity to *Schistosoma mansoni*: Observation on mechanism, and implications for control. *Immunological Investigation*, 21, 391-396.

Butterworth, A.E. & Richardson, B.A. (1985). Factors affecting the levels of antibody- and complement-dependent eosinophil-mediated damage to schistosomula of *Schistosoma mansoni in vitro. Parasite Immunology*, 7, 119-131.

Butterworth, A.E., Sturrock, R.F., Houba, V., Mahmoud, A.A.F., Sher, A. & Rees, P.H. (1975). Eosinophils as mediators of antibody-dependent damage to schistosomula. *Nature*, 256, 727-729.

Butterworth, A.E., Vadas, M.A., Wassom, D.L., Dessein, A., Hogan, M., Sherry, B., Gleich, D.J. & David, J.R. (1979a). Cytolytic T lymphocytes recognize alloantigens on schistosomula of *Schistosoma mansoni*, but fail to induce damage. *Journal of Immunology*, 122, 1314-1321.

Butterworth, A.E., Vadas, M.A., Wassom, D.L., Dessein, A., Hogan, M., Sherry, B., Gleich, G.J. & David, J.R. (1979b). Interaction between human eosinophils and schistosomula of *Schistosoma mansoni* ii. The mechanism of irreversible eosinophil adherence. *Journal of Experimental Medicine*, 150, 1456-1471.

Butterworth, A.E., Wassom, D.L., Gleich, G.J., Loegering, D.A. & David, J.R. (1979c). Damage to schistosomula of *Schistosoma mansoni* induced directly by eosinoiphil major basic protein. *Journal of Immunology*, 122, 221-229.

Viana, I.R., Sher, A., Carvalho, O.S., Massara, C.L., Eloi-Santos, S.M., Peatce, E.J., Colley, D.G., Gazzinelli, G. & Correa-Oliveira, R. (1994). Interferon-gamma production by peripheral blood mononuclear cells from residents of an area endemic for *Schistosoma mansoni*. *Transactions of Royal Society of Tropical Medicine and Hygiene*, 88, 466-470.

Bentley, A.G., Phillips, S.M., Kaner, R.J., Theodorides, V.J., Linette, G,P., Doughty, B.L. (1985). *In vitro* delayed hypersensitivity granuloma formation development of an antigen-coated bead model. *Journal of Immunology*: 134, 4163- 4169.

Bickle, Q.D., Ford, M.J. (1982). Studies on the surface antigenicity and susceptibility to antibody-dependent killing of developing schistosomula using sera from chronically infected mice and mice vaccinated with irradiated cercariae. *Journal of Immunology*: 128, 2101-2106.

Bradley, D.J., McCullough, F.S. (1973). Egg output stability and the epidemiology of *Schistosoma haematobium*. Part II. An analysis of the epidemiology of endemic *S. haematobium*. *Transactions of the Royal Society of Tropical Medicine and Hygiene*: 67, 491-500.

Butterworth, A.E., Capron, M., Cordingley, J.S., Dalton, P.R., Dunne, D.W., Kariuki, H.C., Koech, D., Mugambi, M., Ouma, J.H., Prentice, M.A., Richardson, B.A., Siongok, T.K., Sturrock, R.F. (1985). Immunity after treatment of human schistosomiasis mansoni II. Identification of resistant individuals and analysis of their immune responses. *Transactions of the Royal Society of Tropical Medicine and Hygiene*: 79, 393-408.

Butterworth, A.E., Bensted-Smith, R., Capron, A., Capron, M., Dalton, P.R., Dunne, D.W., Grzych, J.M., Kariuki, H.C., Khalife, J., Koech, D., Mugambi, M., Ouma, J.H., Arap Siongok, T.K., Sturrock, R.F. (1987). Immunity in human schistosomiasis mansoni: prevention by blocking antibodies of the expression of immunity in young children. *Parasitology*: 94, 281-300.

Butterworth, A.E., Dunne, D.W., Gachuhi, K., Ouma, J.H., Sturrock, R.F. (1992). Human immunity to *Schistosoma mansoni*. Observations on mechanisms and implications for control. *Immunological Investigations*: 21, 391-407.

Capron, A., Dessaint, J.P., Joseph, M., Rousseaux, R., Capron M., Bazin, H. (1977). Interaction between IgE complexes and macrophages in the rat: a new mechanism of macrophage activation. *European Journal of Immunology*: 7, 315-322.

Capron, M., Capron, A., Torpier, G., Bazin, H., Bout, D., Joseph, M. (1978). Eosinophil-dependent cytotoxicity in rat schistosomiasis. Involvement of IgG2a, antibody and role of mast cells. *European Journal of Immunology*: 8, 127-133.

Capron, M., Torpier, G., Capron, A. (1979). *In vitro* killing of *Schistosoma mansoni* schistosomula by eosinophils from infected rats: role of cytophilic antibodies. *Journal of Immunology*: 126, 2087-2092.

Capron, A., Dessaint, J.P., Capron, M., Joseph, M., Pestel, J. (1980). Role of anaphylactic antibodies in immunity to schistosomes. *American Journal of Tropical Medicine and Hygiene*: 29, 847-857.

Capron, M., Spiegelberg, H., Prin, L., Bennich, M., Butterworth, A.E., Pierce, R.J., Ouaissi, M.A., Capron, A. (1984). Role of IgE receptors in effector induction of human eosinophils. *Journal of Immunology*: 132, 48-56.

Capron, M., Capron. A. (1986). Rats, mice and men. Models for immune effector mechanisms against schistosomiasis. *Parasitology Today*: 2, 69-75.

Capron, A., Dessaint, J.P., Capron, M., Ouma, J.H., Butterworth, A.E. (1987). Immunity to schistosomes: progress towards vaccine. *Science*: 238, 1065.

Capron, A. (1993). Vaccine strategies: Background and update. *IIIrd CEC Meeting on Schistosomiasis Research. Life Sciences and Technologies for Developing Countries. (STD)* 59- 61.

Capron, A., Capron, M., Grangette, C. & Dessaint, J. (1989). IgE and inflammatory cells. *Ciba Foundation Symposium* , 147, 153-170.

Capron, A., Dessaint, J.P., Capron, M., Joseph, M. & Torpier, G. (1982). Effector mechanisms of immunity to schistosomes and their regulation. *Immunological Reviews*, 61, 41-66.

Capron, A., Dessaint, J.P., Capron, M. & Bazin, H. (1975). Specific IgE antibodies in immune adherence of normal macrophages to *Schistosoma mansoni* schistosomules. *Nature*, 253, 474-475.

Capron, A., Dessaint, J.P., Capron, M., Ouma, J.H. & Butterworth, A.E. (1987). Immunity to Schistosomes: Progress Toward Vaccine. *Science*, 238, 1065-1072.

Capron, A., Dessaint, J.P., Joseph, M., Rousseaux, R., Capron, M. & Bazim, H. (1977). Interaction between antigen complexes and macrophages in the rat: a new mechanism of macrophage activation. *European Journal of Immunology*, 7, 315-322.

Capron, A. & Dessaint, J.P. (1992). Immunologic aspects of schistosomaisis. *Annual Review of Medicine*, 43, 209-218.

Capron, A., Riveau, G., Grzych, J.M., Boulanger, D., Capron, M. & Pierce, R. (1994). Development of vaccine strategy against human and bovine schistosomiasis: Background and update. *Tropical and Geographical Medicine*, 46, 242-246.

Capron, M., Camus, D., Carlier, Y. & Figueiredo, J.F.M.C. (1977). Immunological studies in human schistosomiasis. II. Antibody cytotoxicity for *S.mansoni* schistosomules. *American Journal of Tropical Medicine and Hygiene*, 26, 248-253.

Capron, M. & Capron, A. (1986). Rat, mice and men models for immune effector mechanisms against schistosomiasis. *Parasitology Today*, 2, 69-75.

Capron, M., Capron, A., Torpier, G., Bazin, H., Bout, D. & Joseph, M. (1978). Eosinophil-dependent cytotoxicity in rats schistosomiasis. Involvement of IgG2a antibody and role of mast cells. *European Journal of Immunology*, 8, 127-133.

Capron, M., Grezel, D., Grzych, J.M., Capron, A. (1993). Vaccine induced effector mechanisms in rodents. *IIIrd CEC Meeting on Schistosomiasis Research. Life Sciences and Technologies for Developing Countries: (STD)* 64.

Capron, M., Spiegelberg, H., Prin, L., Bennich, M., Butterworth, A.E., Pierce, R.J., Ouaissi, M.A. & Capron, A. (1984). Role of IgE receptors in effector induction of human eosinophils. *Journal of Immunology*, 232, 28-36.

Chandiwana, S.K. (1987b). Community water contact patterns and the transmission of *Schistosoma haematobium* in the highveld region of Zimbabwe. *Social Science Medicine*, 25, 495-505.

Chandiwana, S.K. (1987a). Seasonal patterns in water contact and the influence of water availability on contact activities in two schistosomiasis endemic areas in Zimbabwe. *Central African Journal of Medicine*, 33, 8-15.

Chandiwana, S.K. & Christensen, N.O. (1988). Analysis of the dynamics of transmission of human schistosomiasis in the highveld region of Zimbabwe. A review. *Tropical Medicine and Parasitology*, 39, 187-193.

Chandiwana, S.K., Christensen, N.O. & Frandsen, F. (1987). Seasonal patterns in the transmission of *Schistosoma haematobium*, *S.mattheei* and *S.mansoni* in the highveld region of Zimbabwe. *Acta Tropical*, 44, 433-444.

Chandiwana, S.K., Makaza, D. & Taputaira, A. (1987). Variation in incidence of schistosomiasis in the highveld region of Zimbabwe. *Tropical Medicine and Parasitology*, 39, 313-319.

Chandiwana, S.K., Woolhouse, M.E.J. & Bradley, M. (1991). Factors affecting the intensity of reinfection with *S.haematobium* following treatment with Praziquantel. *Parasitology*, 102, 73-83.

Cheever, A.W., Williams, M.E., Wynn, T.A., Finkelman, F.D., Seder, R.A., Cox, T.M., Hieney, S., Caspar, P. & Sher, A. (1994). Anti-IL-4 treatment of *Schistosoma mansoni*-infected mice inhibits development of T cells and non-B, non-T cells expressing Th2 cytokines while decreasing egg-induced hepatic fibrosis. *Journal of Immunology*, 170, 753-759.

Cheever, A.W., Xu, Y., Macedonia, J.G. & Cox, T. (1992). The role of cytokines in the pathogenesis of hepatric granulomatous disease in *Schistosoma mansoni* infected mice. *Mem. Inst. Oswaldo Cruz*, 87, 81-85.

Cher, D.J. & Mosmann, T.R. (1987). Two types of murine hepler T cell clones. Delayed type hypersensitivity is mediated by Th1 clones. *Journal of Immunology*, 138, 3688-3694.

Chikunguwo, S.M., Harris, T.S., Brodeur, P.H., Harn, D.A. & Stadecker, M.J. (1992). The cell-mediated response to schistosomal antigens at the clonal level: development and characterisation of a panel of egg antigen-specific murine T cell clones. *European Journal of Immunology*, 22, 917-922.

Chikunguwo, S.M., Kanazaw, T., Dayal, Y. & Stadecker, M.J. (1991). The cell-mediated response to schistosomal antigens at the clonal level. *In vivo* functions of the cloned murine egg antigen-specific CD4$^+$ T helper type 1 lymphocytes. *Journal of Immunology*, 147, 3921-3925.

Clarke, V. de (1966). The influence of acquired resistance in the epidemiology of bilharziasis. *The Central African Journal of Medicine*: 12, 1-30.

Clegg, J.A., Smithers, S.R. (1972). The effects of immune rhesus monkey serum on schistosomula of *Schistosoma mansoni* during cultivation in vitro. *International Journal for Parasitology*: 2, 79-98.

Clegg, J.A., Smith, S.R. & Terry, R.J. (1971). Acquisition of human antigens by *S. mansoni* during cultivation *in vitro*. *Nature*, 232, 653-654.

Coffman, R.L. & Carty, J. (1986). A T cell activity that enhances polyclonal IgE production and its inhibition by interferon-gamma. *Journal of Immunology*, 136, 949-54.

Coffman, R.L., Seymour, B.W.P., Lebman, D.A., Hiraki, D.D., Christiansen, J.A., Shrader, B., Cherwisnki, H.M., Savelkoul, H.F.J., Finkelman, F.D., Bond, M.W. & Mosmann, T.R. (1988). The role of helper T cell products in Mouse B cell differentiation and isotype regulation. *Immunological Reviews*, 102, 5-28.

Coffman, R.L., Varikila, K., Scott, P. & Chatelain, R. (1991). Role of cytokines in the differentiation of CD4$^+$ T-cell subsets *in vivo*. *Immunological Reviews*, 123, 189-194.

Coffman, R.L. & von der Weid, T. (1997). Multiple pathways for the initiation of T helper 2 (Th2) responses. *Journal of Experimental Medicine*, 185, 373-375.

Colley, D.G. (1981a). Immune responses and immunoregulation in experimental and clinical schistosomiasis. *Parasitic Diseases:* 1, 1-83.

Colley, D.G. (1981b). T lymphocytes that contribute to the immunoregulation of granuloma formation in chronic schistosomiasis. *Journal of Immunology:* 126, 1465-1468.

Colley, D.G., Barsoum, I.S., Dahawi, M.S.S., Gamil, M.F., Habib, M. & Alamy, M.A. (1986). Immune response and immunoregulation in relation to human schistosomiasis in Egypt. III. Immunity and longitudinal studies of *in vitro* responsiveness after treatment. *Transactions of Royal Society of Tropical Medicine and Hygiene*, 80, 952-958.

Colley, D.G., Cook, J.A., Freeman, G.L., Bartholomew, R.K. & Jordan, P. (1977a). Immune responses during human schistosomiasis mansoni. I. *In vitro* lymphocyte blastogenic response to heterogenous antigenic preparations from schistosome eggs, worms and cercaraie. *International Archives of Allergy and Applied Immunology*, 53, 420-437.

Colley, D.G., Hieney, S.E., Bartholomew, R.K. & Cook, J.A. (1977b). Immune response during human schistosomiasis mansoni. III. Regulatory effect of patient sera on human lymphocyte blastogenic responses to schistosome antigen preparations. *American Journal of Tropical Medicine and Hygiene*, 26, 917-925.

Colley, D.G., Katz, N., Rocha, R.S., Abrantes, W., da Silva, A.L. & Gazzinelli, G. (1983). Immune responses during human schistosomiasis mansoni: ix. T-lymphocytes subset analysis by monoclonal antibodies in hepatosplenic disease. *Scandanavian Journal of Immunology*, 17, 297-302.

Cot, M., Le Hesran, J.Y., Miailhes, P., Esveld, M., Etya'ale, D. & Breart, G. (1995). Increase in birth weight following chloroquine chemoprophylaxis during the first pregnancy: results of a randomised trial in Cameroon. *American Journal of Tropical Medicine and Hygiene*, 53, 581-585.

Couissinier-Paris, P. and Dessein, A.J. (1995). Schistosoma-specific helper T cell clones from subjects resistant to infection by *Schistosoma mansoni* are Th0/2. *European Journal of Immunology*, 25, 2295-2302.

Dalton, P.R., Pole, D. (1978). Water-contact patterns in relation to *Schistosoma haematobium* infection. *Bulletin of the World Health Organisation:* 56, 417 - 426.

Dean, D.A. (1977). Decreased binding of cytotoxic antibody by developing *Schistosoma mansoni*. Evidence for a surface change independent of host antigen absorption and membrane turnover. *Journal of Parasitology*, 63, 418-426.

Deelder, A.M., De Jonge, N., Boerman, O.C., Fillie, Y.E., Hilberath, G.W., Rotmans, J.P., Gerriste, M.J. & Schut, D.W.O.A. (1989). Sensitivity determination of circulating anodic antigen in *Schistosoma mansoni* infected individuals by enzyme-linked immuno-sorbent assay using monoclonal antibodies. *American Journal of Tropical Medicine and Hygiene*, 40, 268-272.

De Cock, K.M., Awadh, S., Raja, R.S., Wankya, B.M., Jupp, R.A., Slavin, B., Arap Siongok, T.K., Rees, P.H., Bertrand, J. & Lucas, S.B. (1987). *Transactions of Royal Society of Tropical Medicine and Hygiene*, 81, 107-110.

Deelder, A.M., Klappe, H.T.M., Van den Aardweg, G.J.M.J., Van Meerbeke, E.H.E.M. (1976). *Schistosoma mansoni*: Demonstration of two circulating antigens in infected hamsters. *Experimental Parasitology:* 40, 189-197.

Deelder, A.M., De Jonge, N., Boerman, O.C., Fillie, Y.E, Hilberath, G.W., Rotmans, J.P., Gerriste, M.J., Schut, D.W.O.A. (1989). Sensitivity determination of circulating anodic antigen in *Schistosoma mansoni* infected individuals by an enzyme-linked immuno-sorbent assay using monoclonal antibodies. *American Journal of Tropical Medicine and Hygiene*: 40, 268-272.

De Jonge, N., De Caluwe, P., Hilberath, G.W., Krijger, F.W., Polderman, A.M. & Deelder, A.M. (1989a). Circulating anodic antigen levels in serum before and after chemotherapy with praziquantel in schistosomiasis mansoni. *Transactions of Royal Society of Tropical Medicine and Hygiene*, 83, 368-372.

De Jonge, N., Fillie, Y.E., Hilberath, G.W., Krijger, F.W., Lengeler, C., de Savigny, D.H., van Vliet, N.G. & Deelder, A.M. (1989b). Presence of the schistosome circulating anodic antigen (CAA) in urine of patients with *Schistosoma mansoni* or *S.haematobium* infections. *American Journal of Tropical Medicine and Hygiene*, 41, 563-569.

De Jonge, N., Polderman, A.M., Hilberath, G.W., Krijger, F.W. & Deelder, A.M. (1990). Immunodiagnosis of schistosomiasis patients in the Netherlands: Comparison of antibody and antigen detection before and after chemotherapy. *Tropical Medicine and Parasitology*, 41, 257-261.

De Jonge, N., Gryseels, B.M., Hilberath. G.W., Polderman. A.M., Deelder, A.M. (1988). Detection of circulating anodic antigen by ELISA for sero-epidemiology of schistosomiasis mansoni. *Transactions of the Royal Society of Tropical Medicine and Hygiene*: 82, 591-594.

De Vries, J.E., de Waal Malefyt, R., Yssel, H., Roncarolo, M.G. and Spits, H. (1991). Do human Th1 and Th2 CD4+ clones exist? *Research in Immunology*, 142, 59-63.

Del Pozo, V., De Andres, B., Martin, E., Cardaba, B., Fernandez, J.C., Gallardo, S., Tramon, P., Leyva-Cobian, F., Palomino, P. & Lahoz, C. (1992). Eosinophil as antigen-presenting cell: activation of T cell clones and T cell hybridoma by eosinophils after antigen processing. *European Journal of Immunology*, 22, 1919-1925.

Del Prete, G., De Carli, M., Almerigogna, F., Giudizi, M.G., Biagiotti, R. & Romagnani, S. (1993). Human IL-10 is produced by both type 1 helper (Th1). and type 2 helper (Th2). T cell clones and inhibits their antigen-specific proliferation and cytokine production. *Journal of Immunology*, 150, 353-360.

Del Prete, G., Maggi, E., Parronchi, P., Chretien, I., Tiri, A., Macchia, D., Ricci, M., Banchereau, J., de Vries, J.E. & Romagnani, S. (1988). IL-4 is essential factor for the IgE synthesis induced *in vitro* by human T cell clones and their supernatants. *The Journal of Immunology*, 140, 4193-4198.

Del Prete, G., Maggi, E. & Romagnani, S. (1994). Biology of Disease: Human Th1 and Th2 cells: Functional properties, mechanism of regulation, and role in disease. *Laboratory Investigation*, 70, 299-306.

Demeure, C.E., Rihet, P., Abel, L., Quattara, M., Bourgois, A. & Dessein, A.J. (1993). Resistance to *S.mansoni* in humans - Influence of the IgE/IgG4 balance and IgG2 in immunity to reinfection after chemotherapy. *Journal of Infectious Diseases*, 168, 1000-1008.

Dessein, A.J., Demeure, C.E., Rihet, P., Kohlstaedt, S., Carneiro-Carvalho, D., Ouattara, M., Goudot-Crozel, V., Dessein, H., Bourgois, A., Abel, L., Carvallo, E.M. & Prata, A. (1992). Environmental, genetic and immunological factors in human resistance to *Schistosoma mansoni*. *Immunological Investigations*, 21, 423-453.

Dessein, A.J., Samuelson, J.C., Butterworth, A.E., Hogan, M., Sherry, B., Vadas, M.A. & David, J.R. (1981). Immune evasion by *Schistosoma mansoni*: loss of susceptibility to antibody or complement-dependent eosinophil attack by schistosomula cultured in medium free of macromolecules. *Parasitology*, 82, 357-360.

Doenhoff, M.J., Butterworth, A.E., Hayes, R.J., Sturrock, R.F., Ouma, J.H., Koech, D., Prentice, M.A. & Bain, J. (1993). Sero-epidemiology and serodiagnosis of schistosomiasis in Kenya using crude and purified egg antigens of *Schistosoma mansoni* in ELISA. *Transaction of the Royal Society of Tropical Medicine and Hygiene*, 87, 42-48.

Doenhoff, M.J., Pearson, S., Dunne, D.W., Bickle, Q.L.S., Bain, J., Musalliam, R., Hassounah, O. (1981). Immunological control of hepatotoxicity and parasite egg excretion in Schistosoma mansoni infections: Stage specificity of the reactivity of immune serum in T-cell deprived mice. *Transactions of the Royal Society of Tropical Medicine and Hygiene*: 75, 41 - 53.

Doenhoff, M.J., Butterworth, A.E., Hayes, R.J., Sturrock, R.F., Ouma, J.H., Koech, D., Prentice, M., Bain, J. (1993). Sero-epidemiology and serodiagnosis of schistosomiasis in Kenya using crude and purified egg antigens of *Schistosoma mansoni* in ELISA. *Transactions of the Royal Society of Tropical Medicine and Hygiene*: 87, 42-48.

Domingo, E.O. & Warren, K.S. (1968). Endogenous desensitization. Changing host granulomatous response to schistosome eggs at different stages of infection with *Schistosoma mansoni*. *American Journal of Pathology*, 52, 369-380.

Doughty, B.L., Ottesen, E.A., Nash, T.E., Phillips, S.M. (1984). Delayed hypersensitivity granuloma formation around *Schistosoma mansoni* eggs *in vitro*. III. Granuloma formation and modulations in human schistosomiasis. *Journal of Immunology*: 133, 993-997.

Doumenge, J.P., Mott, K.E., Cheung, C., Villenave, D., Chapuis, O., Perrin, M.F., Reaud-Thomas, G. (1987). *Atlas de la re-partition mondiale des schistosomiases/Atlas of the global distribution of schistosomiasis*: Geneva, OMS/WHO; Talence,PUB.:3.

Dresden, M.H., Payne, D.C. (1981). A sieving method for the collection of schistosome eggs from mouse intestines. *Journal of Parasitology:* 67, 450-452.

Dunne, D.W. (1990). Schistosome carbohydrates. *Parasitology Today*, 6, 45-48.

Dunne, D.W., Bickle, Q.D. (1987). Identification and characterization of a polysaccharide-containing antigen from *Schistosoma mansoni* eggs which cross-react with the surface of schistosomula. *Parasitology:* 94, 255-268.

Dunne, D.W., Bickle, Q.D., Butterworth, A.E. & Richardson, B.A. (1987). The blocking of human antibody-dependent, eosinophil-mediated killing of *Schistosoma mansoni* schistosomula by monoclonal antibodies which cross-react with a polysaccharide-containing egg antigen. *Parasitology*, 94, 269-280.

Dunne, D.W., Butterworth, A.E., Fulford, A.J.C., Kariuki, H.C., Langley, J.G., Ouma, J.H., Capron, A., Pierce, R.J. & Sturrock, R.F. (1992). Immunity after treatment of schistosomiasis mansoni: association between IgE antibodies to adult worm antigens and resistance to reinfection. *European Journal of Immunology*, 22, 1483-1494.

Dunne, D.W., Doenhoff, M.J. (1983). *Schistosoma mansoni* egg antigens and hepatocyte damage in infected T-cell deprived mice. *Contributions to Microbiology and Immunology*: 7, 22-29.

Dunne, D.W., Grabowska, A.M., Fulford, A.J.C., Butterworth, A.E., Sturrock, R.F., Koech, D. & Ouma, J.H. (1988). Human antibody responses to *S.mansoni*: the influence of epitopes shared between different life-cycle stages on the response to the schistosomulum. *European Journal of Immunology*, 18, 123-131.

Dunne, D.W., Hagan, P. & Abath, F.G.C. (1995). Prospects for immunological control of schistosomiasis. *The Lancet*, 345, 1488-1492.

Dunne, D.W., Hassounah, O., Musallam, R., Lucas, S., Pepys, M.B., Baltz. M., Doenhoff, M.J. (1983). Mechanisms of *Schistosoma mansoni* egg excretion: Parasitological observations in immunosuppressed mice reconstituted with immune serum. *Parasite Immunology*: 5, 47-60.

Dunne, D.W., Lucas., Bickle, Q., Pearson, S., Madgwick, L., Bain,J., Doenhoff, M.J. (1981). Identification and partial purification of an antigen (WI) from *Schistosoma mansoni* eggs which is putatively hepatotoxic in T cell deprived mice. *Transactions of the Royal Society of Tropical Medicine and Hygien*e: 75, 54 - 71.

Ellner, J.J. & Mahmoud, A.A.F. (1979). Killing of schistosomula of *Schistosoma mansoni* by normal human monocytes. *Journal of Immunology*, 123, 1483-1494.

Ellner, J.J., Olds, G.R., Lee, C.W., Kleinherz, M.E. & Edwards, K.L. (1982). Destruction of the multicellular parasite *Schistosoma mansoni* by T lymphocytes. *Journal of Clinical Investigations*, 70, 369-378.

Ellner, J.J., Tweardy, D.J., Osman, G.S., Wilson, C., El Kholy, A. & Rocklin, R.E. (1985). Increased blastogenic responses to worm antigen and loss of adherent suppresor cell activity after treatment for human infection with *Schistosoma mansoni*. *Journal of Infectious disease*, 151, 320-324.

Evans, A.C. (1983). Control of schistosomiasis in large irrigation schemes by use of niclosamide. A ten year study in Zimbabwe. *American Journal of Tropical Medicine and Hygiene*, 32, 1029-1039.

Feldmeier, H., Gastl, G.A. & Poggensee, U. (1985a). Relationship between intensity of infection and immunomodulation in human schistosomiasis. I. Lymphocyte sub-populations and specific antibody responses. *Clinical and Experimental Immunology*, 60, 225-233.

Feldmeier, H., Gastl, G.A. & Poggensee, U. (1985b). Relationship between intensity of infection and immunomodulation in human schistosomiasis. II. NK cell activity and in vitro lymphocyte proliferation. *Clinical and Experimental Immunology*, 60, 234-240.

Feldmeier, H., Kern, P. & Niel, G. (1981). Modulation of *in vitro* lymphocyte proliferation in patients with schistosomiasis haematobium, schistosomiasis mansoni and mixed infections. *Tropenmedizin and Parasitologie*, 32, 237-242.

Feldmeier, H., Nogueirao-Queiroz, J.A. & Peixoto-Queiroz, M.A. (1986). Detection and quantification of circulating antigen in schistosomiasis by monoclonal antibody. II. The quantification of circulating antigens in human schistosomiasis mansoni and haematobium: relationship to intensity of infection and disease status. *Clinical and Experimental Immunology*, 65, 232-243.

Finkelman, F.D., Katona, I.M., Urban, J.J., Snapper, C.M., Ohara, J. & Paul, W.E. (1986). Suppression of *in vivo* polyclonal IgE responses by monoclonal antibody to the lymphokine B-cell stimulatory factor 1. *Proceeding National Academy of Science, USA*, 83, 9675-9678.

Finkelman, F.D. & Urban, J.F. (1992). Cytokines: making the right choice. *Parasitology*, 8, 311-314.

Fiorentino, D.F., Zlotnik, A., Mosmann, T.R., Howard, M. & O'Garra, A. (1991). IL-10 acts on the antigen-presenting cell to inhibit cytokine production by Th1 cells. *Journal of Immunology*, 146, 3444-3449.

Fisher, A.C. (1934). A study of the schistosomiasis of the Stanleyville district of the Belgian Congo. *Transactions of the Royal Society of Tropical Medicine and Hygiene:* 28, 277-306.

Fulford, A.C.J., Butterworth, A.E., Sturrock, R.F. & Ouma, J.H. (1992). On the use of age intensity data to detect immunity to parasite infections with special reference to *Schistosoma mansoni* in Kenya. *Parasitology*, 105, 219-227

Gaafar, T., Ismail, S., Helmy, M., Afifi, A., Guirguis, N., EL Ridi, R. (1993). Identification of the *Schistosoma haematobium* soluble egg antigens that elicit human granuloma formation *in vitro*. *Parasitology Research:* 79, 103-108.

Gazzinelli, G., Katz, N., Rocha, R.S. & Colley, D.G. (1983). Immune responses during human schistosomiasis mansoni. x. Production and standardization of an antigen-induced mitogenic activity by peripheral blood mononuclear cells from treated, but not active cases of schistosomiasis. *Journal of Immunology*, 130, 2891-2895.

Gazzinelli, G., Lambertucci, J.R., Katz, N., Rocha, R.S., Lima, M.S. & Colley, D.G. (1985). Immune responses during human schistosomiasis mansoni, XI. Immunological status of patients with acute infections and after treatment. *Journal of Immunology*, 135, 2121-2127.

Goll, P.H., Wilkins, H.A., Marshall, T.F.de C. (1984). Dynamics of *Schistosoma haematobium* infection in a Gambian community. II. The effect of transmission of the control of *Bulinus senegalensis* by the use of niclosamide. *Transactions of the Royal Society of Tropical Medicine and Hygiene:* 78, 222-226.

Goudot-Crozel, V., Caillol, D., Djabali, M. & Dessein, A.J. (1989). The major parasite surface antigen associated with human resistance to schistosomiasis is a 37-kDa glyceraldehyde-3-P dehydrogenase. *Journal of experimental Medecine*, 170, 2065-2075.

Greenwood, B.M., David, P.H., Gtoo-Forbes, L.N., Allen, S.J., Alonso, P.L., Armstrong Schellenberg, J.R., Byass, P., Hurrwitz, M., Menon, A. & Snow, R.W. (1995). Mortality and morbidity from Malaria after stopping malaria chemoprophylaxis. *Transaction of the Royal Society of Tropical Medicine and Hygiene*, 89, 629-633.

Grencis, R.K. (1996). T cell and cytokine basis of host variability in response to intestinal nematode infections. *Parasitology*, 112, S31-S37.

Grencis, R.K., Hultner, L. & Else, K.J. (1991). Host protective immunity to *Trichinella spiralis* in mice: activation of Th cell subsets and lymphokine secretion in mice expressing different response phenotype. *Immunology*, 74, 329-332.

Grogan, J.L., Kremsner, P.G., Deelder, A.M. & Yazdanbakhsh, M. (1996). Elevated proliferation and interleukin-4 release from CD4[+] cells after chemotherapy in human *Schistosoma haematobium* infection. *European Journal of Immunology*, 26, 1365-1370.

Grogan, J.L., Kremsner, P.G., van Dam, G.J., Metzger, W., Mordmuller, B., Deelder, A.M. & Yazdanbakhsh, M. (1996). Antischistosome IgG4 and IgE responses are affected differentially by chemotherapy in children versus adults. *Journal of Infectious disease*, 173, 1242-1247.

Gryseels, B. & de Vlas, S.J. (1996). Worm burdens in schistosome infections. *Parasitology Today*, 12, 115-118.

Gryseels, B., Nkulikyinka, L. & Engels, D. (1994). Impact of repeated community-based selective chemotherapy on morbidity due to Schistosomiasis mansoni. *American Journal of Tropical Medicine and Hygiene*, 51, 634-641.

Gryseels, B., Stelma, F., Talla, I., Polman, K., Van Dam, G., Sow, S., Diaw, M., Sturrock, R.F., Decam, C., Niang, M., Doehring-Schwerdtfeger, E. & Kardorff, R. (1995). Epidemiology, immunology and chemotherapy of *Schistosoma mansoni* infections in a recently exposed community in Senegal. *Tropical and Geographical Medicine*, 46, 209-219.

Gryzch, J., Grezel, D., Xu, C.B., Neyrinck, J., Capron, M., Ouma, J.H., Butterworth, A.E. & Capron, A. (1993). IgA antibodies to protective antigen in human schistosomiasis mansoni. *Journal of Immunology*, 150, 527-535.

Grzych, J.M., Capron, M., Dissous, C. & Capron, A. (1984). Blocking activity of rat monoclonal antibodies in experimental schistosomiasis. *Journal of Immunology*, 133, 998-1004.

Grzych, J.M., Capron, M., Lambert, P.H., Dissous, C., Torres, S. & Capron, A. (1985). An anti-idiotype vaccine against experimental schistosomiasis. *Nature, London*, 316, 74-76.

Gryzch, J.-M., Capron, M., Bazin, H., Capron, A. (1982). *In vitro* and *in vivo* effector function of rat IgG2a monoclonal anti-*S. mansoni* antibodies. *Journal of Immunology*: 129, 2739-2743.

Grzych, J.-M., Neyrinck, J.L., Grezel, D., BoXu, C., Renom, G., Capron, A., (1993). *In vitro* human antibody responses to recombinant *Schistosoma mansoni* antigen. *IIIrd CEC Meeting on Schistosomiasis Research. Life Sciences and Technologies for Developing Countries. (STD)* 66.

Gundersen, S. G., Haagensen, I., Jonassen, T. O., Figenschau, K. J., De Jonge, N., Deelder, A. M. (1992). Development of a Magnetic Bead Antigen Capture Enzyme Linked Immuno Assay (MBAC-EIA) for the rapid detection of schistosomal circulating anodic antigen. *Journal of Immunological Methods*: 148, 1-8.

Hackett, F., Simpson, A.J.G., Omer- Ali, P., Smithers, S.R. (1987). Surface antigens of and cross-protection between two geographical isolates of *Schistosoma mansoni*. *Parasitology*: 94, 301-312.

Hagan, P., Moore, P.J., Adjukiewicz, A.B., Greenwood, B.M., Wilkins, H.A. (1985a). *In vitro* antibody-dependent killing of schistosomula of *Schistosoma haematobium* by human eosinophiles. *Parasite Immunology*: 7, 617-624.

Hagan, P., Wilkins, H.A., Blumenthal, U.J., Hayes, R.J., Greenwood, B.M. (1985b): Eosinophilia and resistance to *Schistosoma haematobium* in man. *Parasite Immunology:* 7, 625-632.

Hagan, P., Blumenthal, U.J., Chaudri, M., Greenwood, B.M., Hayes, R.J., Hodgson, J., Kelly, C., Knight, M. Simpson, A.J.G., Smithers, S.R., Wilkins, H.A. (1987). Resistance to reinfection with *Schistosoma haematobium* in Gambian children: analysis of their immune responses. *Transactions of The Royal Society of Tropical Medicine and Hygiene:* 81, 938-946.

Hagan, P., Blumenthal, U.J., Dunne, D., Simpson, A.J.G., Wilkins, H.A., (1991). Human IgE, IgG4 and resistance to reinfection with *Schistosoma haematobium. Nature*: 349, 243-245.

Hagan, P. (1987). The human immune response to schistosome infection: In: Rollinson, D and Simpson, A J G. eds. *The Biology of Schistosome. Academic Press, London*, 295-320.

Hagan, P. (1992). Reinfection, exposure and immunity in human schistosomiasis. *Parasitology Today*, 8, 12-16.

Hagan, P. (1993). Points of view: IgE and protective immunity to helminth infections. *Parasite Immunology*, 15, 1-4.

Hagan, P., Chandiwana, S.K., Ndhlovu, P., Woolhouse, M. & Dessein, A.J. (1994). The epidemiology, immunology and morbidity of *Schistosoma haematobium* infections in diverse communities in Zimbabwe. The study design. *Tropical Geographical Medicine*, 46, 227

Haggi, H., Huldt, G., Loftenius, A. & Schroder, H. (1990). Antibody responses in schistosomiasis haematobium in Somalia. Relation to age and infection intensity. *Annals of Tropical Medicine and Parasitology*, 84, 171-179.

Hamburger, J., Pelley, R.P., Warren, K.S. (1976). *Schistosoma mansoni* soluble egg antigens: determination of the stage and species specificity of their serologic reactivity by radio-immunoassay. *Journal of Immunology*: 117, 1561-1566.

Hamburger, J., Lustigman, S., Siongok, T.K.A., Ouma, J.H., Mahmoud, A.A.F. (1982). Characterization of a purified glycoprotein from *Schistosoma mansoni* eggs: specificity, stability, and the involvement of carbohydrate and peptide moieties in its serologic activity. *Journal of Immunology:* 128, 1864-1869.

Harn, D.A., Reynolds, S.R., Chikunguwo, S., Furlong, S. & Dahl, C. (1995). Synthetic peptide vaccines for schistosomiasis. *Pharmacological Biotechnology*, 6, 891-905.

Harris, A.R.C., Russel, R.J. & Charters, A.D. (1984). A review of schistosomiasis in immigrants in Western Australia demonstrating the unusual longevity of *Schistosoma mansoni. Transaction of the Royal Society of Tropical Medicine and Hygiene*, 78, 385-388.

Hayunga, E.G., Mollegard, I., Duncan, J.F., Sumner, M.P., Stek, M.Jr. & Hunter, K.W.Jr. (1986). Development of circulating antigen assay for rapid detection of acute schistosomiasis. *Lancet*, 2, 716-717.

Hayunga, E.G., Vannier, W.E., Chesnut, R.Y. (1981). Partial characterization of radio-labelled antigens from adult *Schistosoma haematobium*. *Journal of Parasitology*: 67, 589-591.

Henkle, K.L., Davern, K.M., Wright, M.D., Ramos, A.J. & Mitchell, G.F. (1990). Comparison of the cloned genes of the 26- and 280 kilodalton glutathione S-transferases of *Schistosoma japonicum* and *Schistosoma mansoni*. *Molecular and Boichemical Parasitology*. 40, 23-34.

Hillyer, G.V., Pacheco, E. (1986). Isolation and characterization of *Schistosoma haematobium* egg antigens. *American Journal of Tropical Medicine and Hygiene:* 35, 777-785.

Hillyer, G.V. & Gomez de Rios, I. (1979). The enzyme-linked immunosorbent assay (ELISA) for the immunodiagnosis of schistosomiasis. *American Journal of Tropical Medicine and Hygiene*, 28, 237-241.

Hofstetter, M., Fasano, M.S. & Ottesen, E.A. (1983). Modulation of the host response in human schistosomiasis. IV. Parasite antigen induced release of histamine that inhibits lymphocyte responsiveness in vitro. *Journal of Immunology*, 130, 1376-1380.

Hofstetter, M., Poindexter, R.W., Ruiz-Tiben, E. & Ottesen, E.A. (1982). Modulation of the host response in human schistosomiasis. III. Blocking antibodies specifically inhibit immediate hypersensitivity responses to parasite antigens. *Immunology*, 46, 777-785.

Hustings, E.L. (1983). Human water contact activities related to the transmission of bilharziasis (schistosomiasis). *Journal of Tropical Medicine and Hygiene*, 86, 23-35.

Iarotski, L.S. & Davis, A. (1981). The schistosomiasis problem in the world: Results of a W.H.O. questionnaire survey. *Bulletin of the World Health Organisation*, 59, 115-127.

Jassim, A., Hassan, K., Catty, D. (1987). Antibody isotypes in human schistosomiasis mansoni. *Parasite Immunology*: 9, 627-650.

James, S.L., Glaven, J., Goldenberg, S., Meltzer, M.S., Pearce, E. (1990). Tumour necrosis factor (TNF) as a mediator of macrophage helminthotoxic activity. *Parasite Immunology:* 12, 1-13.

James, S.L. & Sher, A. (1990). Cell-mediated immune response to schistosomiasis. *Current Topics in Microbiological Immunology*, 155, 21-31.

Janeway, J.C.A. & Bottomley, K. (1994). Signals and signs for lymphocyte responses. *Cell*, 76, 275-285.

Jarret, E.E.E. & Haig, D.M. (1984). Mucosal mast cells *in vivo* and *in vitro*. *Immunology Today*, 5, 115-120.

Joseph, M., Auriault, C., Capron, A., Vorng, H. & Veins, P. (1983). A new function for platelets: IgE-dependent killing of schistosomes. *Nature*, 303, 810-812.

Joseph, M., Capron, A., Butterworth, A.E., Sturrock, R.F. & Houba, V. (1978). Cytotoxicity of human and baboon mononuclear phagocytes against schistosomula *in vitro*: induction by immune complexes containing IgE and *S.mansoni* antigens. *Clinical Experimental Immunology*, 33, 48-56.

Joseph, M., Auriault, C., Capron, A., Vorng, H., Viens, P. (1983). A new function for platelets: IgE-dependent killing of schistosomes. *Nature: London* 303, 810-812.

Jordan, P., Webbe, G. 1982. Schistosomiasis. Epidemiology, treatment and control. *William Heinemann Medical Books LTD, London.*

Kamogawa, Y., Minasi, L.E., Carding, S.R., Bottomly, K. & Flavell, R.A. (1993). The relationship of IL-4- and IFN-γ-producing T cell studied by ablation of IL-4-producing cells. *Cell*, 75, 985-995.

Kelly, C. (1987). Molecular studies of schistosome immunity. In: Rollinson, D and Simpson, A J G. eds. *The biology of schistosome. Academic Press, London*, 265-293

Kelly, C., Simpson, A.J.G., Fox, E., Phillips, S.M. & Smithers, S.R. (1986). The identification of *Schistosoma mansoni* surface antigen recognized by protective monoclonal antibodies. *Parasite Immunology*, 8, 193-198.

Kelly, C., Hagan, P., Knight, M., Hodgson, J., Simpson, A.J.G., Hackett, F., Wilkins, H.A., Smithers, S.R. (1987). Surface and species-specific antigens of *Schistosoma haematobium*. *Parasitology:* 95, 253-266.

Khalife, J., Capron, M., Capron, A., Gryzch, J., Butterworth, A.E., Dunne, D.W. & Ouma, J.H. (1986). Immunity in human schistosomiasis mansoni. Regulation of protective immune mechanisms by IgM blocking antibodies. *Journal of Experimental Medicine*, 164, 1626-1640.

Khalife, J., Dunne, D.W., Richardson, B.A., Mazza, G., Thorne, K.J.L., Capron, A. & Butterworth, A.E. (1989). Functional role of IgG subclasses in eosinophil mediated killing of schistosomula of *Schistosoma mansoni*. *Journal of Immunology*, 142, 4422-4427.

Kigoni, E.P., Elsas, P.P.X., Lenzi, H.L., Dessein, A.J. (1986). IgE antibody and resistance to infection. II. Effect of IgE suppression on the early and late skin reactions and resistance of rats to *Schistosoma mansoni* infection. *European Journal of Immunology:* 16, 589-595.

King, C.H., Spagnuolo, P.J., Ellner, J.J. (1986). Differential contribution of chemotoxins and opsonins to neutrophil-mediated killing of *Schistosoma mansoni* larvae. *Infection and Immunity:* 52, 748-755.

King, C.L., Low, C.C. & Nutman, T.B. (1993). IgE production in human helminth infection: reciprocal interrelationship between interleukin-4 and interferon-γ. *Journal of Immunology*, 150, 1873-1878.

King, C.L., Medhat, A., Malhotra, I., Nafeh, M., Helmy, A., Khaudary, J., Ibrahim, S., El-Sherbiny, M., Zaky, S., Stupi, R.J., Brustoski, K., Shehata, M. & Shata, M.T. (1996). Cytokine control of parasite-specific anergy in human urinary schistosomiasis. IL-10 modulates lymphocytes reactivity. *Journal of Immunology*, 156, 4715-4721.

King, C.L. & Nutman, T.B. (1993). IgE and IgG subclass regulation by interleukin-4 and interferon-y in human helminth infections: assessment by B cell precursor frequencies. *Journal of Immunology*, 151, 458-463.

Kloetzel, K. (1967). A rationale for the treatment of *Schistosoma mansoni* even when reinfection is expected. *Transaction of the Royal Society of Tropical Medicine and Hygiene*, 61, 609-610.

Kloetzel, K., da Silva, J.R. (1967). Schistosomiasis mansoni acquired in adulthood: behaviour of egg counts and the intradermal test. *American Journal of Tropical Medicine and Hygiene*: 16: 167-169.

Klumpp, P.K. & Chu, K.Y. (1987). Fical Mollusciciding: an effective way to augment chemotherapy of schistosomaisis. *Parasitology Today*, 3, 74-76.

Kusel, J.R., Mackenzie, P.E. & McLaren, D.J. (1975). The release of membrane antigens into culture by adult *Schistosoma mansoni*. *Parasitology*, 71, 247-259.

Lammie, P.J., Linette, G.P., Phillips, S.M. (1985). Characterization of *S. mansoni* derived T-clones that form granulomas *in vitro*. *Journal of Immunology*: 134, 4170-4175.

Lazdins, J.R., Stein, M.J., David, J.R., Sher, A. (1982). *Schistosoma mansoni*: Rapid isolation and purification of schistosomula of different developmental stages by centrifugation on discontinuous density gradient of Percoll. *Experimental Parasitology*: 53, 39-44.

Liew, F.Y., Millot, S., Li, Y., Lelchuk, R., Chan, W.L. & Ziltener, H. (1989). Macrophage activation interferon-gamma from host-protective T cells is inhibited by interleukin (IL)-3 and IL-4 produced by disease-promoting T cells in Leishmaniasis. *European Journal of Immunology*, 19, 1227-1232.

Linsley, P.S., Brady, W., Grosmaire, L., Arufo, A., Damle, N.K. & Ledbetter, J.A. (1991). Binding of the B cell activation antigen B7 to CD28 costimulates T cell proliferation and interleukin-2 mRNA accumulation. *Journal of Experimental Medicine*, 173, 721-730.

Lopez, A.F., Sanderson, C.J., Gamble, J.R., Campbell, H.D., Young, I.G. & Vadas, M.A. (1988). Recombinant human interleukin 5 is a selective activator of human eosinophil function. *Journal of Experimental Medicine*, 167, 219-224.

Lunde, M., Ottesen, E.A., Cheever, W. (1979). Serological differences between acute and chronic schistosomiasis mansoni detected by enzyme-linked immunosorbent assay (ELISA). *American Journal of Tropical Medicine and Hygiene*: 28, 87-91.

Magnusson, C.G.M., Johansson, S.G.O. (1989). Anti-IgE antibodies: characteristics and isotypic distribution, IgM. *Clinical Review in Allergy*: 7, 74-100.

Mahanty, S., Abrams, J.S., King, C.L., Limaye, A.P. & Nutman, T.B. (1992). Parallel regulation of IL-4 and IL-5 in human helminth infections. *Journal of Immunology*, 148, 3567-3571.

Mahanty, S., King, C.L., Kumaraswami, V., Regunathan, J., Maya, A., Jayaraman, K., Abrams, J.S., Ottesen, E.A. & Nutman, T.B. (1993). IL-4 and IL-5 secreting lymphocyte populations are preferentially stimulated by parasite-derived antigens in human tissue invasive nematode infections. *Journal of Immunology*, 151, 3704-3711.

Mahmoud, A.A.F., Arap Siongok, T.K., Ouma, J.H., Houser, H.B. & Warren, K.S. (1983). Effects of targeted mass treatment on intensity of infection and morbidity in schistosomiasis mansoni. *Lancet*, 1, 849-851.

Maizels, M.R., Bundy, D.A.P., Selkirl, E.M., Smith, D.F. & Anderson, M.R. (1993). Immunological modulation and evasion by helminth parasites in human populations. *Nature*, 365, 797-805.

Mansour, M.M., Ali, P.O., Farid, Z., Simpson, A.J.G., Woody, J.W. (1988). Serological differentiation of acute and chronic schistosomiasis mansoni by antibody responses to keyhole limpet haemocyanin. *American Journal of Tropical Medicine and Hygiene*: 41, 338-344.

Marquet, S., Abel, L., Hillaire, D., Dessein, H., Kalil, J., Feingold, J., Weissenbach, J. & Dessein, A.J. (1996). Genetic localization of a locus controlling the intensity of infection by *Schistosoma mansoni* on chromosome 5q31-q33. *Nature Genetics*, 14, 181-184.

Marrack, P. & Kappler, J. (1994). Subversion of the immune system by pathogens. *Cell*, 76, 323-332.

Mazza, G., Dunne, D.W. & Butterworth, A.E. (1990). Antibody isotype responses to the *Schistosoma mansoni* schistosomulum in the CBA/N mouse induced by different stages of the parasite life cycle. *Parasite Immunology*, 12, 529-543.

McCullough, F.S., Bradley, D.J. (1973). Egg output stability and the epidemiology of *Schistosoma haematobium*. Part I. Variation and stability of *S. haematobium* egg counts. *Transactions of the Royal Society of Tropical Medicine and Hygiene:* 67, 475-490.

Mclaren, D.J., Clegg, J.A., Smithers, S.R. (1975). Acquisition of host antigens by young *Schistosoma mansoni* in mice: correlation with failure to bind antibody *in vitro*. *Parasitology:* 70, 67-75.

McLaren, M.L., Lilleywhite, J.E., Dunne, D.W., Doenhoff, M.J. (1981). Sero-diagnosis of human *Schistosoma mansoni* infection enhanced sensitivity and specificity in ELISA using a fraction containing *Schistosoma mansoni* egg antigens W1. *Transaction the Royal Society of Tropical Medicine and Hygiene*: 75, 72-79.

McLaren, D.J., Terry, R.J. (1982). The protective role of acquired host antigens during schistosome maturation. *Parasite Immunology*: 4, 129-148.

McLaren, D.J. (1985). Parasite defence mechanisms. In "Inflammation: Basic Mechanisms, Tissue injuring Principles and Clinical Models" (Venge, p. and Lindbom, A., eds) 219-254. Almquist and Wiksell, Sweden.

McLaren, D.J., Clegg, J.A. & Smithers, S.R. (1975). Acquisition of host antigens by young *Schistosoma mansoni* in mice: correlation with failure to bind antibody *in vitro*. *Parasitology*, 70, 67-75.

McLaren, D.J. & Incani, R.N. (1982). *S.mansoni*: acquired resistance of developing schistosomula to immune attack *in vitro*. *Experimental Parasitology*, 53, 285-298.

McLaren, D.J. & James, S.C. (1985). Ultrastructural studies of the killing of schistosomula of *S.mansoni* by activated macrophages *in vitro*. *Parasite Immunology*, 7, 315-331.

McLaren, D.J., Ramalho-Pinto, F.J. & Smithers, R.S. (1978). Ultrastructural evidence for complement and antibody-dependent damage to schistosomula of *S.mansoni* rat eosinophils in vitro. *Parasitology*, 77, 313-324.

McLaren, D.J. & Terry, R.J. (1982). The protective role of acquired host antigens during schistosome maturation. *Parasite Immunology*, 4, 129-148.

McLaren, M.L., Drapper, C.C., Roberts, R.M., Minter-Goedbloed, E., Lighthart, G.S., Teesdale, C.H., Amin, M.A., Omer, A.H., Bartlett, A. & Voller, A. (1978). Studies on the Enzyme-Linked Immunosorbent Assay (ELISA). Test for *Schistosoma mansoni* infection. *Annals of Tropical Medicine and Parasitology*, 72, 243-253.

McLaren, M.L., Lillywhite, J.E., Dunne, D.W. & Doenhoff, M.J. (1981). Serodiagnosis of human schistosoma infections: enhanced sensitivity and specificity in ELISA using a fraction containing *S.mansoni* egg antigens w1 and a1. *Transactions of the Royal Society of Tropical Medicine and Hygiene*, 75, 72-79.

Mduluza, T., PD Ndhlovu, N Midzi, C Mary, CP Paris, CMR Turner, SK Chandiwana, MEJ Woolhouse, AJ Dessein & P Hagan (**2001**). T cell clones from *Schistosoma haematobium* infected and exposed individuals lacking distinct cytokine profiles for Th1/Th2 polarisation. *Mem Inst Oswaldo Cruz*, Vol. 96: 89-101

Mduluza, T., PD Ndhlovu, N Midzi, Spicer, J.T., Mutapi, F; C Mary, CP Paris, CMR Turner, SK Chandiwana, MEJ Woolhouse, AJ Dessein & P Hagan (**2003**). Contrasting cellular responses in Schistosoma haematobium infected and exposed individuals from areas of high and low transmission in Zimbabwe: *Immunology Letters*. 88/3, 249-256.

Ministry of Health. (1992). Secretary for health report at the annual debriefing of Malaria and Schistosomiasis in Zimbabwe.

Moser, G., Sher, A. (1981). Studies of the antibody-dependent killing of schistosomula of *Schistosoma mansoni* employing haptenic target antigens. ii. *In vitro* killing of TNP-schistosomula by human eosinophils and neutrophils. *Journal of Immunology:* 126, 1025-1029.

Mosmann, T.R., Cherwisnki, H.M., Bond, M.W., Giedlin, M.A. & Coffman, R.L. (1986). Two types of murine helper T cell clone. I. Definition according to profiles of lymphokine activities and secreted proteins. *Journal of Immunology*, 136, 2348-2357.

Mosmann, T.R. & Coffman, R.L. (1989). Th1 and Th2 cells: different patterns of lymphokine secretion lead to different functional properties. *Annual Reviews of Immunology*, 7, 145-173.

Mosmann, T.R. & Sad, S. (1996). The expanding universe of T-cell subsets: Th1, Th2 and more. *Immunology Today*, 17, 138-146.

Mott, K.E., Baltes, R., Bambagha, J. & Baldassini, B. (1982). Field studies of the reusable polyamide filter for detection of *S.haematobium* eggs by urine filtration. *Propernmedlizin and Parasitologie*, 33, 227-228.

Mott, K.E. & Cline, B.L. (1980). Advances in epidemiology survey methodology and techniques in schistosomiasis. *Bulletin of the World Health Organisation*, 58, 639-647.

Mott, K.E., Dixon, H. (1982). Collaborative study on antigens for immunodiagnosis of schistosomiasis. *Bulletin of the World Health Organization*: 60, 729-753.

Murphy, E., Shibuya, K., Hosken, N., Openshaw, P., Maino, V., Davis, K., Murphy, K., and O'Garra, A. (1996). Reversibility of T helper 1 and 2 populations is lost after long-term stimulation. *Journal of Experimental Medicine*, 183, 901-913.

Mutapi, F., Ndhlovu, P.D., Hagan, P. & Woolhouse, M.E.J. (1997). A comparison of humoral responses to *Schistosoma haematobium* in areas with low and high levels of infection. Parasite Immunology, 19, 255-263.

Mutapi, F., Ndhlovu, P.D., Hagan, P., Spicer, J.T., Mduluza, T., Turner, C.M.R., Chandiwana, S.K. & Woolhouse, M.E.J. (1997). Chemotherapy accelerates the development of acquired immune responses to *Schistosoma haematobium* infection. (submitted).

Ndhlovu, P.D., Chandiwana, S.K. & Makura, O (1992). Progress in the control of Schistosomiasis in Zimbabwe since 1984. *The Central African Journal of Medicine*, 38, 316-321.

Ndhlovu, P., Cadman, H., Gundersen, S.G., Vennervald, B.J., Friis, H., Christensen, N., Mutasa, G., Haagensen, I., Chandiwana, S.K. & Deelder, A.M. (1995). Optimization of the Magnetic Bead Antigen Capture Assay for the detection of circulating anodic antigens in mixed *Schistosoma* infections. *Acta Tropica*, 59, 223-235.

Ndhlovu, P., Cadman, H., Vennervald, B.J., Christensen, N.O., Chidimu, M. & Chandiwana, S.K. (1996). Age-related antibody profiles in *Schistosoma haematobium* infections in a rural community in Zimbabwe. *Parasite Immunology*, 18, 181-191.

Ndhlovu P., H. Cadman, S. Gundersen, B. Vennervald, H. Friis, N. Christensen, G. Mutasa, K.Kaondera, G. Mandaza and A.M.Deelder (1996). Circulating anodic antigen (CAA) levels in different age groups in a Zimbabwean rural community endemic for *Schistosoma haematobium* determined using the Magnetic Beads Antigen Capture Enzyme Linked Immuno Assay. *American Journal Of Tropical Medicine And Hygiene* 54: 5; *537-542.*

Ndhlovu P., Woolhouse M.E.J (1996). Correlations between antibody levels and urine egg counts for *Schistosoma haematobium* , *Transactions of the Royal Society of Tropical Medicine and Hygiene (1996)* 90: *324-325.*

Olgilvie, B.M., Smithers, S.R. & Terry, R.J. (1966). Reagin-like antibodies in experimental infections of *Schistosoma mansoni* and the passive transfer of resistance. *Nature (London)*, 209, 1221-1223.

Openshaw, P., Murphy, E.E., Hosken, N.A., Maino, V., Davis, K., Murphy, K. & O'Garra, A. (1995). Heterogeneity of intracellular cytokine synthesis at the single-cell level in polarized T helper 1 and T helper 2 populations. *Journal of Experimental Medicine*, 182, 1357-1367.

Ottesen, E.A. (1979). Modulation of the host immune response in human schistosomiasis. I. Adherent suppressor cells that inhibit lymphocyte proliferative response to parasite antigens. *Journal of Immunology*, 123, 1639-1644.

Ottesen, E.A., Hiatt, J.A., Freeman, G.C., Bartholomew, R.K. & Jordan, P. (1977). Immune responses during human schistosomiasis mansoni. I. *In vitro* lymphocyte blastogenesis in response to heterogenous antigenic preparations from schistosome eggs, worms and cercaria. *International Archives of Allergy and Applied Immunology*, 53, 420-437.

Ottesen, E.A., Hiatt, R.A., Cheever, A.W., Sotomayor, R.Z. & Neva, F.A. (1978). The acquisition and loss of antigen-specific cellular immune responsiveness in acute and chronic schistosomiasis in man. *Clinical and Experimental Immunology*, 33, 38-47.

Ottesen, E.A. & Poindexter, R.W. (1980). Modulation of the host response in human schistosomiasis. ii. Humoral factors which inhibit proliferative responses to parasite antigens. *American Journal of Tropical Medicine and Hygiene*, 29, 592-597.

Parish, C.R. & Liew, F.Y. (1972). Immune response to chemically modified flagellin. III. Enhanced cell-mediated immunity during high and low zone antibody tolerance to flagellin. *Journal of Experimental Medicine*, 135, 298-311.

Paul, W.E. & Seder, R.A. (1994). Lymphocyte responses and cytokines. *Cell*, 76, 241-251.

Payares, G., McLaren, D.J., Evans, W.H. & Smither, S.R. (1985). Changes in the surface antigens profile of *Schistosoma mansoni* during maturation from cercaria to adult worm. *Parasitology*, 91, 83-99.

Pearce, E.J., Caspar, P., Grzych, J., Lewis, F.A. & Sher, A. (1991). Downregulation of Th1 cytokine production accompanies induction of Th2 responses by a parasitic helminth, *Schistosoma mansoni. Journal of Experimental Medicine*, 173, 159-166.

Pearce, E.J. & James, S.L. (1986). Post lung stage schistosomula of *Schistosoma mansoni* exhibit transient susceptibility to macrophages-mediated cytotoxicity *in vitro* that may relate to late phase killing *in vivo. Parasite Immunology*, 8, 513-527.

Pearce, E.J., James, S.L., Dalton, J., Barrall, A., Ramos, C., Strand, M., Sher, A. (1986). Immunochemical characterisation and purification of Sm-97, a *Schistosoma mansoni* antigen mono-specifically recognised by antibodies from mice protectively immunized with a non-living vaccine. *Journal of Immunology:* 137, 3593-3600.

Pene, J., Rousset, F., Briere, F., Chretien, I., Paliard, X., Banchereau, J., de Vries, J.E. (1988a). IgE production by normal human B cells induced by alloreactive T cell clones is mediated by IL-4 and suppressed by IFN-γ. *Journal of Immunology*: 141, 1218-1229.

Pene, J., Rousset, F., Briere, F., Chretien, I., Bonnefoy, J-L., Spits, H., Yokota,T., Arai, N., Arai, K-L., Banchereau, J and de Vries, J.E. (1988b). IgE production by normal human lymphocytes is induced by interleukin 4 and suppressed by interferon gamma and alpha and prostaglandin E2. *Proceedings of the National academy of Science:* USA 85, 6880-6884.

Pemberton, R.M., Coulson, P.S., Smythies, L.E. & Wilson, R.A. (1993). Phenotypic and functional properties of Th lines and clones recognizing larval antigens of *Schistosoma mansoni. Parasite Immunology*, 15, 373-382.

Pemberton, R.M., Malaquias, L.C.C., Falcao, P.L., Silveira, A.M.S., Rabello, A.L.T., Katz, N., Amorim, M., Mountford, A.P., Coffman, R.L., Correa-Oliveira, R. & Wilson, R.A. (1994). Cell mediated immunity to schistosomes: Evaluation of the mechanism operating against lung-stage parasites which might be exploited in a vaccine. *Tropical Geographical Medicine*, 46, 247-254.

Pemberton, R.M. & Wilson, R.A. (1995). T-helper type-1-dominated lymph node responses induced in C57BL/6 mice by optimally irradiated cercariae of *Schistosoma mansoni* are downregulated after challenge infection. *Immunology*, 84, 310-316.

Peters, P.A., El Alany, M., Warren, K.S. & Mahmoud, F.a. (1980). Quick Kato smear for field quantification of *S.mansoni* eggs. *American Journal of Tropical Medicine and Hygiene*, 29, 217-219.

Peters, P.A., Mahmoud, A.A., Warren, K.S., Ouma, J.H. & Arap-Siongok, T.K. (1976). Field studies of a rapid, accurate means of quantifying *Schistosoma haematobium* eggs in urine samples. *Bull World Health Organisation*, 54, 159-162.

Phillips, S.M. (1992). Schistosomiasis: An immunological diseases. Aquaculture and Schistosomiasis. *Proceedings of a network meeting held in Manila Philippines Development National Research Council National Academy Press Washington D C:* 161-165.

Phillips, S.M. & Lammie, P.J. (1986). Immunopathology of granuloma formation and fibrosis in schistosomiasis. *Parasitology Today*, 2, 296-303.

Phillips, S.M., Reid, W.A., Doughty, B., Khoury, P.B. (1978). The cellular and humoral immune response to *Schistosoma mansoni* infections in inbred rats. III. Development of optimal protective immunity following natural infections and artificial immunizations. *Cellular Immunology*: 38, 225-238.

Pretolani, M. & Goldman, M. (1997). IL-10: A potential therapy for allergic inflammation? *Immunology Today*, 18, 277-280.

Punnonen, J., Aversa, G., Cocks, B.G., McKenzie, A.N., Menon, S., Zurawski, G., De Waal Malefyt, R. & de Vries, J.E. (1993). Interleukin-13 induces interleukin 4-independent IgG4 and IgE synthesis and CD23 expression by human B cells. *Proceedings of National Academy Science, USA*, 90, 3730-3734.

Ribeiro de Jesus, A.M., Almeoda, R.P., Bacellar, O., Araujo, M.I., Demuere, C., Bina, J.C., Dessein, A.J. & Carvalho, E.M. (1993). Correlation between cell-mediated immunity and degree of infection in subjects living in an endemic area of schistosomiasis. *European Journal of Immunology*, 23, 152-158.

Rihet, P., Demeure, C.E., Bourgois, A., Prata, A., Dessein, A.J. (1991). Evidence for an association between resistance to *Schistosoma mansoni* and high anti-larval IgE levels. *European Journal of Immunology*: 21, 2679-2686.

Rihet, P., Demeure, C.E., Dessein, A.J. & Bourgois, A. (1992). Strong serum inhibition of specific IgE correlated to competing IgG4, revealed by a new methodology in subjects from a *S.mansoni* endemic area. *European Journal of Immunology*, 22, 2063-2070.

Roberts, M., Butterworth, A.E., Kimani, G., Kamau, T., Fulford, A.J.C., Dunne, D.W., Ouma, J.H. & Sturrock, R.F. (1993). Immunity after treatment of human schistosomiasis: Association between cellular responses and resistance to reinfection. *Infection and Immunity*, 61, 4984-4993.

Rollinson, D. & Simpson, A.J.G. (1987a). The human immune response to schistosome infection: In: Rollinson, D. and Simpson, A.J.G. eds. *The biology of schistosome. Academic Press, London*, 295-320.

Romagnani, S. (1991). Human Th1 and Th2 subsets: doubts no more. *Immunology Today*, 12, 256-260.

Romagnani, S. (1992). Induction of Th1 and Th2 responses: a key role for the natural immune response? *Immunology Today*, 13, 379-381.

Romagnani, S. (1997). The Th1/Th2 paradigm. *Immunology Today*, 12, 263-265.

Romagnani, S. (1994). Lymphokine production by human T cells in diseases states. *Annual Reviews of Immunology*, 12, 227-233.

Rousset, F., Garcia, E., Defrance, T., Peronne, C., Vezzio, N., Hsu, D.H., Kastelein, R., Moore, K.W. & Banchereau, J. (1992). Interleukin-10 is a potent growth and differentiation factor for activated human B lymphocytes. *Proceedings of National Academy of Science, USA*, 89, 1890-1893.

Sathe, B.D., Pandit, C.H., Chanderkar, N.G., Badade, D.C., Sengupta, S.R., Renapurkar, D.M. (1991). Sero-diagnosis of schistosomiasis by ELISA test in an endemic area of Gimvi village, India. *Journal of Tropical Medicine and Hygiene*: 94, 76-78

Scott, P. (1993). IL-12: initiation cytokine for cell-mediated immunity. *Science*, 260, 496-497.

Seder, R.A., Gazzinelli, G., Sher, A. & Paul, W.E. (1993). IL-12 acts directly on CD4+ T cells to enhance priming for IFN-γ production and diminishes IL-4 inhibition of such priming. *Proceedings of National Academy of Science, USA*, 90, 10188-10192.

Seder, R.A., Le Gros, G., Ben-Sasson, S.Z., Urban, J.J., Finkelman, F.D. & Paul, W.E. (1991). Increased frequency of interleukin 4-producing T cells as a result of polyclonal priming. Use of single-cell assay to detect interleukin 4-producing cells. *European Journal of Immunology*, 21, 1241-1247.

Seder, R.A. and Paul, W.E. (1994). Acquisition of lymphokine-producing phenotype by CD4[+] T cells. *Annual Review of Immunology*, 12, 635-674.

Seder, R.A., Paul, W.E., Davis, M.M. & Fazekas de St.Groth, B. (1992). The presence of interleukin-4 during *in vitro* priming determines the lymphokine-producing potential of CD4[+] T cells from T cell receptor transgenic mice. *Journal of Experimental Medicine*, 176, 1091-1098.

Sher, A. (1992). Schistosomiasis. Parasitizing the cytokine system [news; comment]. *Nature:* 356, 565-566..

Sher, A. & Coffman, R.L. (1992). Regulation of immunity to parasites by T cells and T cell-derived cytokines. *Annals Review of Immunology*, 10, 385-409.

Sher, A., Coffman, R.L., Hieny, S., Scott, P. & Cheever, A.W. (1990). Interleukin 5 is required for the blood and tissue eosinophilia but not granuloma formation induced by infection with *Schistosoma mansoni*. *Proceedings of National Academy of Science, USA*, 10, 61-65.

Sher, A., Fiorentino, D., Caspar, P., Pearce, E.J. & Mosmann, T.R. (1991). Production of IL-10 by CD4$^+$ T lymphocytes correlates with downregulation of Th-1 cytokine synthesis in helminth infection. *Journal of Immunology*, 147, 2713-2716.

Shiff, C.J., Coults, W.G.C., Yiannakis, C. & Holmes, R.W. (1979). Seasonal patterns in the transmission of *Schistosoma haematobium* in Rhodesia and its control by winter application of molluscide. *Transaction of the Royal Society of Tropical Medicine and Hygiene*, 73, 375-380.

Shiff, C.J., Evans, A., Yiannakis, C. & Eardley, M. (1975). Seasonal influence on the production of *Schistosoma haematobium* and *Schistosoma mansoni* cercariae in Rhodesia. *International Journal of Parasitology*, 5, 119-123.

Simpson, A.J.G., Knight, M., Hagan, P., Hodgson, J., Wilkins, H.A., Smithers, S.R. (1985). The schistosomulum surface antigens of *Schistosoma haematobium*. *Parasitology*: 90, 499-508.

Simpson, A.J.G., Hackett, F., Kelly, C., Knight, M., Payares, G., Omer-Ali, P., Lillywhite, J., Fleck, S.L. & Smithers, S.R. (1986). The recognition of *Schistosoma mansoni* surface antigens by antibodies from patients infected with *S.mansoni* and *S.haematobium*. *Transactions of the Royal Society of Tropical Medicine and Hygiene*, 80, 261-270.

Smithers, S.R., Terry, R.J. (1969a). Immunity in schistosomiasis. *Annals of the New York Academy of Sciences*: 160, 826-840.

Smithers, S.R. Terry, R.J., Hockley, D.J. (1969b). Host antigens in schistosomiasis. *Proceedings of the Royal Society London:* B171, 483-494.

Smithers, S.R. & Terry, R.J. (1967a). Naturally acquired resistance to experimental infections of *Schistosoma mansoni* in rhesus monkey (*Macaca mulatta*). *Parasitology*, 55, 701-710.

Smithers, S.R. & Gammage, K. (1980). The recovery of *S.mansoni* from the skin, lungs and hepatic portal system of naive mice previously exposed to *S.mansoni*. Evidence for two phases of parasite attrition in immune mice. *Parasitology*, 80, 289-297.

Smithers, S.R., McLaren, D.J. & Clegg, J.A. (1985). Acquisition of host antigens by young *Schistosoma mansoni* in mice: correlation with failure to bind antibody *in vitro*. *Parasitology*, 70, 67-75.

Smithers, S.R. & Terry, R.J. (1967b). Resistance to experimental infection with *S.mansoni* in rhesus monkeys induced by the transfer of adult worms. *Transactions of the Royal Society of Tropical Medicine and Hygiene*, 61, 517-523.

Smithers, S.R., Terry, R.J. (1965). Infection of laboratory hosts with cercariae of *Schistosoma mansoni* and the recovery of the adult worms. *Parasitology*: 55, 695-700.

Snapper, C.M., Finkelman, F.D. & Paul, W.E. (1988). Regulation of IgG1 and IgE production by interleukin 4. *Immunological Reviews*, 102, 51-73.

Snapper, C.M. & Paul, W.E. (1987). Interferon gamma and B cell stimulatory factor-1 reciprocally regulate Ig isotype production. *Science*, 236, 944-947.

Soisson, L.A., Masterson, C.P., Tom, T.D., McNally, M.T., Lowell, G.H. & Strand, M. (1992). Induction of protective immunity in mice using a 62 kDa recombinant fragment of *S.mansoni* surface antigen. *Journal of Immunology*, 149,3612-3620.

Soisson, L.A., Reid, G.D.F., Farah, I.O., Nyindo, M. & Strand, M. (1993). Protective immunity in baboons vaccinated with a recombinant antigen or radiation-attenuated cercariae of *Schistosoma mansoni* are antibody-dependent. *Journal of Immunology*, 151, 4782-4789.

Sontoro, F., Lachmann, P.J., Capron, A. and Capron, M. (1979). Activation of complement by *Schistosoma mansoni* schistosomula: Killing of parasite by the alternative pathway and requirement of IgG for classical pathway activation. *Journal of Immunology*, 123, 1551-1557.

Sprent, J. (1994). T and B memory cells. *Cell*, 76, 315-322.

Street, N.E., Schumacher, J.H., Fong, T.A., Bass, H., Fiorentino, D.F., Leverah, J.A. & Mosmann, T.R. (1990). Heterogeneity of mouse helper T cells. Evidence from bulk cultures and limiting dilution cloning for precursors of Th1 and Th2 cells. *Journal of Immunology*, 144, 1629-1639.

Sturrock, R.F. (1987). Immunity after treatment of human schistosomiasis mansoni. III. Long-term effects of treatment and retreatment. *Transaction of the Royal Society of Tropical Medicine and Hygiene*, 81, 303-314.

Sturrock, R.F., Kimani, T., Joseph, M., Butterworth, A.E., David, J.R., Houba, V. (1981). Heat-labile and heat-stable anti-schistosomula antibodies in Kenyan school children infected with *Schistosoma mansoni*. *Transactions of the Royal Society of Tropical Medicine and Hygiene*: 75, 219-227.

Sturrock, R.F., Kimani, T., Cottrell, B.J., Butterworth, A.E., Seitz, H.M., Arap Siongok, T.K., Houba, V. (1983). Observations on possible immunity to reinfection among Kenyan school children after treatment for *Schistosoma mansoni*. *Transactions of the Royal Society of Tropical Medicine and Hygiene*: 77, 363-371.

Sturrock, R.F., Kimani, G., Joseph, M., Butterworth, A.E., David, J.R., Capron, A. & Houba, V. (1981). Heat-labile and heat-stable anti-schistosomular antibodies in Kenyan schoolchildren infected with *Schistosoma mansoni*. *Transaction of the Royal Society of Tropical Medicine and Hygiene*, 75, 219-227.

Swain, S.L., Bradley, L.M., Croft, M., Tonkonogy, S., Atkins, G., Weinberg, A.D., Duncan, D.D., Hedrick, S.M., Dutton, R.W. & Huston, G. (1991). Helper T-cell subsets: phenotype, function and the role of lymphokines in regulating their development. *Immunological Reviews*, 123, 115-144.

Swain, S.L., Weinberg, A.D., English, M. & Huston, G. (1990). IL-4 directs the development of Th2-like helper effectors. *Journal of Immunology*, 145, 3796-3806.

Taylor, P. & Makura, O. (1985). Prevalence and distribution of schistosomiasis in Zimbabwe. *Annals of Tropical Medicine and Parasitology*, 79, 287-299.

Trottein, F., Godin, C., Pierce, R.J., Sellin, B., Taylor, M.G., Gorillot, I., Silva, M.S., Lecocq, J.P. & Capron, A. (1992). Inter-species variation of schistosome 28-kDa glutathione S-transferases. *Molecular and Biochemical Parasitology*, 54, 63-72.

Tiu, U.W., Davern, K.M., Wright, M.D., Board, P., Mitchell, G.M., (1988). Molecular and serological characteristics of the glutathione S-transferase of *Schistosoma japonicum* and *Schistosoma mansoni. Parasite Immunology*: 10, 693-706.

Urban, J.J., Katona, I.M., Paul, W.E. & Finkelman, F.D. (1991). Interleukin-4 is important in protective immunity to gastrointestinal nematode infection in mice. *Proceedings of National Academy of Science, USA*, 88, 5513-5517.

Urban, J.J., Madden, K.B., Svetic, A., Cheever, A., Trotta, P.P., Gause, W.C., Katona, I.M. & Finkelman, F.D. (1992). The importance of Th2 cytokines in protective immunity to nematodes. *Immunological Reviews*, 127, 205-220.

Vella, A.T. & Pearce, E.J. (1992). CD4$^+$ Th2 response induced by *S.mansoni* eggs develops rapidly through an early transient, Th0-like stage. *Journal of Immunology*, 148, 2283-2290.

Vermund, S.H., Bradley, D.J. & Ruiz-Tiben, E. (1983). Survival of *Schistosoma mansoni* in the human host: estimates from a community-based prospective study in Peurto Rico. *American Journal of Tropical Medicine and Hygiene*, 32, 1040-1048.

von Lichtenberg, F. (1964). Studies on granuloma formation: III. Antigen sequestration and destruction in the schistosome pseudotubercle. *American Journal of Pathology:* 45, 75-94.

Warren, K.S. (1972). The immunopathogenesis of schistosomiasis: A multidisciplinary approach. *Transactions of the Royal Society of Tropical Medicine and Hygiene:* 66, 417-434.

Warren, K.S.(1973). Regulation of the prevalence and intensity of schistosomiasis in man. Immunology or ecology? *Journal of Infectious Diseases*: 127, 595-609.

Warren, M.K. & Vogel, S.N. (1985). Bone marrow-derived macrophages: development and regulation of differentiation markers by colony-stimulating factor and interferon. *Journal of Immunology*, 134, 982-989.

Warren, K.S., Arap Siongok, T.K., Hanser, H.B., Ouma, J.H. & Peters, P.A. (1978). Quantitation of infection with *S.haematobium* in relation to epidemiology and selective population chemotherapy. 1. Minimal number of daily egg counts in urine necessary to establish intensity of infection. *Journal of Infectious Diseases*, 138, 849-855.

Webster, M., Fallon, P.G., Fulford, A.J.C., Butterworth, A.E., Ouma, J.H., Kimani, G. & Dunne, D.W. (1997). Effect of praziquantel and oxamniquine treatment on human isotype responses to *Schistosoma mansoni*: elevated IgE to adult worm. *Parasite Immunology*, 19, 333-335.

Webster, M., Fallon, P.G., Fulford, A.J.C., Butterworth, A.E., Ouma, J.H., Kimani, G. & Dunne, D.W. (1997). IgG4 and IgE responses to *Schistosoma mansoni* adult worm after treatment. *Journal of Infectious Diseases*, 175, 493-494.

Weinstock, J.V. (1992). The pathogenesis of granulomatous inflammation and organ injury in schistosomiasis: Interactions between the schistosome ova and the host. *Immunological Investigations*, 21, 455-475.

WHO (1993). The control of *Schistosomias*: Second report of the WHO Expert Committee 830.

Wiest, P.M. (1996). The epidemiology of morbidity of schistosomiasis. *Parasitology Today*, 12, 215-220.

Wilkins, H.A., Goll, P.H., Marshall, T.F. de C., Moore, P.J. (1984a). Dynamics of *Schistosoma haematobium* infection in a Gambian community. The pattern of human infection in the study area. *Transactions of the Royal Society of Tropical Medicine and Hygiene*: 78, 216 - 221.

Wilkins, H.A., Goll, P.H., Marshall, T.F. de C., Moore, P.J. (1984b). Dynamics of *Schistosoma haematobium* infection in a Gambian community. III. The acquisition and loss of infection. *Transactions of the Royal Society of Tropical Medicine and Hygiene*: 78, 227-232.

Wilkins, H.A., Blumenthal, U.J., Hagan, P., Tulloch, S., Hayes, R.J. (1987). Resistance to reinfection after treatment for urinary schistosomiasis. *Transactions of the Royal Society of Tropical Medicine and Hygiene:* 81, 29-35.

Wilkins, H.A. (1977). *Schistosoma haematobium* in a Gambian community I. The intensity and prevalence of infection. *Annals of Tropical Medicine and Parasitology*, 71, 53-58.

Wilkins, H.A. & Scott, A. (1978). Variation and stability in *Schistosoma haematobium* egg counts: a four-year study of Gambian children. *Transactions of the Royal Society of Tropical Medicine and Hygiene*, 72, 397-404.

Williams, M.E., Montenegro, S., Domingues, A.L., Wynn, T.A., Teixerira, K., Mahanty, S., Coutinho, A. & Sher, A. (1994). Leukocytes of patients with *Schistosoma mansoni* respond with a Th2 pattern of cytokine production to mitogen or egg antigens but with a Th0 pattern to worm antigens. *Journal of Infectious Diseases*, 170, 946-954.

Wilson, R.A., Coulson, P.S., Betts, C., Dowling, M.A. & Smythies, L.E. (1996). Impaired immunity and altered pulmonary responses in mice with a disrupted interferon-gamma receptor gene exposed to the irradiated *Schistosoma mansoni* vaccine. *Immunology*, 87, 275-282.

Woolhouse, M., Taylor, P., Matanhire, D. & Chandiwana, S.K. (1991). Acquired immunity and epidemiology of *Schistosoma haematobium*. *Nature*, 351, 757-759.

Woolhouse, M.E.J. (1991). On the application of mathematical models of schistosome transmission dynamics I: natural transmission. *Acta Tropica*. 49, 241-250.

Woolhouse, M.E.J. (1992). On the application of mathematical models of schistosome transmission dynamics, II: control. *Acta Tropica*. 50, 189-204.

World Health Organisation. (1985). The control of schistosomiasis. *Technical Report Series*, 728, 113.

Wyler, D.J. (1992). Why does liver fibrosis occur in schistosomiasis? *Parasitology Today*, 8, 277-279.

Wynn, T.A., Eltoum, I., Oswald, I.P., Cheever, A.W. & Sher, A. (1994). Endogenous interleukin 12 (IL-12). regulates granuloma formation induced by eggs of *Schistosoma mansoni* and exogenous IL-12 both inhibits and prophylatically immunizes against egg pathology. *Journal of Experimental Medicine*, 170, 1551-1561.

Wynn, T.A., Jankovic, D., Hieny, S., Cheever, A.W. & Sher, A. (1995). IL-12 enhances vaccine-induced immunity to *Schistosoma mansoni* in mice and decreases T helper 2 cytokine expression, IgE production, and tissue eosinophilia. *Journal of Immunology*, , 4701-4709.

Wynn, T.A., Reynolds, A., James, S., Cheever, A.W., Caspar, P., Hieney, S., Jankovic, D., Strand, M. & Sher, A. (1996). IL-12 enhances vaccine-induced immunity to schistosomes by augmenting both humoral and cell-mediated immune responses against the parasite. *Journal of Immunology*, 157, 4068-4078.

In: Progress in Immunology Research
Editor: Barbara A. Veskler, pp. 207-230
ISBN 1-59454-380-1
©2005 Nova Science Publishers, Inc.

Chapter IX

Molecular and Cellular Mechanisms Leading to Pathogenesis of *Trypanosoma Cruzi*, The Agent of Chagas'Disease

Ali Ouaissi[1], Margarida Borges[1,2]
and Anabela Cordeiro-Da-Silva[2,3]

[1]Institut de la Recherche pour le Développement, Unité de Recherche n° 008
« Pathogénie des Trypanosomatidae », Montpellier, France
[2]Biochemical Laboratory, Faculty of Pharmacy, University of Porto, Portugal
[3]Institut of Molecular and Cellular Biology, University of Porto, Portugal

Abstract

Trypanosoma cruzi, the causative agent of Chagas'disease is a parasitic protozoa that infects more than eighteen million people in South and Central America. Chagas'disease is characterized by a heart defect and megaviscera in a proportion of patients, and these clinical signs are associated with extensive destruction of parasympathetic, enteric and other neurons and degeneration of cardiac muscle. Chagas'disease is often associated with the presence of autoantibodies (autoAb) against host tissues, including specialized components of striated muscle, neurons and connective tissue, which make autoimmune reactions likely. Interestingly, a remarkable polyspecificity of the autoAb was observed. Indeed, the autoAb have been found to cross-react with animal erythrocytes and distinct structural basement membrane proteins such as laminin and collagen. The purpose of the present review is to summarize some of the current data on the pathophysiology of *T. cruzi* infection. Special attention is given to recent data mainly from our own laboratory illustrating the important role of an immunomodulatory factor released by the parasite, the Tc52 protein, in the induction and perpetuation of chronic disease.

Introduction

Chagas'disease, or American trypanosomiasis due to the protozoan parasite *Trypanosoma cruzi* (*T. cruzi*), is endemic in Latin America. Eighteen million individuals are infected and 100 million of people are at risk (WHO, 2002). The life cycle of *T. cruzi* involves obligatory passage through a definitive vertebrate host (mammals, including man) and an intermediate invertebrate host (hematophagous triatomine bugs). The trypomastigote, ingested by the insect, differentiates into the proliferative epimastigote in the midgut of the insect vector and transforms into an infective metacyclic trypomastigote in the posterior intestine (hindgut) and released in the faeces of the bug during or just after the blood meal (Bonaldo, 1988). This latter form, following invasion of vertebrate host cells, undergoes differentiation into amastigotes, which after several reproductive cycles, transform into the infective trypomastigote forms released in the blood circulation after the rupture of the host cell.

The two available drugs (nifurtimox and benznidazole) present some toxic effects and cure a very low percentage of chronic cases, likely due to the resistance of some strains (Rodrigues Coura & Castro, 2002). Thus, the search of an effective vaccine against the disease is still needed.

Different strategies were developed in order to identify parasite components which could be used as a vaccine candidate. The use of inactivated parasites (Andrews *et al.,* 1985), fractionated parasite material (Gonzales Cappa *et al.,* 1981) and irradiated non infectious *T. cruzi* trypomastigotes (Zweerink *et al.,* 1984) have been shown to induce partial protection of mice against a lethal *T. cruzi* challenge. Due to the cross reactivity between some parasite and host components (Khoury *et al.,* 1979) and toxicity of some parasite molecules, identification and isolation of antigens that induce protective immune mechanisms is needed for the development of an effective convenient vaccine. In this way, the use of defined parasite surface antigens has already been shown to induce partial protection against lethal infection with reduced parasite levels (Scott & Snary, 1979; Snary, 1983).

Moreover, several experiments performed using the mouse model, such as passive transfer of specific anti-*T. cruzi* polyclonal antibodies (Kretteli & Brener, 1976), different anti-*T. cruzi* antibody classes (Takehara *et al.,* 1981), or monoclonal antibodies against the *T. cruzi* flagellar fraction (Segura *et al.,* 1986), have already demonstrated the importance of the humoral response in host's defense against the *T. cruzi* challenge. Fischer rats have also been shown to be susceptible to *T. cruzi* infection (Rivera-Vanderpoas *et al.,* 1983), and could be considered as an experimental model for studies of the immune response involved in resistance against the *T. cruzi* infection. Using this model, Rodriguez *et al.* (1981) showed that the neonatally initiated injections of anti-μ antibodies in rats resulted in a loss of antibody production and an increase in rat susceptibility to the acute infection.

Important immune dysfunction is observed during the acute phase of Chagas'disease, namely polyclonal activation of B and T cell populations, whose autoreactive sub-populations may lead to severe chronic tissue damage (Minoprio *et al.,* 1986; Engman & Leon, 2002). Paradoxically, a state of immunosuppression could be observed during the acute phase of the disease (Tarleton, 1988; Santos Lima & Minoprio, 1996).

Molecular cloning of relevant *T. cruzi* genes involved in the host-parasite relationships has produced a whole series of significant details on the mechanisms of the action of certain *T. cruzi* virulence factors and their interaction with the cells of the immune system (Ouaissi *et al.* 2004). The purpose of this review is to summarize some of the current data related to the *T.cruzi*-host cell interplay.

Immunology of *T. Cruzi* Infection

Trypomastigotes are able to invade and multiply within different host cells, including macrophages, endothelial cells, smooth and striated muscles, fibroblasts and even neurons (Ouaissi, 1993). After the recognition phase and adhesion, the trypomastigote is internalized, evades the phagolysosome, undergoes transformation into amastigotes and multiplies within the cytoplasm of the infected host cell (Ouaissi, 1993; see Figure 1).

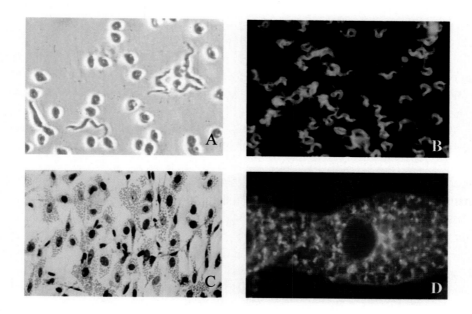

Figure 1. Different *T. cruzi* developmental stages. A) phase contrast microscopy showing an heterogenous population comprising trypomastigotes (slender forms with flagellum) and extracellular amastigotes (round forms without flagellum); B) trypomastigotes reacting with a monoclonal antibody (mAb VG3/G11, Ouaissi *et al.*, 1991) directed against a carbohydrate epitope carried by an 85 kDa surface polypeptide; C) heavily *in vitro*-infected L929 mouse fibroblast cell line showing high density of intracellular amastigotes; D) VG3/G11 immunofluorescence staining of L929-infected fibroblasts showing intracellular trypomastigotes originating from amastigotes after a step of differentiation reacting with the mAb.

In man, the immune system is able to induce a reduction in parasitaemia during acute phase and to actively control the infection for years, but the infection persists for life. In the last several years, a number of studies have provided a better understanding of some characteristics of the anti-*T. cruzi* immune response in man and experimental models. A number of humoral and cellular mediators have been shown to be important in resistance (Krettli & Brener, 1982; Tarleton, 1991). However, the mechanisms by which this machinery interacts with the parasite and vice-versa still needs to be better clarified.

The immune control of *T. cruzi* is mediated by different cell populations: T cells, B cells, macrophages and NK cells (Kierszenbaum & Pienlowsky, 1979; Schmunis *et al.*, 1971; Kierszenbaum & Howard, 1976; Kierszenbaum *et al.*, 1974, Rottenberg *at al.*, 1988; Trishmann, 1983).

There are several mechanisms by which T cells contribute to the eradication of *T. cruzi*: (i) direct destruction of infected cells (Nickell *et al.*, 1993); (ii) stimulation of antigen specific antibody production (Rottenberg *et al.*, 1992); (iii) secretion of lymphokines which in turn modulate different parasiticidal mechanisms (Nogueira & Cohn, 1978; Frosch *et al.*, 1997). It is now clear, that both $CD4^+$ and $CD8^+$ T cells mediate effector functions that control parasite growth and dissemination into the host leading therefore to protection against Chagas'disease (Tarleton, 1991).

It has been shown that mice treated with anti-$CD4^+$ (Araujo, 1989; Rottenberg *et al.*, 1992) or anti-$CD8^+$ antibodies (Tarleton, 1990) exhibited a higher susceptibility to infection. Similar observations were obtained by infecting gene knockout mice lacking CD4+ or CD8+ molecules (Rottenberg *et al.*, 1993) or α_2-microglobulin (Tarleton *et al.*, 1992). Moreover, it has also been reported that mice depleted of $CD4^+$ and $CD8^+$ T cells have increased susceptibility to infection and exacerbation of chronic pathology, and also enhanced myocardial parasitism (Tarleton *et al.*, 1994). Further, mice lacking the major histocompatibility complex (MHC) class I and class II molecules exhibited a high rate of mortality compared to the control mice (Tarleton, 1996), demonstrating the critical importance of both CD8+ and CD4+ T cell-mediated immune responses in the control of *T. cruzi* infection. Given that *T. cruzi* lives free within the cytoplasm of the infected host cell and that the parasite antigens are processed and presented on MHC class I, inducing $CD8^+$ T cytolytic cells. This induction mechanism would be an important arm of the immune system to control *T. cruzi* infection (Buckner *et al.*, 1997).

There are two distinct subsets of $CD4^+$ T cells, T helper type 1 (Th1) and T helper type 2 cells (Th2), which produce distinct and restricted patterns of cytokines that cross regulate each other and thus mediate different types of immune responses (Townsend & McKenzie, 2000). Th1 subset produces abundant interleukin 2 (IL-2) and interferon γ (IFN-γ). These cytokines are known to be involved in classic cell-mediated functions such as clonal expansion of cytotoxic T lymphocytes (CTLs), macrophage activation and class-switching to IgG isotypes that mediate complement lysis of sensitized cells (reviewed by Murray, 1998). Th2 subset produces the functionally opposite cytokines, as IL-4 and IL-5, which are known to activate B cells to switch to neutralizing antibodies (IgG1 in the mouse) and IgE, the initiator of immediate hypersensitivity (Murray, 1998).

Unlike the situation observed in murine infections of *Leishmania major* in which the resistant and susceptible mouse strains exhibit a dominance of either Th1 (IL-2, IFN-γ, IL-12) or Th2 (IL-4, IL-10) type cytokine responses (Sher & Coffman, 1992) respectively. During the acute and chronic phases of infection by *T. cruzi* both type 1 and type 2 patterns of cytokine are developed in both resistant and susceptible mouse strains (Eksi *et al.*, 1996). Nevertheless, recent studies have shown that antigen-specific Th1 (but not Th2) cells provide protection against lethal *T. cruzi* infection in mice (Kumar & Tarleton, 2001).

Macrophage activation by lymphokines, secreted by T lymphocytes, leads to increased trypanocidal activity (Roitt, 1985a). The production of nitric oxide (NO) by IFN-γ-activated macrophages and also in synergism with TNF-α, is a major effector mechanism during an experimental *T. cruzi* infection (Muñoz-Fernández *et al.*, 1992). Indeed, NO-mediated trypanocidal activity can be blocked by L-arginine analogs that inhibit the induced nitric oxide synthase (iNOS) pathway (Hölscher *et al.*, 1998). Moreover, chemoattractant molecules, such as leukotriene B (4) or platelet-activating factor, may also activate macrophages to induce NO-mediated killing of *T. cruzi* (Talvani *et al.*, 2002). Moreover, studies have shown that other factors such as myoglobin may facilitate a *T. cruzi* infection by acting as a scavenger of NO (Ascenzi *et al.*, 2001). Chemokines such as a macrophage inflammatory protein 1 alpha (MIP-1α), has been shown to play a role in macrophage/monocyte influx in infected organs during infection by *T. cruzi* (Petray *et al.*, 2002). Others have demonstrated that MIP-1α, RANTES and MIP-1β, released by macrophages following *T. cruzi* infection, are able to increase the *T. cruzi* trypomastigote uptake and destruction *in vitro* and in mouse inflammatory macrophages both *in vivo* and *in vitro*, through the cytotoxic NO-dependent mechanism (Brenier-Pinchart *et al.*, 2001).

Dendritic cells (DC) are crucial in the initiation of the immune response and are distinguishable from the other antigen presenting cells by their high efficient antigen presentation. DCs are specialized to acquire and process antigen in peripheral non-lymphoid sites, and to transport the antigen to the secondary lymphoid organs where the stimulation of naïve lymphocytes occurs. During their migration, DCs enter a process of maturation that determines whether adaptative immune response occurs and the nature of that immune response. Moreover, DCs are early producers of IL-12 for the initiation of Th1 responses (Banchereau *et al.*, 2000). It is only in recent years that investigations were developed to explore the relationship between *T. cruzi* and DC. In fact, it has been shown that *T. cruzi* released soluble factors could prevent DC maturation and secretion of IL-12 and TNF-α (Van Overtvelt *et al.*, 1999). Further, it was demonstrated that *T. cruzi* is able to inhibit the LPS-induced up-regulation of MHC class I molecules at the surface of human DC (Overtvelt *et al.*, 2002). Studies with the glycoinositolphospholipid (GIPL) from *T. cruzi* have demonstrated that this molecule leads to a down-regulation of human DC surface antigens, such as CD80, CD86, HLA-DR, CD40 and CD57 that are important for T cell activation (Brodskyn *et al.*, 2002). Taken together, these observations led investigators to propose a novel efficient mechanism leading to the alteration of DC function and maturation, that may be used by *T. cruzi* to escape the host immune response. However, although these investigations are interesting and provocative, it is important to remind that the GPI anchors express biological activities similar to those of LPS. Given the fact that LPS induces the maturation of dendritic cells, one would expect that *T. cruzi*-derived LPS-like substances could activate rather than inhibit DC maturation. Therefore, the observations showing DC-inhibition await further explanation. In this regard, it is noteworthy that GPI anchors and

GIPLs from *T. cruzi* are potent activators of the human and mouse macrophage Toll-like receptor 2 (TLR2) (Campos *et al.*, 2001). Interestingly, the TLR2 activation by GPI led to the synthesis of IL-12 and TNF-α when using LPS as a triggering agent. Moreover, it has been shown that *trans*-sialidase (TS) from *T. cruzi*, a parasite-secreted protein, was able to activate mouse DC (Todeschini *et al.*, 2000). Furthermore, in recent studies we found that a *T. cruzi* released protein related to the thiol-disulfide oxidoreductase family, called Tc52, which is crucial for parasite survival and virulence, induces human DC maturation. Tc52-treated immature DC acquire CD83 and CD86 expression, produce inflammatory chemokines (IL-8, MCP-1, and MIP-1α, and present potent costimulatory properties. Tc52 binds to DC by a mechanism with the characteristics of a saturable receptor system and signals via TLR2. While Tc52-mediated signaling involves its GSH-binding site, another portion of the molecule is involved in Tc52 binding to DC (Figure 2). In contrast to GPI and GIPLs-induced DC activation via TLR2, TLR2 activation by Tc52 resulted in the secretion of IL-8, MCP-1 and MIP-1α no increased production of TNF-α occurred. This provides the first evidence, to our knowledge, that a protozoan parasite-derived molecule which belongs to the thiol-disulfide oxidoreductase family, by interacting with TLR2 leads to selective release of inflammatory chemokines, a pattern distinct from the classical profile observed in the case of LPS and related molecules such as GPI and GIPLs. Together these data evidence complex molecular interactions between the *T. cruzi*-derived molecule, Tc52, and DC, and suggest that Tc52 and a related class of proteins might constitute a new type of pathogen-associated molecular patterns (PAMPs).

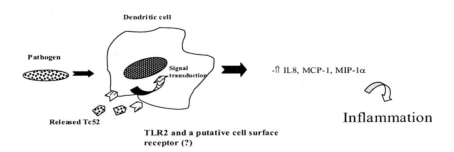

Figure 2. A model for *T. cruzi* Tc52-DC interaction and signaling pathway. The binding of as yet uncharacterized DC surface structure (Tc52R) followed by the interaction of the glutathion binding domain of Tc52 to the Toll-like receptor two (TLR2) led to the activation of the signaling cascade and NF-κB nuclear translocation and regulation of nuclear gene expression.

Mechanisms Leading to the State of Immunosuppression

In mice and humans, the acute phase of Chagas'disease is marked by a state of immunosuppression (Ouaissi *et al.*, 2001). This phenomena is characterized by a decrease of interleukin-2 (IL-2) production, increase of suppressive activity by splenic T cells and macrophages, down regulation of CD3, CD4 and CD8 T cell markers and inhibition of IL-2 receptor (IL-2R) expression in the case of human peripheral blood mononuclear cells (Sztein & Kierszenbaum, 1993). *In vitro* activated T cells from infected mice showed low levels of IL-2 production and proliferative response to mitogens, characteristic of the immunosuppressive response (Tarleton, 1988). In mice, the immunosuppressive activity has been directly involved in the immunodeficiency. In fact, several investigators have documented the notion of suppressive cell populations (T cells, NK cells or macrophages) being involved in the immunosuppression phenomena (Tarleton, 1988; Cerrone *et al.*, 1992; Takle & Snary, 1993). Investigators have demonstrated that prostaglandins were also able to mediate suppression of lymphocyte proliferation and cytokine synthesis during acute *T. cruzi* infection (Pinge-Filho *et al.*, 1999).

Furthermore, several parasite suppressive factors contained in, or released by *T. cruzi* could be responsible for immunological abnormalities (Liew *et al.*, 1988; Serrano & O'Daly, 1990). Indeed, a trypanosomal immunosuppressive factor (TIF), released by the parasite, has been found to suppress the lymphoproliferation and the expression of several molecules involved in lymphocyte activation such as CD3, CD4, CD8, CD25 and the transferrin receptor. Furthermore, this factor (s) with a molecular mass between 30 and 100 kDa was able to curtail a mechanism controlling cell progression through G1 phase (Kierszenbaum *et al.*, 1990, 1998).

Among these parasite-released molecules, a *T.cruzi* protein termed Tc52 (a polypeptide of molecular mass 52 kDa, sharing structural and functional properties with the thioredoxin and glutaredoxin family involved in thiol-disulphide redox reactions), expressed by all *T. cruzi* developemental stages (Figure 3), was shown to suppress T cell proliferation induced by anti-CD3 stimulation (Ouaissi *et al.*, 1995), and exerted several cytokine and chemokine-like activities able to synergize with IFN-γ to stimulate NO production by macrophages (Figure 4) and to modulate IL-1α, IL-12 and IL-10 encoding genes (Fernandez-Gomez *et al.*, 1998) as well as DC chemokine secretion pattern (see above and also Figure 2). NO has a number of other physiological effects that may play a role in the complex host-parasite relationships (James, 1995). Indeed, it has been shown that NO production during Toxoplasmosis in C57BL/6 mice has two opposite effects: being protective against *Toxoplasma gondii* and down-regulating the immune response, suggesting its possible contribution in the establishment of chronic infections (Hayashi *et al.*, 1996).

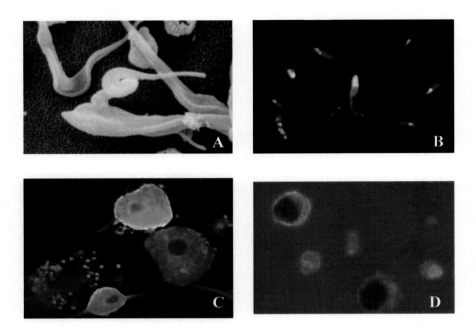

Figure 3. A) scanning electron microscopy analysis of a *T. cruzi* trypomastigotes preparation; B) immunofluorescence staining of *T. cruzi* epimastigotes with anti-Tc52 mouse immune serum showing high density intracellular vesicles containing the reactive Tc52 molecule; C) amastigotes released from ruptured-infected J774 mouse macrophages and the macrophage surface exibited positive immune reactivity with anti-Tc52 immune serum; D) immunofluorescence staining of Tc52-treated J774 macrophages with anti-Tc52 immune serum showing positive labelling on the surface as well as inside the cell.

In the case of *T. cruzi*, previous studies have shown that experimental infection induces NO production, and suggest that IFN-γ and TNF-α are involved in this phenomenon (Munoz-Fernandez *et al.*, 1992a; Petray *et al.*, 1994). Furthermore, independent experiments have shown that IFN-γ and TNF-α mediated activation of macrophages leads to increased production of NO which in turn suppress T cell activation (Abrahamson & Coffman, 1995). Moreover, the participation of NO in the suppression of T cell activation has been reported in a number of other biological systems (James, 1995). Furthermore, NO markedly inhibited the induction of IL-2 promoter, which could account for most of the reduction in IL-2 production, and weakly increased the activation of IL-4 promoter (Chang *et al.*, 1997). This mechanism could be involved in the down-regulation of IL-2 gene expression observed during *T.cruzi* infection (Soong & Tarleton, 1994). Therefore, it is likely that NO production during the initial phase of acute *T.cruzi* infection might participate in the clearance of parasites by macrophages whereas its overproduction during the late phase of acute infection would account for the immunosuppression observed.

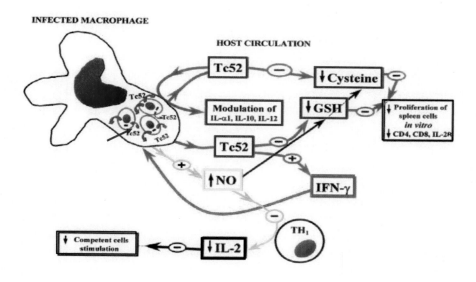

Figure 4. Schematic representation of Tc52 released protein -macrophage interplay.

Another interesting observation is that Tc52 acts directly on macrophages to modulate IL1-α expression. The proinflammatory cytokine IL-1α has potential deleterious effects. Indeed, inappropriate expression of IL-1α in the blood during sepsis correlates with hypotension, shock and death (Dinarello & Wolff, 1993; Ferreira, 1993). Moreover, evidences showing that NO stimulates the release of IL-1α from activated macrophages has beeen recently reported (Hill *et al.*, 1996). Taken together, these observations and our data may suggest that Tc52 could trigger a feed-back loop amplification for the production of the major proinflammatory cytokine IL-1α. However, due to the potential deleterious effects of IL-1α, its activity is regulated at the levels of synthesis, processing and release, and by a naturally occurring receptor antagonist (IL-1Rα) and serum proteins such as α2-macroglobulin (Dinarello, 1993; Arend *et al.*, 1994) whose production has been shown to increase during *T.cruzi* infection (Araujo-Jorge *et al.*, 1992).

These observations may have some implications *in vivo*. Indeed, we showed that elevated levels of circulating Tc52 in the blood of *T. cruzi* experimentally infected mice occurred during the acute phase of the disease and were associated with decreased responsiveness of T cells to mitogen or anti-CD3 stimulation (Ouaissi *et al.*, 1995). Thus, it is reasonable to assume that exogenous Tc52 might participate directly or indirectly at least via NO production in the immunosuppression observed during *T. cruzi* infection. However, it is noteworthy that Chagas'disease almost exclusively involves intracellular amastigotes, which also expressed Tc52 protein. Experiments carried out with murine macrophages harboring a eukaryotic plasmid carrying the *Tc52* gene, showed increased IL-10 mRNA levels (Borges *et* al., 2001). It is tempting to speculate that Tc52-induced increased IL-10 secretion might participate in the downregulation of IL-2 production. This is in agreement with previous studies showing that murine IL-10 can downregulate the host immune response by decreasing the production of IL-2 (Fiorentino et al., 1991) and inhibiting mitogen driven T-cell proliferation (Ding & Shevach, 1992).

Although the basic molecular and immunological approaches have revealed interesting features regarding the cytokine and chemokine-like activities of Tc52, we thought that further in vivo functional studies were needed to ascertain its biological function (s). Thus, genetic manipulation of T. cruzi allowed us to produce parasite clones lacking a Tc52 protein-encoding allele (Tc52$^{+/-}$) (Allaoui et al., 1999). Subsequently, the disease phenotype in Tc52$^{+/-}$- infected BALB/c mice, during the acute and chronic phases of the disease was examined. The results obtained are in agreement with the observations made when using in vitro experimental models. Indeed, a lack of suppression of IL-2 production and of T-cell proliferation inhibition was observed in the case of spleen cells from Tc52$^{+/-}$-infected mice when compared to the wild-type (WT) parasite infected ones. Also, increased production of IL-10 was observed in the case of spleen cells from WT-infected mice, whereas the levels measured in the case of Tc52$^{+/-}$-infected mice were comparable to those of normal mice spleen cells, suggesting therefore, that Tc52 play a role in IL-10 cytokine regulation during in vivo T. cruzi infection (Garzon et al., 2003). Therefore, it is reasonable to suggest that the reduction of Tc52 production by gene targeting which in turn downregulates the IL-10 synthesis could be among the mechanisms participating in the immunoregulatory mechanisms leading to the control of IL-2 production.

We have already shown that Tc52, under conditions of experimental infections, appeared relatively immunologically silent during the early acute phase, failing to elicit significant levels of antibodies and lymphocyte proliferation (Ouaissi et al., 1995). This observation allowed us to make the hypothesis that the analysis of structure-function relationship in the Tc52 molecule could reveal discrete domains, which either contribute to minimize its antigenicity and/or act as an immunoregulatory factor. Studies conducted showed that, indeed, a major peptide fragment of 28 kDa molecular mass (Tc28k) localized in the carboxy-terminal portion of Tc52 carry the inhibitory capacity in T cell activation (Borges et al., 2003). Synthetic peptides spanning the amino terminal or carboxy-terminal domain of Tc52 protein indicated that the activity mapped to the 432-445 aa residues of Tc52 sequence. Moreover, the peptide when coupled to a carrier protein, exhibited increased inhibitory activity on T lymphocyte activation. Interestingly, the coupled peptide significantly down regulated IFN-γ and IL-2 secretion. Likewise, in immunized mice, the coupled peptide 432-445 was a very poor B and T cell antigen compared to the other Tc52 derived peptides. Therefore, the immunomodulatory portion of T.cruzi Tc52 virulent factor seems to reside, at least in part, in a conserved sequence within its carboxyl-terminal domain which could minimize its antigenicity (Borges et al., 2003). Such molecules may permit parasites to escape immune surveillance and to grow unimpeded by normal immune responses. Moreover, the impairment of multiple immune effector functions by blocking the signal transduction pathways utilized by cytokines such as IL-2 and IFN-γ, the host may become more susceptible to opportunistic infections as well.

Autoimmune Reactions

A number of investigators have proposed autoimmunity as the major contributor to Chagas'diease pathology (Kierszenbaum, 1999). *T. cruzi* infection causes lysis of cardiac myocytes, releasing cardiac antigens, such as myosin, considered to be the major autoantigen. Moreover, the pro-inflammatory environment induced by the presence of inflammatory factors as chemokines and cytokines are also fundamental conditions to this process (Cunha-Neto *et al.*, 1995). Indeed, the pro-inflammatory reactions may induce proliferation of autoreactive T cells, in response to self antigens presented on the host APC. Further, the elevated levels of myocardial antigens, in the presence of pro-inflammatory cytokines, lead to increased presentation of self antigens and stimulation and expansion of autoreactive cells (Fedoseyeva *et al.*, 1999). Furthermore, in *T. cruzi* infection, the tissue inflammation may cause new cryptic epitopes to be presented by APC. Because the circulating T cells are not tolerant to these "novel" epitopes, they became activated and initiate autoimmunity (Leon & Engman, 2001).

The "molecular mimicry" is another possibility leading to the autoimmune reactions. Indeed, *T. cruzi* antigens may share amino-acid (aa) sequences or three-dimensional epitopes with host components (Table 1). These shared peptides may initiate a cross-reactive T cell response leading to activation of autoreactive T cells. However, others argue that this mechanism is less than convincing (Benoist & Mathis, 2001).

Table 1. Molecular mimicry in Chagas'disease

T. cruzi molecule	Host component
α-Gal residues	α-Gal, EVI [1]
Sulphated glycolipids	Neurons [1]
FI-160	47 kDa neuronal protein [1]
Ribosomal P protein	Ribosomal P protein (Hela cell line) [1]
23 kDa ribosomal protein	23 kDa ribosomal protein (Hela cells reticulocytes) [1]
Microtubule-associated proteins	Microtubule-associated proteins (brain) [1]
Ribosomal P0 protein	β1-adrenoreceptor [1, 2]
Carboxyl-terminal end of ribosomal P proteins	β1-adrenoreceptor [3, 4]
B13 protein	Human cardiac myosin heavy chain [1, 5]

[1] Kalil & Cunha-Neto, 1996. [2] Skeiky *et al.*, 1992. [3] Motrán *et al.*, 1998. [4] Lopez Bergami *et al.*, 2001. [5] Cunha-Neto *et al.*, 1995.

The precise mechanism responsible for myocardial and conducting damages in Chagas'disease is not completely understood. The existence of autoantibodies directed against host tissues during *T. cruzi* infection is well documented. On the other hand, several observations suggested the possible involvement of autoreactive cell-mediated immunity in the production of Chagasic cardiomyopathy (Ouaissi *et al.*, 2001). The presence of antoanti-Idiotype (Id) T cells in Chagas'patients has been demonstrated (Gazzinelli *et al.*, 1988). The authors suggested that Id-anti-Id interactions may play a role in determinning the pathogenesis of Chagas, cardiomayopathy. In a previous report we have shown that acetylcholinesterase (AChE) exhibited immunological cross-reactivity with *T. cruzi* (Ouaissi *et al.*, 1988). Moreover, antibodies to AChE and anti-idiotypic antibodies were detected in sera of patients presenting the chronic cardiac form of the disease. The data provide a biochemical basis supporting, in part, the denervation hypersensitivity in Chagas'disease. In addition, it provided the notion of an Id-anti-Id regulation of conducting tissue damage during the course of *T. cruzi* infection.

Induction of effector functions by T cells requires two signals provided by antigen-presenting cells (APC) (Mueller *et al.*, 1989): signal 1 is delivered by T-cell receptor (TCR), which recognizes specific peptides presented on MHC molecules; signal 2 is provided by some integral membrane proteins: APC-like lymphocyte functional antigen 3 (LFA-3), intercellular adhesion molecule 1 (ICAM-1), ICAM-2 and ICAM-3, or B7-1/2 and co-receptor molecules such as CD2, LFA-1, and CD28 (Damle *et al.*, 1992; Freeman *et al.*, 1993). It was shown that infection with *T. cruzi* was able to up-regulate B7-2 molecules on macrophages and, in this way, enhance their costimulatory activity (Frosch *et al.*, 1997) and increase the expression and secretion of ICAM-1 in inflammatory cells, as well as cardiac myocytes in infected mice (Laucella *et al.*, 1996). Moreover, an increase in the expression of ICAM-1, LFA-1, integrin VLA-4 and vascular cellular adhesion molecules (VCAM-1) in chagasic hearts from infected experimental animals has been demonstrated and associated with a persistent production of inflammatory and anti-inflammatory cytokines from the early acute stage through the late chronic stage of infection (Zhang & Tarleton, 1996).

Polyclonal Activation

T. cruzi infection is associated with a poly-isotypic production of non-specific immunoglobulins as well as parasite specific response. This response is characterized essentially by a predominance of IgM, IgG2a and IgG2b that persists during all the chronic phase (El Cheikh *et al.*, 1992). During the acute phase of infection, the parasitaemia level decreases quickly coinciding with increased antibody levels, suggesting the protective role of antibodies against blood forms of the parasite (Krettli & Brener, 1976; Brener, 1980). Moreover, animals producing weak levels of antibodies exhibited a high susceptibility to *T. cruzi* infection (Kierszenbaum & Howard, 1976). Indeed, the transfer of immune sera to athymic mice delayed the appearance of parasites and increased mouse survival (Kierszenbaum, 1980). Indirect evidence supported the implication of specific antibodies in the control of infection. For example, *in vivo* transfer of spleen cells from mice which had recovered from the acute phase of infection to naïve mice induce protection against challenge infection, removal of B lymphocytes abolished the immune protection (Scott, 1981). Nevertheless, studies during *T. cruzi* mouse infection have demonstrated clearly an important non-specific B and T lymphocytes polyclonal activation (Ortiz-Ortiz *et al.*, 1980; D'Imperio Lima *et al.*, 1985, 1986).

The non-specific polyclonal lymphocyte activation that occurs during infection plays a role in the pathogenesis of Chagas' disease. Extensive polyclonal B and T cell activation occurred during the acute and chronic phases of *T. cruzi* infection (Minoprio *et al.*, 1986a, 1986b). Simultaneously, massive polyclonal activation also occurred in minor lymphocyte subsets, such as CD5B and γδ T lymphocytes, which are associated with autoimmune reactivity and pathological processes (El Cheikh *et al.*, 1992; Santos Lima & Minoprio, 1996). In the early stages of infection, the number of immunoglobulin-secreting cells in the spleen and peripheral lymph nodes is very high and the majority of activated B cells secrete antibodies nonspecific for the parasite antigens (Reina-San-Martin *et al.*, 2000; el Bouhdidi *et al.*, 1994). Isolation of lipopolysaccharide (LPS)-like substances from *T. cruzi* has been reported (Goldenberg *et al.*, 1983). An octadecapeptide derived from the sequence of an 85-kDa surface protein from *T. cruzi* trypomastigotes has been shown to be able to act as a comitogenic molecule (Pestel, *et al.*, 1992). Recent studies have also shown that *T. cruzi* Tc24 released protein induced an increase in the number of B cell secreted immunoglobulins mainly of IgM isotype in Tc24 treated mice (Da Silva *et al.*, 1998). It is well known that the antigenic complex composition of *T. cruzi* parasites can contribute to the polyclonal activation resulting in a "hyperstimulation" of lymphocyte clones directed against a multitude of challenging antigens. As a consequence, a panclonal activation of the immune system develops resulting in a nonspecific B cell response during experimental infections (Cordeiro-da-Silva *et al.*, 2002; Requena *et al.*, 2000). These responses lead to the expansion of antiself clones that may be responsible for the killing of parasitised and nonparasitised cells. High cellularity (60-70% of lymphocytes) is concomitant with very large lymphocyte activation (Minoprio, 1986a). In this phase of infection, the B cell polyclonal proliferation is characterised by a typical isotypic profile, IgG2a and IgG2b, in spleen and lymph nodes. The hybridomas produced by spleen B cells or lymph node T cells of infected mice fail to bind parasite Ag but are autoreactive clones, in such a way that the antibodies recognize natural structures like actin, tubulin, myosin, keratin, myoglobulin, thyroglobulin or myelin (Reina-San-Martin *et al.*, 2000 and Ouaissi *et al.*, 2001). Moreover, a recent study has demonstrated that a protein from *T. cruzi* named TcPA45 induced polyclonal activation. Moreover, the TcPA45 was shown to be a eucaryotic proline racemase hortologue, allowing the parasite to synthesize and express itself on its surface proteins containing D-proline, an aminoacid resistant to host-induced proteolytic mechanisms, and that this enzymatic activity was necessary for mitogenic activity (Reina-San-Martin *et al.*, 2000).

Apoptosis as a Possible Process Involved in the Immune System Dysegulation

Programmed cell death (PCD) seems of great importance in the field of immunology and cell biology of protozoan parasites. Indeed, in the last few years, a large amount of evidence supported the notion that apoptosis of immune cells could be among factors which help the parasite to avoid the elimination favoring, therefore, its persistence in the immune host. Apoptosis can occur in a variety of cell types during the acute phase of *T. cruzi* infection (Zhang *et al.*, 1999; Ameisen *et al.*, 1995). However, apoptosis observed in immune cells, such as macrophages, lymphocytes and interstitial dendritic cells, may have functional significance different from that found in cardiac myocytes and endothelial cells, such as regulation and modulation of inflammatory responses (Zhang *et al.*, 1999). Lopez and co-workers have observed that splenic CD4+ T cells from mice infected with *T. cruzi* undergo apoptosis *ex-vivo* upon stimulation (Lopes *et al.*, 1995), a fact that allowed the authors to suggest that this phenomenon may participate in the immune deficiency *in vivo*, helping, the parasite in this way to persist during the chronic phase of the disease. Moreover, spleen CD4+ T cells apoptosis in *T. cruzi*-infected BALB/c mice, was due to increased expression of Fas and Fas ligands (FasL), two markers known to be essential to the PCD. It was also demonstrated that, Fas: FasL pathway controls parasite replication *in vitro* and prevents an exacerbated Th2 type immune response to the parasite (Lopes *et al.*, 1999). Furthermore, it was demonstrated that IFN- γ is able to modulate Fas and Fas-L expression and nitric oxide-induced apoptosis during *T. cruzi* infection (Martins *et al.*, 1998). Furthermore, parasite or host-derived factors could be involved in apoptosis of immune cells.

During the last few years, a number of investigations have focused on the study of an important family of proteins from *T. cruzi*: Trans-sialidse (TS) and mucin families (Frasch, 2000). *T. cruzi* is unable to synthesize *de novo* the monosaccharide sialic acid, but can incorporate sialic acid derived from the host (Frash, 1994). The parasite expresses TS able to catalyse the transfer of sialic acid from host glycoconjugates to mucin-like molecules located on the parasite surface membrane. TS is released by the parasite into the external milieu as a soluble factor (Frasch, 2000). TS is expressed in the invasive trypomastigote stage and is defined by two regions: a globular amino-terminus of about 640 amino acids containing the catalytic activity and a variable number of repeated highly antigenic motifs of 12 amino acid named SAPA located at the C terminus (Frasch, 2000). Members of the TS-like family in the intracellular amastigote stage stimulate a CD8+ cytotoxic T-cell response (Wizel *et al.*, 1997) demonstrating that these antigens enter the class I pathway in infected cells. Moreover, it was found that members of TS and TS-like families induced an antibody response against the parasite (Cazzulo & Frasch, 1992). However, it was demonstrated that TS, through its C-terminal of long tandem repeats induced an abnormal polyclonal B cell activation and Ig secretion (Gao *et al.*, 2002). SA85-1.1 surface protein belonging to TS family of *T. cruzi*, has been shown to stimulate a polarized Th1 response to become anergic (Millar *et al.*, 1999). Further study has identified an epitope of SA85-1.1, named epitope 1, able to protect mice and stimulating 4% to 6% of the splenic CD4+ cells during *T. cruzi* infection (Millar *et al.*, 2000).

Furthermore, independent investigators have shown that TS was able to induce apoptosis *in vivo* in spleen, thymus and peripheral ganglia (Leguizamon *et al.*, 1999). In contrast, another study showed that TS blocked activation-induced cell death in CD4+ T cells from *T. cruzi*-infected mice requiring CD43 signaling (Todeschini *et al.*, 2002). Further, depending of the host cell type infected by *T. cruzi*, TS could exert an anti or pro-apoptotic function. For example, TS is able to enter into synergism with neurocytokines to prevent neuron and Schwann cell apoptosis (Chuenkova & Pereira, 2000; Chuenkova *et al.*, 2001). Alternatively, TS was responsible for the early thymic alterations via apoptosis induction in the "nurse cell complex" (Mucci *et al.*, 2002).

Further, GIPL-derived ceramide from *T. cruzi* has been reported to be able to synergize with IFN- γ to induce intense macrophage apoptosis (Freire-de-Lima *et al.*, 1998). Moreover, the demonstration of increased parasite replication inside macropahges as a result of apoptotic cells uptake, confirms the important contribution of apoptosis in the maintenance of *T. cruzi* in the immune host (Freire-de-Lima *et al.,* 2000). However, it has also been postulated that infected cells were still able to counteract the parasite by either initiating their own death by apoptosis and/or being more easily recognized and phagocytosed by macrophages. This led to the parasite elimination from the infected cell. It has also been reported that the protozoan parasite, by activating the expression of heat shock proteins or NF-kappa B, could interfere with molecules of cellular death machinery, thus regulating the transcription of anti-apoptotic molecules (Heussler *et al.*, 2001).

Conclusion

T. cruzi can elicit a complex series of cellular interactions which result in specific and non-specific immune responses. The immunosuppression appears to predominate at certain phases of the infection. Relieving these immunological alterations could provide the means to restore the efficacy of the immune protection mechanisms. It is hoped that the identification of target genes that induce immune dysfunction and their mode of action will help to develop new therapeutic interventions.

References

Abrahamson, I.A. & Coffman, R.L. (1995). Cytokine and nitric oxide regulation of the immunosuppression in *Trypanosoma cruzi* infection. *J Immunol* 15: 3955-63.

Allaoui, A., Francois, C., Zemzoumi, K., Guilvard, E. & Ouaissi, A. (1999). Intracellular growth and metacyclogenesis defects in *Trypanosoma cruzi* carrying a targeted deletion of a Tc52 protein-encoding allele. *Mol Microbiol* 32: 1273-86.

Ameisen, J.C., Idziorek, T., Billaut-Mulot, O., Loyens, M., Tissier, J., Potentier, A. & Ouaissi, A. (1995). Apoptosis in a unicellular eukaryote (*Trypanosoma cruzi*): implications for the evolutionary origin and role of programmed cell death in the control of cell proliferation, differentiation and survival. *Death and Differentiation* 2: 285-300.

ANDREWS, N.W., ALVES, M.J.M., SCHUMACHER, R.I. & COLLI, W. (1985) *Trypanosoma cruzi:* protection in mice immunized with 8-Methoxyproralen-inactivated trypomastigotes. *Exp Parasitol* 60: 255- 62

Araujo, F.G. (1989). Development of resistance to *Trypanosoma cruzi* in mice depends on a viable population of L3T4+ (CD4+) T lymphocytes. *Infect Immun* 57: 2246-8.

ARAUJO-JORGE, T.C., LAGE, M.J.F., RIVERA, M.T. et al (1992). *Trypanosoma cruzi:* enhanced alpha-macroglobin levels correlate with the resistance of BALB/cj mice to acute infection. *Parasitol Res* 78 : 215-221.

AREND, W. P., MALYAK, M., SMITH, M. J. et al (1994). Binding of IL-1 □, IL-1 □, and IL-1 receptor antagonist by soluble IL-1 receptors and levels of soluble IL-1 receptors in synovial fluids. *J Immunol* 153 : 4766-4774.

Ascenzi, P., Salvati, L. & Brunori, M. (2001). Does myoglobin protect *Trypanosoma cruzi* from the antiparasitic effects of nitric oxide? *FEBS Lett* 501: 103-5.

Banchereau, J., Briere, F., Caux, C., Davoust, J., Lebecque, S., Liu, Y.J., Pulendran, B. & Palucka, K. (2000). Immunobiology of dendritic cells *Annu Rev Immunol* 18:767-811.

Benoist, C. & Mathis, D. (2001). Autoimmunity provoked by infection: how good is the case for T cell epitope mimicry? *Nat Immunol* 2: 797-801.

Bonaldo, M.C., Souto Padron, T., De Souza, W. & Goldenberg, S. (1988). Cell-substract adhesion during *Trypanosoma cruzi* differentiation. *J Cell Biol* 106: 1349-1358.

BORGES, M., GUILVARD, E., CORDEIRO DA SILVA, A. & OUAISSI, A. (2001). Endogenous *Trypanosoma cruzi* Tc52 protein expression upregulates the growth of murine macrophages and fibroblasts and cytokine gene expression. *Immunol Lett* 39 : 127-134.

Brener Z. (1980). Immunity to *Trypanosoma cruzi*. *Adv Parasitol* 18:247-92.

Brenier-Pinchart, M., Pellox, H., Derouich-Guergour, D. & Ambroise-Thomas, P. (2001). Chemokines in host-protozoan-parasite-interactions. *Trends in Parasitol* 17: 292-296.

Brodskyn, C., Patricio, J., Oliveira, R., Lobo, L., Arnholdt, A., Mendonca-Previato, L., Barral, A. & Barral-Neto, M. (2002). Glycoinositolphospholipid from *Trypanosoma cruzi* interfere with macrophages and dendritic cell responses. *Infect Immun* 70: 3736-43.

Buckner, F.S., Wipke, B.T. & Van Voorhis, W.C. (1997) *Trypanosoma cruzi* infection does not impair major histocompatibility complex class I presentation of antigen to cytotoxic T lymphocytes. *Eur J Immunol* 27: 2541-8.

CAMPOS, M.A., ALMEIDA, I.C., TAKEUCHI, O., AKIRA, S., VALENTE, E.P., PROCOPIO, D.O., TRAVASSOS, L.R., SMITH, J.A., GOLENBOCK, D.T. & GAZZINELLI, R.T. (2001). Activation of Toll-like receptor-2 by glycosylphosphatidylinositol anchors from a protozoan parasite. *J Immunol* 167: 416-23.

Cazzulo, J.J. & Frasch, A.C.C. (1992). SAPA/trans-sialidase and cruzipain: two antigens from *Trypanosoma cruzi* contain immunodominant but enzymatically inactive domains. *FASEB J* 6: 3259-64.

Cerrone, M.C., Ritter, D.M. & Kuhn, R.E. (1992). Effect of antigen-specific T helper cells or interleukin-2 on suppressive ability of macrophage subsets detected in spleens of *Trypanosoma cruzi*-infected mice as determined by limiting dilution-partition analysis. *Infect Immun* 60: 1489-98.

CHANG, R.H., LIN FENG, M.H. & LAI, M.Z. (1997). Nitric oxide increased interleukin-4 expression in T lymphocytes. *Immunology* 90 : 364-369.

Chuenkova, M. V. & Pereira M. A. (2000). A trypanosomal protein synergizes with the cytokines ciliary neurotrophic factor and leukemia inhibitory factor to prevent apoptosis of neuronal cells. *Mol Biol Cell* 11: 1487-98.

Chuenkova, M.V., Furnari, F.B., Cavenee, W.K. & Pereira, M.A. (2001). *Trypanosoma cruzi* trans-sialidase: a potent and specific survival factor for human Schwann cells by means of phosphatidylinositol 3-kinase/Akt signaling. *Proc Natl Acad Sci U S A* 98: 9936-41.

Cordeiro-da-Silva, A., Lemesre, J-L., Sereno, D. & Ouaissi, A. (2002). Leishmaniasis: from infection to autoimmunity. *Recent Res Devel Immunology* 4:43-52.

Cunha-Neto, E., Gruber, A., Zingales, B., De Messias, I., Stolf, N., Belloti, G., Patarroyo, M.E., Pillegi, F. & Kalil, J. (1995). Autoimmunity in Chagas Disease cardiopathy: biological relevance of a cardiac myosin-specific epitope crossreactive to an immunodominant *Trypanosoma cruzi* antigen. *Proc Natl Acad Sci USA* 92: 3541-5.

Da Silva, A.C., Espinoza, A.G., Taibi, A., Ouaissi, A. & Minoprio, P. (1998). A 24,000 MW *Trypanosoma cruzi* antigen is a B-cell activator. *Immunology* 94: 189-96.

Damle N.K., Klussman K., Linsley P.S. & Aruffo A. (1992). Differential costimulatory effects of adhesion molecules B7, ICAM-1, LFA-3, and VCAM-1 on resting and antigen-primed CD4+ T lymphocytes. *J Immunol* 148: 1989-1992.

D'Imperio Lima, M.R., Eisen, H., Minoprio, P., Joskowicz, M. & Coutinho, A. (1986). Persistence of polyclonal B cell activation with undetectable parasitemia in late stages of experimental Chagas' disease. *J Immunol* 137: 353-6.

D'Imperio Lima, M.R., Joskowicz, M., Coutinho, A., Kipnis, T. & Eisen, H. (1985). Very large and isotypically atypical polyclonal plaque-forming cell responses in mice infected with *Trypanosoma cruzi*. *Eur J Immunol* 15: 201-3.

DINARELLO, C. A. (1993). Modalities for reducing IL-1 activity in disease. *Trends Pharmacol Sci* 14 : 155-159.

DINARELLO, C. A., WOLFF, S. M. (1993). The role of IL-1 in disease. *N Engl J Med* 328 : 106-113.

DING, L. & SHEVACH, E.M. (1992). IL-10 inhibits mitogen-induced T cell proliferation by selectively inhibiting macrophage costimulatory function. *J Immunol* 148: 3133-3139.

Eksi, S., Wassom, D.L. & Powell, M.R. (1996). Host genetics and resistance to acute *Trypanosoma cruzi* infection in mice: profiles and compartmentalization of IL-2-, -4-, -5-, -10-, and IFN-gamma-producing cells. *J Parasitol* 82: 59-65.

el Bouhdidi, A., Truyens, C., Rivera, M.T., Bazin, H. & Carlier, Y. (1994). *Trypanosoma cruzi* infection in mice induces a polyisotypic hypergammaglobulinaemia and parasite-specific response involving high IgG2a concentrations and highly IgG1 antibodies. *Parasite Immunol* 16: 69-76.

El Cheikh, M.C., Hontebeyrie-Joskowicz, M., Coutinho, A. & Minoprio, P. (1992). CD5 B cells. Potential role in the (auto) immune responses to *Trypanosoma cruzi* infection. *Ann N Y Acad Sci* 651: 557-63.

Engman, D.M. & Leon, J.S. (2002). Pathogenesis of Chagas heart disease: role of autoimmunity. *Acta Tropica*. 81: 123-132.

Fedoseyeva E.V., Zhang F., Orr P.L., Levin D., Buncke H.J. & Benichou G. (1999). *De novo* autoimmunity to cardiac myosin after hearth transplantation and its contribution to the rejection process. *J Immunol* 162: 6836-42.

Fernandez-Gomez, R., Esteban, S., Gomez-Corvera, R., Zoulika, K. & Ouaissi, A. (1998). *Trypanosoma cruzi*: Tc52 released protein-induced increased expression of nitric oxide synthase and nitric oxide production by macrophages. *J Immunol* 160: 3471-9.

FERREIRA, S.H. (1993). The role of interleukins and nitric oxide in the mediation of inflammatory pain and its control by peripheral analgesics. *Drugs* 46 : 1-9.

FIORENTINO, D.F., ZLOTNIK, A., VIEIRA, P. et al (1991). IL-10 acts on the antigen-presenting cell to inhibit cytokine production by Th1. *J Immunol* 146: 3444-51.

Frash, A.C.C. (1994). Trans-sialidase, SAPA amino acid repeats and the relation between *Trypanosoma cruzi* and the mammalian host. *Parasitology* 108: S37-S44.

Frasch, A.C.C. (2000). Functional diversity in the trans-sialidase and mucin families in *Trypanosoma cruzi*. *Parasitol Today* 16 : 282-286.

Freeman G.J., Borriello F., Hodes R.J., Reiser H., Gribben J.G., Ng J.W., Kim J., Goldberg J.M., Hathcock K., Laszlo G., Lombard L.A., Wang S., Gray G.S., Nadler L.M. & Sharpe A.H. (1993). Murine B7-2, an alternative CTLA4 counter-receptor that costimulates T cell proliferation and interleukin 2 production. *J Exp Med* 178: 2185-2192.

Freire-de-Lima, C.G., Nunes, M.P., Corte-Real, S., Soares, M.P., Previato, J. O., Mendonca-Previato, L. & Dos Reis, G.A. (1998). Proapoptotic activity of a *Trypanosoma cruzi* ceramide-containing glycolipid turned on in host macrophages by IFN-gamma. *J Immunol* 161: 4909-16.

Freire-de-Lima, C.G., Nascimento D.O., Soares, M.B, Bozza, P.T, Castro-Faria-Neto, H.C, de Mello, F.G, Dos Reis, G.A. & Lopes, M.F. (2000). Uptake of apoptotic cells drives the growth of a pathogenic trypanosome in macrophages. *Nature* 403: 199-203.

Frosch, S., Kuntzlin, D. & Fleischer, B. (1997). Infection with *Trypanosoma cruzi* selectively upregulates B7-2 molecules on macrophages and enhances their costimulatory activity. *Infect Immun* 65: 971-7.

Gao, W., Wortis, H.H. & Pereira, M.A. (2002). The *Trypanosoma cruzi* trans-sialidase is a T cell-independent B cell mitogen and an inducer of non-specific Ig secretion. *Int Immunol* 14: 299-308.

GARZON, E., COUTINHO BORGES, M., CORDEIRO DA SILVA, A., NACIFE, V., MEIRELLES, NAZARETH, M., GUILVARD, E., BOSSENO, M.F., GUEVARA, A., BRENIERE, S.F. & OUAISSI, A. (2003) *Trypanosoma cruzi* carrying a targeted deletion of a *Tc52* protein-encoding allele elicits attenuated Chagas' disease in mice. *Immunol Lett* 89 : 67-80.

GAZZINELLI, R.I., MARATO, M.J., NUNES, R.M.B., CANCADO, J.R., BRENER, Z. & GAZZINELLI, G. (1988). Idiotype stimulation of T lymphocytes from Trypanosoma cruzi-infected patients. *J Immunol* 140: 3167-72.

Goldenberg, S., Cordeiro, M.N., Silva-Pereira, A.A. & Mares-Guida, L. (1983). Release of lipopolysaccharide (LPS) from cell surface of *Trypanosoma cruzi* by EDTA. *Int J Parasitol* 13:11-17.

GONZALEZ CAPPA, S.M., BRONZINA, A., KATZIN, A.M., GOLFERA, H., De MARTINI, G.W. & SEGURA, E. (1981). Antigens of subcellular fractions of T. cruzi. III. Humoral immune response and histopathology of immunized mice. *J Protozool* 27: 467-474.

HAYASHI, S., CHAN, C., GAZZINELI, R., ROBERGE, F. (1996). Contribution of nitric oxide to the host parasite equilibrium in Toxoplasmosis. *J Immunol* 156: 1476-1481.

Heussler, V.T., Kuenzi, P. & Rottenberg, S. (2001). Inhibition of apoptosis by intracellular protozoan parasites. *Int J Parasitol* 31: 1166-76.

HILL, J.R., CORBETT, J.A., KWON, G. et al (1996). Nitric oxide regulates interleukin 1 bioactivity released from murine macrophages. *J Biol Chem* 271 : 22672-22678.

Hölscher, C., Köhler, G., Müller, U., Mossmann, H., Schaub, G.A. & Brombacher, F. (1998). Defective nitric oxide effector functions lead to extreme susceptibility of *Trypanosoma cruzi*-infected mice deficient in gamma interferon receptor or inducible nitric oxide synthase. *Inf Immun* 66: 1208-15.

JAMES, S. L. (1995). Role of nitric oxide in parasitic infections. *Microbiol Rev* 59 : 533-547.

Kalil, J. & Cunha-Neto E. (1996). Autoimmunity in Chagas Disease cardiomyopathy: Fulfilling the criteria at last? *Parasitol Today* 12: 396-398.

KHOURY, E.L., RITACCO, V., COSSIO, P.M., LAGUENS, R.P., SZARFMAN, A., DIEZ, C., ARANA, R.M. (1979). Circulating antibodies to peripheral nerve in American trypanosomiasis (Chagas' disease). *Clin Exp Immunol* 36 :8-15.

Kierszenbaum F. (1980). Protection of congenitally athymic mice against *Trypanosoma cruzi* infection by passive antibody transfer. *J Parasitol* 66: 673-5.

Kierszenbaum F. (1999). Chagas disease and the autoimmunity hypothesis. *Clin Microbiol Rev* 12: 210-23.

Kierszenbaum F., Knecht E., Budzko D.B. & Pizzimenti M.C. (1974). Phagocytosis: a defense mechanism against infection with *Trypanosoma cruzi. J Immunol* 112: 1839-44.

Kierszenbaum, F. & Howard, J.G. (1976). Mechanisms of resistance against experimental *Trypanosoma cruzi* infection: the importance of antibodies and antibody-forming capacity in the Biozzi high and low responder mice. *J Immunol* 116: 1208-11.

Kierszenbaum, F., Cuna, W.R., Belrz, L.A. & Sztein, M.B. (1990). A secretion product(s) of *Trypanosoma cruzi* that inhibits proliferation and IL-2 receptor expression by activated human peripheral blood mononuclear cells. *J Immunol* 144: 4000-4004.

Kierszenbaum, F., Majumder, S., Paredes, P., Tanner, M.K. & Sztein, M.B. (1998). The *Trypanosoma cruzi* immunosuppressive factor (TIF) targets a lymphocyte activation event subsequent to increased intracellular calcium ion concentration and translocation of protein kinase C but previous to cyclin D2 and cdk4 mRNA accumulation. *Molec Biochem Parasitol* 92: 133-145.

Kierszenbaum, F. & Pienlowsky, M.M. (1979). Thymus dependent control of host defense mechanisms against *Trypanosoma cruzi* infection. *Infect Immun* 24: 117-20.

Krettli, A.U. & Brener, Z. (1976). Protective effects of specific antibodies in *Trypanosoma cruzi* infections. *J Immunol* 116: 755-60.

Krettli, A.U. & Brener, Z. (1982). Resistance against *Trypanosoma cruzi* associated to anti-living trypomastigote antibodies. *J Immunol* 128: 2009-12.

Kumar, S. & Tarleton, R.L. (2001). Antigen-specific Th1 but not Th2 cells provide protection from lethal *Trypanosoma cruzi* infection in mice. *J Immunol* 166: 4596-603.

Laucella, S., Salcedo, R., Castanos-Velez, E., Riarte, A., De Titto, E.H., Patarroyo, M., Orn, A. & Rottenberg, M.E. (1996). Increased expression and secretion of ICAM-1 during experimental infection with *Trypanosoma cruzi. Parasite Immunol* 18: 227-239.

Leguizamon, M.S., Mocetti, E., Garcia Rivello, H., Argibay, P. & Campetella, O. (1999). *Trans*-sialidase from *Trypanosoma cruzi* induces apoptosis in cells from the immune system *in vivo. J Infect Dis* 180: 1398-402.

Leon, J.S. & Engman, D.M. (2001). Autoimmunity in Chagas heart disease. *Internat J Parasitol* 31:555-561.

Liew, F.Y., Schmidt, J.A., Liu, D.S., Millott, S.M., Scott, M.T., Dhaliwal, J.S. & Croft, S.L. (1988). Suppressive substance produced by T cells from mice chronically infected by *Trypanosoma cruzi* II. Partial Biochemical Characterization. *J Immunol* 140:969-973.

Lopes, M.F., da Veiga, V.F., Santos, A.R., Fonseca, M.E. & Dos Reis, G.A. (1995). Activation-induced CD4+ T cell death by apoptosis in experimental Chagas' disease. *J Immunol* 154: 744-52.

Lopes, M.F., Nunes M.P., Henriques-Pons, A., Giese, N., Morse, H.C., Davidson, W.F., Araujo-Jorge, T.C. & Dos Reis, G.A. (1999). Increased susceptibility of Fas ligand-deficient gld mice to *Trypanosoma cruzi* infection due to a Th2-biased host immune response. *Eur J Immunol* 29: 81-9.

Lopez Bergami, P., Scaglione, J. & Levin, M.J. (2001). Antibodies against the carboxy-terminal end of *Trypanosoma cruzi* ribosomal P proteins are pathogenic. *FASEB J* 15: 2602-12.

Martins, G.A., Cardoso M.A., Aliberti, J.C & Silva, J.S. (1998). Nitric oxide-induced apoptotic cell death in the acute phase of *Trypanosoma cruzi* infection in mice. *Immunol Lett* 63: 113-20.

Millar, A.E. & Kahn, S.J. (2000). The SA85-1.1 protein of the *Trypanosoma cruzi* trans-sialidase superfamily is a dominant T-cell antigen. *Inf Immun* 68: 3574-80.

Millar, A.E., Wleklinski-Lee, M & Kahn, S.J. (1999). The surface protein superfamily of *Trypanosoma cruzi* stimulates a polarized Th1 response that becomes anergic. *J Immunol* 162: 6092-9.

Minoprio, P.M., Coutinho, A., Joskowicz, M., D'Imperio Lima, M.R. & Eisen, H. (1986b). Polyclonal lymphocyte responses to murine *Trypanosoma cruzi* infection. II. Cytotoxic T lymphocytes. *Scand J Immunol* 24: 669-679.

Minoprio, P.M., Eisen, H., Forni, L., D'Imperio Lima, M.R., Joskowicz, M. & Coutinho, A. (1986a). Polyclonal lymphocyte responses to murine *Trypanosoma cruzi* infection. I. Quantification of both T- and B-cell responses. *Scand J Immunol* 24: 661-668.

Motrán, C.C., Cerbán, F.M., Rivarola, W., Iosa, D. & Vottero de Cima, E. (1998). *Trypanosoma cruzi*: immune responses and functional heart damage induced in mice by the main linear B-cell epitope of parasite ribosomal proteins. *Exp Parasitol* 88: 223-230.

Mucci, J., Hidalgo, A., Mocetti, E., Argibay, P.F., Leguizamon, M.S. & Campetella, O. (2002). Thymocyte depletion in *Trypanosoma cruzi* infection is mediated by trans-sialidase-induced apoptosis in nurse cell complex. *Proc Natl Acad Sci U S A* 99 : 3896-901.

Mueller D.L., Jenkins M.K. & Schwartz R.H. (1989). Clonal expansion versus functional clonal inactivation: a costimulatory signalling pathway determines the outcome of T cell antigen receptor occupancy. *Annu Rev Immunol* 7: 445-480.

MUNOZ-FERNANDEZ, M., FERNANDEZ, M.A., FRESNO, M. (1992a). Activation of human macrophages for the killing of intracellular *Trypanosoma cruzi* by TNF-□ and IFN-□ through a nitric oxide-dependent mechanism. *Immunol Lett* 33 : 35-40.

Muñoz-Fernández, M.A., Fernández, M.A. & Fresno (1992). Synergism between tumor necrosis factor-alpha and interferon- gamma on macrophage activation for the killing of intracellular *Trypanosoma cruzi* through a nitric oxide-dependent mechanism. *Eur J Immunol* 22: 301-7.

Murray, J.S. (1998). How the MHC selects Th1/Th2 immunity. *Immunol Today* 19: 157-162.

Nickell, S.P., Stryker, G.A. & Arevalo, C. (1993). Isolation from *Trypanosoma cruzi*-infected mice of CD8+, MHC-restricted cytotoxic T cells that lyse parasite-infected target cells. *J Immunol* 150: 1446-57.

Nogueira, N. & Cohn, Z.A. (1978). *Trypanosoma cruzi*: *in vitro* induction of macrophage microbicidal activity. *J Exp Med* 148: 288-300.

Ortiz-Ortiz, L., Parks, D.E., Rodriguez, M. & Weigle, W.O. (1980). Polyclonal B lymphocyte activation during *Trypanosoma cruzi* infection. *J Immunol* 124: 121-6.

OUAISSI, A., CORNETTE, J., VELGE, P. & CAPRON, A. (1988). Identification of anti-acetylcholinesterase and anti-idiotype antibodies in human and experimental Chagas'disease : pathological implications. *Eur J Immunol* 18 : 1889-1894.

Ouaissi MA. (1993). *Trypanosoma cruzi*-maladie de Chagas: biologie de l'agent pathogène et immunité. –Editions Techniques- *Encycl Med Chir* (Paris, France). Maladies infectieuses, 8-505-A-10. 7 p.

Ouaissi, A., Cordeiro da Silva, A., Guevara, A.G., Borges, M. & Guilvard, E. (2001). *Trypanosoma cruzi*-induced host immune dysfunction: a rationale for parasite immunosuppressive factor(s) encoding gene targeting. *J Biomed Biotech* 1: 11-17.

Ouaissi, A., Guevara-Espinoza, A., Chabe, F., Gomez-Corvera, R. & Taibi, A. (1995). A novel and basic mechanism of immunosuppression in Chagas' disease: *Trypanosoma cruzi* releases *in vitro* and *in vivo* a protein which induces T cell unresponsiveness through specific interaction with cysteine and glutathione. *Immunol Lett* 48: 221-4.

Ouaissi, M., Dubremetz, J.F., Schoneck, R., Fernandez-Gomez, R., Gomez-Corvera, R., Billaut-Mulot, O., Taibi, A., Loyens, M., Tartar, A., Sergheraert, C. & Kusnierz, J.P. (1995). *Trypanosoma cruzi*: a 52- kDa protein sharing sequence homology with glutathione S-transferase is localized in parasite organelles morphological resembling reservosomes. *Exp Parasitol* 81: 453-461.

OUAISSI, A., OUAISSI, M., TAVARES, J. & CORDEIRO DA SILVA, A. (2004). Host cell phenotypic variability induced by Trypanosomatid parasite released immunomodulatory factors: physiopathological implications. *J Biomed Biotech* 3: 167-174.

Overtvelt, L.V., Andrieu, M., Verhasselt, V., Connan, F., Vercruysse, V., Goldman, M., Hosmalin, A. & Vray, B. (2002). *Trypanosoma cruzi* down-regulates lipopolysaccharide-induced MHC chass I on human dendritic cells and impairs antigen presentation to specific CD8(+) T lymphocytes. *Int Immunol* 14: 1135-1144.

Pestel, J., Defoort, J.P., Gras-Masse, H., Afchain, D., Capron, A, Tartar, A. & Ouaissi, A. (1992). Polyclonal cell activity of a repeat peptide derived from the sequence of an 85-kilodalton surface protein of *Trypanosoma cruzi* trypomastigotes. *Infect Immun* 60: 715-9.

PETRAY, P., ROTTENBERG, M.E., GRINSTEIN, S. et al. (1994). Release of nitric oxide during the experimental infection with *Trypanosoma cruzi*. *Parasite Immunol* 16 : 193-199.

Petray, P., Corral, R., Meckert, P. & Laguens, R. (2002). Role of macrophage inflammatory protein 1alpha (MIP-1alpha) in macrophage homing in the spleen and heart pathology during experimental infection with *Trypanosoma cruzi*. *Acta Trop* 83: 205-11.

Pinge-Filho, P., Tadokoro, C.E. & Abrahamsohn, I.A. (1999). Prostaglandins mediate suppression of lymphocyte proliferation and cytokine synthesis in acute *Trypanosoma cruzi* infection. *Cell Immunol* 193: 90-8.

Reina-San-Martin, B., Cosson, A. & Minoprio P. (2000). Lymphocyte polyclonal activation: A pitfall for vaccine design against infectious agents. *Parasitol Today* 16: 62-67.

Reina-San-Martin, B., Degrave W., Rougeot C., Cosson A., Chamond N., Cordeiro-da-Silva A., Arala-Chaves M., Coutinho A. & Minoprio P. (2000). A B-cell mitogen from a pathogenic trypanosome is a eukaryotic proline racemase. *Nature Medicine* 6: 890-897.

Requena, J.M., Alonso, C. & Soto, M. (2000). Evolutionary conserved proteins as prominent immunogens during *Leishmania* infections. *Parasitol* Today 16: 246-50.

RIVERA-VANDERPAS, M.T., RODRIGUEZ, A.M., AFCHAIN, D., BAZIN, H., CAPRON, A. (1983). *Trypanosoma cruzi*: variation in susceptibility of inbred strains of rats. *Acta Trop* 40:5-10.

RODRIGUEZ, A.M., SANTORO, F., AFCHAIN, D., BAZIN, H., CAPRON, A. (1981). *Trypanosoma cruzi* infection in B-cell-deficient rats. *Infect Immun* 31:524-9.

Rodriques Coura, J. & de Castro, S.L. (2002). A critical review on Chagas disease chemotherapy. *Mem Inst Oswaldo Cruz* 97: 3-24.

Roitt I, Brostoff J & Male D. (1985a). Immunity to protozoa and Worms. *In* Immunology, Gower medical Publishing Ltd., pp 18.4.

Rottenberg M., Cardoni R.L., Andersson R., Segura E.L. & Orn A. (1988). Role of T helper/ inducer cells as well as natural killer cells in resistance to *Trypanosoma cruzi* infection. *Scand J Immunol* 28: 573-82.

Rottenberg, M.E., Bakhiet, M., Olsson, T., Kristensson, K., Mak, T., Wigzell, H. & Orn, A. (1993). Differential susceptibilities of mice genomically deleted of CD4 and CD8 to infections with *Trypanosoma cruzi* or *Trypanosoma brucei*. *Infect Immun* 61: 5129-33.

Rottenberg, M.E., Rodriguez, D.A. & Orn, A. (1992). Control of *Trypanosoma cruzi* infection in mice deprived of T-cell help. *Scand J Immunol* 36: 261-8.

Santos Lima, E.C. & Minoprio, P. (1996). Chagas'disease is attenuated in mice lacking γδ T cells. *Inf Immun* 64: 215-221.

Schmunis, G.A., Cappa, S.M., Traversa, O.C. & Janovsky, J.F. (1971). The effect of immuno-depression due to neonatal thymectomy on infections with *Trypanosoma* cruzi in mice. *Trans R Soc Trop Med Hyg* 65: 89-94.

Scott, M.T. (1981). The nature of immunity against *Trypanosoma cruzi* in mice recovered from acute infection. *Parasite Immunol* 3: 209-18.

SCOTT, M.T., SNARY, D. (1979). Protective immunisation of mice using cell surface glycoprotein from *Trypanosoma cruzi*. *Nature* 282 :73-4.

SEGURA, E.L., BUA, J., ROSENSTEIN de CAMPANINI, A., SUBIAS, E., ESTEVA, M., MORENO, M., RUIZ, A.M. (1986). Monoclonal antibodies against the flagellar fraction of epimastigotes of *Trypanosoma cruzi*: complement-mediated lytic activity against trypomastigotes and passive immunoprotection in mice. *Immunol Lett* 13:165-71.

Serrano, L.E. & O'Daly, J.A. (1990). Splenocyte membrane changes and immunosuppression during infection and reinfection with *Trypanosoma cruzi*. *Invest Clin* 31: 17-31.

Sher, A. & Coffman, R.L. (1992) Regulation of immunity to parasites by T cells and T cell-derived cytokines. *Annu Rev Immunol* 10: 385-409.

Skeiky, Y.A., Benson, D.R., Parsons, M., Elkon, K.B. & Reed, S.G. (1992). Cloning and expression of *Trypanosoma cruzi* ribosomal protein P0 and epitope analysis of anti-P0 autoantibodies in Chagas'disease patients. *J Exp Med* 176: 201-211.

SOONG, L. & TARLETON, R.L. (1994). *Trypanosoma cruzi* infection suppresses nuclear factors that bind to specific sites on the interleukin-2 enhancer. *Eur J Immunol* 24 : 16-23.

Sztein, M.B. & Kierszenbaum, F. (1993). Mechanisms of development of immunosupression during *Trypanosoma* infections. *Parasitol Today* 9: 424-32.

TAKEHARA, H.A., PERINI, A., da SILVA, M.H., MOTA, I. (1981). *Trypanosoma cruzi*: role of different antibody classes in protection against infection in the mouse. *Exp Parasitol* 52:137-46.

Takle, G.B. & Snary, D. South American trypanosomiasis (Chagas´disease). In: Kenneth S. Warren, ed. Immunology of Parasitic Infections. Blackwell Scientific Publications; 1993:213.

Talvani, A., Machado, F.S., Santana, G.C., Klein, A., Barcelos, L., Silva, J.S. & Teixeira, M.M. (2002). Leukotrine B (4) induces nitric oxide synthesis in *Trypanosoma cruzi*-infected murine macrophages and mediates resistance to infection. *Infect Immun* 70: 4247-53.

Tarleton, R.L. (1991). The role of T-cell subpopulations in experimental Chagas' disease. *Res Immunol* 142: 130-34.

Tarleton, R.L. (1988). *Trypanosoma cruzi*-induced suppression of IL-2 production. I. Evidence for the presence of IL-2-producing cells. *J. Immunol.* 140: 2763-8.

Tarleton, R.L. (1990). Depletion of CD8+ T cells increases susceptibility and reverses vaccine-induced immunity in mice infected with *Trypanosoma cruzi*. *J Immunol* 144: 717-24.

Tarleton, R.L., Grusby, M.J., Postan, M. & Glimcher, L.H. (1996). *Trypanosoma cruzi* infection in MHC-deficient mice: further evidence for the role of both class I- and class II-restricted T cells in immune resistance and disease. *Int Immunol* 8: 13-22.

Tarleton, R.L., Koller, B.H., Latour, A. & Postan, M. (1992). Susceptibility of beta 2-microglobulin-deficient mice to *Trypanosoma cruzi* infection. *Nature* 356: 338-40.

Tarleton, R.L., Sun, J., Zhang, L. & Postan, M. (1994). Depletion of T-cell subpopulations results in exacerbation of myocarditis and parasitism in experimental Chagas' disease. *Infect Immun* 62: 1820-9.

Todeschini, A.R., Nunes, M.P., Pires, R.S., Lopes, M.F., Previato, J.O., Mendonca-Previato, L. & Dos Reis, G.A. (2002). Costimulation of host T lymphocytes by a trypanosomal trans-sialidase: involvement of CD43 sinaling. *J Immunol* 168: 5192-8.

Todeschini, A.R., Nunes, M.P., Previato, J.O., Mendonça-Previato, L. & Dos Reis, G.A. (2000). *Trans*-sialidase from *Trypanosoma cruzi* fully activates host T cell and dendritic cells. *Mem Inst Oswaldo Cruz* 95(Suppl. 2): 194.

Townsend, M.J. & McKenzie, A.N. (2000). Unravelling the net? Cytokines and diseases *J Cell Sci* 113 : 3549-50.

Trischmann T.M. (1983). Non- antibody-mediated control of parasitemia in acute experimental Chagas' disease. *J Immunol* 130: 1953-7.

Van Overtvelt, L., Vanderheyde, N., Verhasselt, V., Ismaili, J., De Vos, L., Goldman, M., Willems, F. & Vray, B. (1999). *Trypanosoma cruzi* infects human dendritic cells and prevents their maturation: inhibition of cytokines, HLA-DR, and costimulatory molecules. *Infect Immun* 67: 4033-40.

WHO-World Health Organization (2002). Control of Chagas Disease. Second Report of the WHO Expert Committee, Geneve. Technical report Series, No. 905.

Wizel, B., Nunes, M. & Tarleton, R.L. (1997). Identification of *Trypanosoma cruzi* transsialidase family members as targets of protective CD8+TC1 responses. *J Immunol* 159: 6120-6130.

Zhang, J., Andrade, Z. A., Yu, Z. X., Andrade, S. G., Takeda, K., Sadirgursky, M. & Ferrans, V. J. (1999). Apoptosis in a canine model of acute Chagasic myocarditis. *J Mol Cell Cardiol* 31: 581-96.

Zhang, L. & Tarleton, R.L. (1996). Persistent production of inflammatory and anti-inflammatory cytokines and associated MHC and adhesion molecule expression at the site of infection and disease in experimental *Trypanosoma cruzi* infections. *Exp Parasitol* 84: 203-13.

Zweerink, H.J., Weston, H.D., Andersen, O.F., Garber, S.S. & Hayes, E.C. (1984). Immunity against infection with *Trypanosoma cruzi* in mice correlates with presence of antibodies against three trypomastigote polypeptides. *Infect Immun* 46:826-30.

In: Progress in Immunology Research
Editor: Barbara A. Veskler, pp. 231-247

ISBN 1-59454-380-1
©2005 Nova Science Publishers, Inc.

Chapter X

Activation of Peritoneal Macrophage and B Cell Subsets by Porin of *Shigella dysenteriae* Type 1 for Mucosal Immune Response

Tapas Biswas[*] and Avijit Ray

Division of Immunology and Vaccine Development, National Institute of Cholera and Enteric Diseases, Kolkata-700 010, India

Abstract

Porin was purified to homogeneity from *Shigella dysenteriae* type 1. The protein formed hydrophilic diffusion pores by incorporation into artificial liposome vesicles and exhibited significant porin activity. The molecular weight of the native porin molecule was 130,000, consisting of 38,000 monomer. Murine anti-porin antibody raised against the purified porin reacted with whole cell preparation of *S. dysenteriae* type 1 suggesting that porin possessed surface component. Porin could also be visualized on the bacterial surface by immunoelectron microscopy. The anti-porin antibody of *S. dysenteriae* cross-reacted with porin preparations of *S. flexneri, S. boydii, and S. sonnei,* indicating that porins are antigenically related among *Shigella* species. Porins are of particular interest because they have been characterized as potent adjuvants and have great potential as a novel component of vaccines. Mouse peritoneal cavity (PerC) macrophages (MΦ) treated with the protein showed an increase of mRNA for Toll-like receptor (TLR)2, TLR6 and myeloid differentiation factor 88 (MyD88), an effector molecule associated with TLR-mediated response, indicating that TLR2 and TLR6 in combination is essential for recognition of porin. The PerC MΦ strongly expressed the mRNA for NF-κB and selectively up-regulated CD80 in presence of porin. The cell-surface expression of CD80 was augmented by gamma interferon (IFN-γ) supporting selective regulation of B7-1

[*] Division of Immunology and Vaccine Development, National Institute of Cholera and Enteric Diseases, P-33, C.I.T. Road, Scheme-XM, Kolkata-700 010, West Bengal, India. E-mail: shigellaporin@yahoo.com

(CD80) member of B7 family. Porin induced MΦ to release nitric oxide, interleukin (IL)-1 and IL-12. The IL-12 release could be increased profoundly in combination with IFN-γ. Porin mediated induction of IL-12 release would therefore influence Th1-type response, known to be preferentially triggered due to up-regulation of CD80 expression. Furthermore, porin increased the mRNA levels for TLR2 and TLR6 of PerC B-1a (CD5$^+$), B-1b (CD11b$^+$) and conventional B (B-2) cells (CD23$^+$), implicating that coexpression of TLR2 and TLR6 is essential as a combinatorial repertoire for recognition of porin by the B cell populations. Among the two key TLRs, TLR2 and TLR4, which are responsible for recognizing most of the bacterial products but unmethylated CpG DNA motifs, TLR2 and not TLR4, participates in porin recognition. As found with PerC MΦ, TLR2 was increased on all the B cell populations whereas the TLR4 expression remained unaffected. Besides TLRs, mRNA for MyD88 got increased that suggests of its involvement in the activity of porin. Both of the B-1 cell populations strongly expressed the mRNA for NF-κB in presence of porin, conforming to the earlier finding that coexpression of TLR2 and TLR6, resulted in robust NF-κB activation for signaling. However, the B-2 cell showed moderate NF-κB expression in comparison to the B-1 cell "sister" populations. The B-1 cell populations selectively up-regulated the expression of the costimulatory molecules in response to porin. CD80 expression got enhanced on B-1a cell whereas CD86 was solely expressed on B-1b cell. B-1 cell "sister" populations might have evolved the mechanism of signaling via a specific costimulatory molecule so that these cells can divide its work load of responding to an antigen. In contrast to the B-1 cells, B-2 cell selectively expressed CD86 on its surface. Porin induced cell-surface IgM and IgA on both PerC B-1 and B-2 cell subsets and preferentially IgG$_{2a}$ on B-2 cell only. The porin mediated induction of IgA on both B-1 and B-2, and IgG$_{2a}$ on B-2 cell could be augmented by IL-6. The IgA expressed on the "sister" populations, B-1a and B-1b cell-surfaces, was found to bind to the 38 kDa monomer of porin confirming it to be anti-porin IgA antibody, the signature molecule of mucosal immune response. Understanding the mechanism of adjuvanticity of porin of *S. dysenteriae* type 1 is a necessary step towards the development of a better adjuvant against shigellosis.

Introduction

Shigellosis

Shigellosis (bacillary dysentery) still remains one of the main enteric infections of the world. The number of cases has been estimated to be some 200 million per year, mostly among infants and children up to 5 years of age. The majority of the cases are caused by members of the *Shigella* genus, a group of nonmotile, gram-negative, enteric bacteria, comprising of four species, *S. dysenteriae, S. flexneri, S. boydii and S. sonnei,* each divided into 12, 13, 18, and 1 serotypes, respectively [1]. Shigellosis caused by *Shigella dysenteriae* type 1 is one of the predominant causes of infant mortality in developing countries [2]. The disease is caused in humans [3, 4] by a multistep invasion process. Following infection, the organisms invade colonic epithelial cells [5], multiply [6], cause cell death [7], and spread intra and intercellularly [8], resulting in extensive degeneration of the epithelium [9]. The clinical management of shigellosis has become difficult due to the rapid emergence of drug-resistant strains of *Shigella* spp. [10].

Porin

The outer membrane of gram-negative bacteria harbors some major proteins that constitute about 60% of the bacterial surface proteins called porins [11] because of their pore-forming ability as trimers and role in cellular permeability [12, 13]. Porins form large, open, water-filled, voltage-gated channels that non-specifically mediate the passive diffusion of ions and small, hydrophilic molecules up to an exclusion limit of about 600 Da. In addition to their pore-forming activity, bacterial porins serve as receptors for bacteriophages and bacteriocins, and in the case of pathogenic bacteria, they also appear to be the targets of the immune system [14-16]. Porins are of particular interest because they have been characterized as potent adjuvants and have great potential as a novel component of vaccines [17, 18]. These proteins are immunogenic without the addition of exogenous adjuvants [19] and known to be able to augment the humoral response to otherwise poorly immunogenic substances, for example, polysaccharides and peptides [20, 21]. In our laboratory, porin was purified to homogeneity from the outer membrane of *S. dysenteriae* type 1 [22]. The protein formed hydrophilic diffusion pores by incorporation into artificial liposome vesicles and exhibited significant porin activity. The molecular weight of the native porin molecule was 130,000, consisting of 38,000 monomer. Murine anti-porin antibody raised against the purified porin reacted with whole cell preparation of *S. dysenteriae* type 1 suggesting that porin possessed surface component. Porin could also be visualized on the bacterial surface by immunoelectron microscopy. The anti-porin antibody of *S. dysenteriae* cross-reacted with porin preparations of *S. flexneri, S. boydii, and S. sonnei,* indicating that porins are antigenically related among *Shigella* species. We also found that the purified porin monomer of 38,000 of *S. dysenteriae* type 1 was recognized by convalescent sera of shigellosis patients, indicating that the protein is a target outer membrane antigen of significance to which antibodies are naturally tailored and sustained by human beings [23], the only natural host of *Shigella* spp. [3, 4]. These immunoregulatory properties of porin of *S. dysenteriae* type 1 [22] and its specific and strong recognition by convalescent sera of patients make the protein attractive as an adjuvant for use in vaccine formulation against shigellosis [24]. In order to study the adjuvanticity of porin, its recognition, involvement in activation and induction of response of mouse peritoneal cavity (PerC) macrophages (MΦ) and B cell subsets, B-1 and B-2, was examined.

MΦ, B-1a, B-1b and B-2 Cells

MΦ

MΦ are a critical component of the innate and adaptive immune response to bacterial pathogen. MΦ can be purified by attachment to plastic Petri dishes and they express CD11b (Mac1) on their surface. They internalize and degrade bacteria and induce inflammatory responses. In addition to their effector function, MΦ also contribute to the host response by producing chemokines and cytokines, and by antigen presentation to the activated T cells. During pathogenesis of the bacteria, pattern recognition receptors such as Toll-like receptors induce a variety of inflammatory responses in the MΦ [25].

B Cell Subsets

Mice and human have phenotypically distinct populations of B cells, termed B-1 and B-2, that have been proposed to represent entirely separate B cell lineages [26, 27]. The B-1 cells originally defined by the surface expression of CD5 and high levels of IgM, arise early during ontogeny, home predominantly to the peritoneal and pleural cavities, have a capacity for self-renewal and display different receptor specificities [27, 28]. B-1 cells are further divided into B-1a that expresses CD5 and B-1b "sister" cell population that are CD5⁻ and express CD11b [29]. These cells can take important part in innate immunity by secreting large amounts of natural antibodies of the IgM class, which can be produced without exposure to any environmental antigens or immunization [30]. The conventional B cells are the B-2 cells on which CD23 (FcεR) are present specifically. These conventional CD23⁺ B-2 cells are predominantly found in the spleen and are also present in the PerC. Thus, in the PerC, these markers alone can be used to distinguish conventional B cells from B-1 cells, i. e. B-2 cells are CD23⁺ and B-1 cells are either CD5⁺ (B-1a) or CD11b⁺ (B-1b) [29]. B-1a cells, that constitute 30% of total mouse PerC B cells [31], exhibit a number of properties that suggests an important role for them in mucosal immunity and defense against bacterial pathogens, particularly in the PerC [32]. T cell-dependent immune responses generally involve conventional B-2 cells. By contrast, B-1 cells appear to produce antibodies in a T-independent manner [30].

Study of Adjuvanticity of Porin

Expression of Toll-Like Receptors and Myeloid Differentiation Factor 88

Toll was first identified as an essential molecule for embryonic patterning in *Drosophila* and was subsequently shown to be key in antifungal immunity [33]. A homologous family of Toll receptors, the so-called Toll-like receptors (TLRs), exists in mammals [34]. Members of the TLR family have been demonstrated to participate specifically in the process by which antigen presenting cells (APCs) recognize pathogen-associated molecular patterns that distinguish the infectious agents from self [35, 36]. Each TLR is a type 1 transmembrane protein, that is evolutionarily conserved between insects and humans [37]. Ten members of the TLR family have been reported that can be found in human and mouse (TLRs 1-10). The extracellular regions of each TLR contain leucine-rich domain that is thought to participate in ligand recognition, and an intracellular cytoplasmic tail that contains a conserved region called the Toll-IL-1R, or TIR domain, having a similarity with IL-1 receptors (IL-1R) [38], that, upon activation, generally results in recruiting of myeloid differentiation factor 88 (MyD88), an adaptor protein, that interacts with the TLRs through its own C–terminal TIR domain [39]. TLR family members are expressed differentially among immune cells [40] and appear to respond to different stimuli. TLR expression is observed in a variety of other cells, including vascular endothelial cells, adipocytes, cardiac myocytes and intestinal epithelial cells. The expression of the various TLRs is also modulated in response to a variety of stimuli. Porin induced enhancement of TLR and MyD88 expression of PerC MΦ, B-1 and B-2 cell subsets revealed how the adjuvant is distinguished specifically by the cells of the mucosal immune system.

Porin Induced Toll-Like Receptors and MyD88 Expression of MΦ

RT-PCR was conducted using specific primers of TLRs and MyD88 with total RNA isolated from PerC MΦ after incubation for 24 h with or without porin. Cells grown in presence of porin showed an increase in the mRNA levels of constitutively expressed TLR2 and TLR6 (Fig. 1). The mRNA expression of TLR1 and TLR4 remained unaffected with porin treatment. The enhancement of TLR2 and TLR6 mRNA levels indicated that the combination of the two TLRs is essential for MΦ to respond to porin of *S. dysenteriae* type 1. Besides TLR2 and TLR6, the mRNA for MyD88, an effector molecule associated with TLR-mediated response in immune cells, was enhanced [23]. This suggests that the presence of the effector molecule MyD88 is involved in the activity of the porin.

The mammalian TLRs play an important role in innate immunity [41] that includes B7 up-regulation [42]. It has been postulated that TLRs are involved in regulating the effect of immune adjuvants [43]. The role of unmethylated bacterial CpG DNA motifs seemed to have immunopotentiating effect through TLR9 [44].

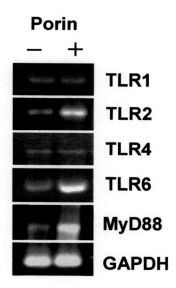

Figure1. Effect of porin on TLR and MyD88 mRNA expression. MΦ were treated with and without 1 μg porin per 1x10⁵ cells for 24 h. Total RNA was extracted from the MΦ after 24 h of culture and subjected to RT-PCR. Ethidium bromide-stained PCR products were photographed, and then images were digitized and analyzed. PCR products were quantified and expressed as the ratio of each product to GAPDH band density.

Porin Induced Toll-Like Receptors and MyD88 Expression of B-1 and B-2 Cell Subsets

RT-PCR was conducted using specific primers of TLRs and MyD88 with total RNA isolated from PerC B-1a, B-1b and B-2 cells after incubation for 72 h with and without porin. The B cell populations grown in the presence of porin showed an increase in TLR2 mRNA expression compared to the untreated cells. Besides the expression of mRNA for TLR2, both B-1 (B-1a and B-1b) and B-2 cell subsets expressed the mRNA for TLR6 [45, 46].

Other than TLR2 and TLR6, there was an increase of mRNA expression for MyD88 by the B-1 cell "sister" and B-2 cell populations. The mRNA expressions for TLR1, known to have homology with TLR6, and TLR4, known to recognize lipopolysaccharide (LPS) primarily, remained unaffected with porin treatment. Similar to that found with PerC MΦ, the enhancement of mRNA levels for TLR2 and TLR6 indicates that the coexpression of TLR2 and TLR6 is essential as a combinatorial repertoire for all the PerC B cell subsets to distinguish the porin of *S. dysenteriae* type 1.

Induction of TLR2 on PerC MΦ and B Cell Repertoire by Porin

Analysis of relative fluorescence intensity showed that porin induced TLR2 strongly and specifically on MΦ (Fig. 2), B-1a, B-1b and B-2 cells. The TLR4 expression on MΦ, B-1 cell "sister" and B-2 populations was found to be absent. This confirms that among the two key TLRs, TLR2 and TLR4, which are primarily responsible other than TLR9, for recognizing majority of the bacterial products, TLR2 and not TLR4, participates in porin recognition. The recognition of porin by TLR2 of PerC MΦ and B cell subsets is in combination with TLR6 as indicated by the study of mRNA expression of TLR6.

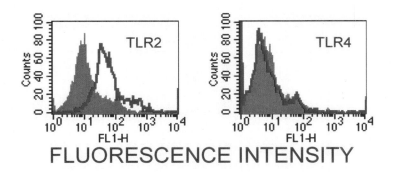

Figure 2. Induction of TLR2 and TLR4 on PerC MΦ in response to porin. MΦ were cultured with complete medium alone (shaded in pink) and porin (blue line) for 24 h, and analyzed by flow cytometry for the expression of TLR2 and TLR4.

Activation of Signaling Molecules: NF-κB and CD80-CD86 (B7 molecules)

Following the recognition by TLRs, the mechanism of adjuvanticity of porin correlates with its ability to initiate signaling pathways that affect the immune response such as activation of the proinflammatory nuclear factor kappa B (NF-κB) or up-regulation of CD80-CD86 (B7 molecules) on the surface of APC and B cells.

NF-κB

A family of proteins, known as NF-κB, binds to specific DNA sequences and activate a specific subset of genes. These genes are involved mainly in the inflammatory response but also a number of genes activated by NF-κB are involved in cell to cell attachment, cell to cell communication, cell survival and cell growth, cell differentiation and development and tissue remodeling. Activation of NF-κB is required for cells to undergo transformation and bypass the cellular death program [47-49]. Since NF-κB activity also play an important role in mucosal immunity, it is likely that its activity is tightly controlled by TLRs. Thereby, porin of *S. dysenteriae* type 1 may activate NF-κB pathway in immune cells via recognition by the TLRs, leading to expression of genes involved in innate host response to deal immediately with pathogens [50-52] and later play an essential role in allowing the host to mount an adaptive immunity (antibody) response.

CD80-CD86

Another feature of these cells that makes them potential candidate for skewing immune responses is their selective expression of the costimulatory molecules [53]. The most fully characterized costimulatory signal is mediated by the binding of CD80 (B7-1) and CD86 (B7-2) on APC to their receptor CD28 on T cells [54]. The induction of B7-2 occurs within 6 h of stimulation and peaks between 18 and 24 h [53]. However, B7-1 expression is not evident until 24 h post stimulation and reaches maximal levels at 48 to 72 h. These time differences in costimulation might provide the mechanism for skewing immunity. In certain situations, T cell or B cell that encounters antigen in absence of costimulatory signals may get anergized or deleted [55].

NF-κB Expression by MΦ and B Cell Subsets in Response to Porin

It was found that PerC MΦ (Fig. 3), and both B-1a and B-1b cells strongly expressed the mRNA for NF-κB in response to porin. It could be due to porin induced coexpression of TLR2 and TLR6 of MΦ and B cell subsets. This corresponds to the earlier finding that stimulation of CHO cells with bacterial lipopeptide resulted in coexpression of TLR2 and TLR6 which was necessary to generate robust activation of NF-κB for signaling [56]. The observation indicates that NF-κB is thereby strongly induced by porin treatment of PerC MΦ and B-1 cell populations. Besides the B-1 cell "sister" populations, B-2 cell also expressed the mRNA for NF-κB in response to the protein. Our result suggests that NF-κB has a dominant role in porin induced mucosal MΦ and B cell response.

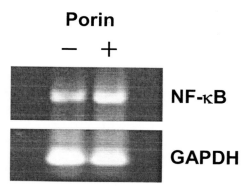

Figure 3. Effect of porin on NF-κB mRNA expression. MΦ were treated with and without porin for 24 h. Total RNA was extracted from the MΦ after 24 h of culture and subjected to RT-PCR. Ethidium bromide stained PCR products were photographed, and then images were digitized and analyzed. PCR products were quantified and expressed as the ratio of each product to GAPDH band density.

Selective Up-Regulation of CD80 on MΦ by Porin

Analysis of relative fluorescence intensity after 24 h incubation of CD11b$^+$ PerC MΦ with porin demonstrated that CD80 expression increased significantly compared to control. The up-regulation of the costimulatory molecule on CD11b$^+$ MΦ was augmented substantially by porin plus IFN-γ. However, the analysis of relative fluorescence intensity showed that porin could not induce the expression of CD86 on CD11b$^+$ PerC MΦ, supporting porin mediated selective up-regulation of the B7-1 (CD80) member of the B7 family. However, the induction of CD80 (B7-1) surface expression on MΦ is unlike B7-2 expression by Neisserial porin on B lymphocytes [57] or cholera toxin (CT) on GM-CSF-activated bone marrow MΦ [58]. Costimulatory molecules, CD80 and CD86, are ligands for the T cell membrane proteins CD28 and CTLA-4. These molecules have been implicated as regulatory determinants on professional APC that participate in murine and human T cell activation or T cell-dependent B cell activation [59]. B7-1 (CD80) overexpression is known to correlate with induction of a dominant Th1 response, and anti-B7-1 antibody drives the immune response toward Th2 pathway while anti-B7-2 antibody drives the immune response to Th1 pathway [60].

Selective Up-Regulation of the Costimulatory Molecules on B-1a, B-1b and B-2 Cells

Analysis of relative fluorescence intensity showed that porin induced CD80 and CD86 on B-1a cells, and CD86 solely on B-1b cells of C57BL/6 mice. Likewise, the B-1a cells of C3H/HeJ mice was found to express CD80 exclusively on their surface, in contrast to their B-1b "sister" population that expressed CD86 solely in response to porin. The cell-surface expression of both the costimulatory molecules was much more with C3H/HeJ mice than found with C57BL/6 mice. The observation indicated that there is a strong bias for the up-regulation of the costimulatory molecules on the B-1 cell populations. It is known that CD86 is expressed as an earlier response to an antigen well ahead of CD80 expression, which takes part in delayed but matured response [53, 61]. B-1 cell "sister" populations might have evolved the mechanism of signaling via a specific costimulatory molecule so that these cells can divide its work load of responding to an antigen. Analysis of relative fluorescence intensity showed that porin strongly induced CD86 on B-2 cells. The inability of the protein to induce the expression of CD80, supports porin-mediated selective up-regulation of CD86 on PerC B-2 cells.

In order to establish porin of *S. dysenteriae* type 1 as an adjuvant, it was necessary to demonstrate that the protein induced the expression of the costimulatory molecules, CD80-CD86, on B cells, since up-regulation of these molecules is one of the essential step for a good adjuvant to proliferate immune cells by skipping tolerance [55]. Neisserial porins are known to act as immune adjuvants by up-regulating the surface expression of the costimulatory molecule B7-2 (CD86) on B cells and probably other APCs [57, 62]. However, we found that the porin induced up-regulation of CD80 and CD86 on murine PerC B cell repertoire was extremely selective perhaps to tightly regulate and control the mucosal immune response.

Response of PerC MΦ to Porin

We found that porin of *S. dysenteriae* type 1 mediated the release of nitric oxide (NO) and interleukin (IL)-1 from PerC MΦ of mouse [63]. The MΦ were further studied for the porin induced release of specific subsets of cytokines, for example IL-12, which promotes Th1 differentiation [53]. The release of IL-12 by the cells were analyzed both *per se* and under the influence of gamma interferon (IFN-γ), the positive feed-back regulator of APC [64].

Porin Mediated Release of IL-12 by MΦ

We found porin induced PerC MΦ to release IL-12 (~ 36 pg/ million MΦ) and not IL-10. The specific induction of IL-12 production and release implicate promotion of Th1-type responses by porin directly. This observation is similar to that reported for LPS [65], both directly [66] and indirectly via the induction of IFN-γ [67]. The release of IL-12 by porin of *S. dysenteriae* treated MΦ was profoundly enhanced by IFN-γ emphasizing the direct synergistic effect of IFN-γ on IL-12 for Th1-type response. However, this is in contrast to the adjuvanticity of CT that has been generally associated with Th2-type cytokine response and immunoglobulin E production [68], which may inhibit Th1 cytokine secretion [69].

Response of PerC B Cell Subsets to Porin

A hallmark of mucosal immunity is the predominance of IgA antibody formation. Mucosal immunologists became interested in PerC B-1 cells when it was shown that these cells reconstitute almost 50% of intestinal IgA plasma cells [70], suggesting that frequent migration of lymphocytes take place between the PerC and gut-associated lymphatic tissues (GALT) [71]. Recent study has shown that B-1-derived IgA plasma cells participate in humoral defense at the mucosal surface [72]. Whether the costimulatory molecules, CD80-CD86, play a regulatory role in mucosal IgA responses has never been adequately explored. Recognition of porin by the TLRs and up-regulation of the costimulatory molecules, CD80-CD86, on the surface of PerC B lymphocytes could be the initiating mechanism behind the immunopotentiating activity of porin towards the generation of IgA. Porin mediated up-regulation of CD80-CD86 and expression of IgA on B-1 cell, and IgG subclasses and IgA on B-2 cell was studied to evaluate if the immunogen could regulate the mucosal immune response. One approach by which the mucosal response may be selectively enhanced is through combination of immunocompetent cytokines. IL-6 is a multifunctional cytokine, which also plays a crucial role in B-cell terminal differentiation and development of secretory IgA responses at the mucosa [73]. Both B-1 and B-2 cell subsets were treated with IL-6 to assess if the cytokine could enhance the porin induced IgA expression on B-1 cell and IgG subclasses and IgA expression on B-2 cell.

Induction of Immunoglobulins on B Cell Subsets by Porin

Flow cytometric analysis of relative fluorescence intensity showed IgM, known to be expressed in high amounts on B-1 cells even without any antigenic provocation, was specifically generated under the influence of porin. Similarly, IgM was expressed on B-2 cells also. We studied whether the B-1 and B-2 cells expressing CD80-CD86 under the influence of porin has the capacity to produce IgA. It was found that PerC B-1a, B-1b and B-2 cells expressed IgA efficiently in response to porin suggesting that the protein is capable of generating IgA, the key immunoglobulin class that participate in mucosal immune response. Although there seemed to be a B-1a and B-1b population specific bias for expression of the costimulatory molecules, there was no correlation between the preference for expression of CD80 or CD86 and induction of IgA. The porin induced IgA on B-1a and B-1b cell-surface, recognized specifically the purified porin monomer of 38 kDa of *S. dysenteriae* type 1 on Western blot, defining it to be anti-porin IgA antibody. Among the IgG subclasses, IgG_{2a} got selectively expressed and the cell surface expression of IgG_1, IgG_{2b} and IgG_3 remained unaffected with porin treatment of B-2 cells. In order to investigate whether the porin mediated IgA response could be enhanced, B cell subsets were treated with the immunogen and IL-6. IL-6 was found to augment the porin induced IgA response of B-1 and B-2 cells, thus highlighting the cytokine to be a positive regulator of the immunogen. Furthermore, porin induced IgG_{2a} on the B-2 cells was also augmented strongly by IL-6 treatment in presence of porin. Previously, IL-6 requirement for IgA secretion by B cell subsets, B-1 and B-2, have been shown. In $IL-6^{-/-}$ mice IgA producing B cell numbers in lamina propria got reduced by more than 60 % compared to wild type control mice ($IL-6^{+/+}$) [74]. The residual intestinal IgA^+ cells in these $IL-6^{-/-}$ mice were B cells coexpressing CD5 suggesting that the cells were of B-1 origin. Murine IL-6- responsive B cells have been found in Peyer's patches also. The IgA^+ B-2 cells of Peyer's patch could be induced to secrete IgA *in vitro* by IL-5 and particularly IL-6 [75].

Concluding Remarks

Porins are of particular interest because they have been characterized as potential adjuvants. The adjuvanticity of porin of pathogenic *Neisseria* spp. has been described [57, 62]. The activation of murine PerC MΦ and B cell populations by porin of *S. dysenteriae* type 1 was studied keeping in view the significance of these cells in regulating mucosal immunity. Mucosal immune response should have a direct role in shigellosis because *Shigella* spp. infects its host by penetration of the colonic mucosa. Immunity can be broadly categorized into innate immunity and adaptive immunity. Innate immunity was formerly thought to be a nonspecific immune response characterized by engulfment and digestion of microorganisms and foreign substances by macrophages and leukocytes. However, innate immunity has considerable specificity and is capable of discriminating between pathogens and self [25, 36, 76]. The initial recognition of these microbial pathogens is mediated by TLRs expressed on APC, for example MΦ, where they may also play critical roles as adjuvant receptors. Adjuvanticity of porin of *S. dysenteriae* type 1 involves the increase of TLR2 and TLR6 of PerC MΦ, which resulted in the induction of NF-κB and cell-surface selective expression of CD80 (Fig. 4). It was followed by the release of proinflammatory cytokine IL-12 that could be augmented by IFN-γ. This may allow the MΦ to become more responsive to porin and favor Th1-type response that in turn could help towards the generation of a certain isotype of antibody. In addition, the activation of the innate immune response can be a prerequisite for the triggering of acquired immunity. Adaptive immunity is mediated by clonally distributed T and B lymphocytes and is characterized by specificity and memory. Adjuvants boost APC signaling to promote immunity by enhancing antigen presenting activity and by inducing cytokine production and costimulatory molecule expression in APCs [41], such as MΦ. APCs use costimulatory molecules, CD80-CD86, in addition to instructive cytokines, to signal T cells and to induce clonal expansion of antigen-specific T cells. Furthermore, the adjuvanticity of porin of *S. dysenteriae* type 1 involves increase of mRNA for TLR2 and TLR6 indicating that coexpression of TLR2 and TLR6 are necessary so that the TLRs may function as a combinatorial repertoire in association with MyD88 for recognition of porin by the PerC B cell populations (Fig. 5). This results in the robust activation of NF-κB and B-1 cell population-specific expression of CD80 and CD86. In B-2 cells the mRNA expression of NF-κB is moderate compared to B-1 cell "sister" populations followed by selective up-regulation of CD86. Induction of these signal molecules may regulate the cell-surface expression of IgM and particularly IgA in all the B cell populations, and preferentially IgG$_{2a}$ expression on B-2 cells which is known to be produced due to Th1-type influence [77]. Analysis of the T cell mediated immune response can further strengthen the correlation between porin-induced IL-12 release by MΦ and IgG$_{2a}$ expression of B-2 cells. Whereas, B-1 cells can participate in innate immunity and appear to produce antibody in a T-independent manner [30]. The IgA and IgG$_{2a}$ expressions on the respective B cells could be augmented by IL-6. Understanding the mechanism of adjuvanticity of porin of *S. dysenteriae* type 1 is essential to develop better adjuvant against shigellosis.

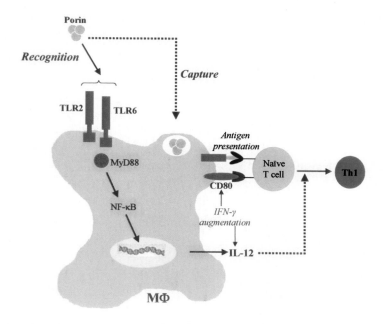

Figure 4. Porin of *S. dysenteriae* induced activation of murine PerC MΦ. Porin induced coexpression of TLR2 and TLR6 along with the effector molecule MyD88, associated with TLR mediated response, in MΦ. It was followed by increased NF-κB activation and selective up-regulation of CD80. This regulation led to the release of proinflammatory cytokine IL-12 by MΦ that favors Th1-type response. The cell-surface expression of CD80 and IL-12 release could be augmented by IFN-γ.

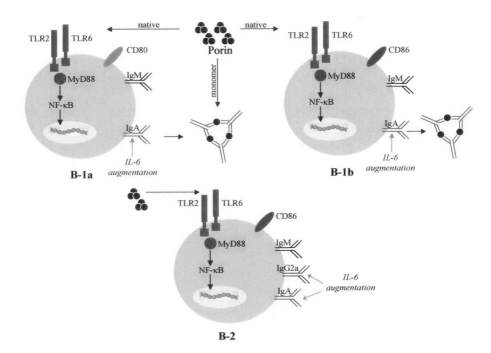

Figure 5. Porin of *S. dysenteriae* induced activation of murine PerC B cell subsets. Porin induced coexpression of TLR2 and TLR6 along with the effector molecule MyD88 in all the B cell populations. It was followed by increased NF-κB activation and selective expression of the costimulatory molecules, CD80-CD86. This led to the cell-surface expression of specific immunoglobulins, particularly IgA, which could be augmented by IL-6.

References

[1] Ewing, W. H. & Lindberg, A. A. (1984). Serology of *Shigella*. *Methods Microbiol., 1,* 113-142.

[2] Keusch, GT. Shigella. In: Gorbach, SL editor. Infectious diarrhea. London: Blackwell; 1986; 31-50.

[3] Hale, T. L. (1991). Genetic basis of virulence in *Shigella* species. *Microbiol. Rev., 55,* 206-224.

[4] Keusch, GT. Shigellosis. In: Evans AS, Feldman HA editors. Bacterial infection of humans: epidemology and control. New York: Plenum Press; 1982; 487-509.

[5] LaBrec, E. H., Schneider, H., Magnani, T. J. & Formal, S. B. (1964). Epethelial cell penetration as an essential step in pathogenesis in bacillary dysentery. *J. Bacteriol., 88,* 1503-1518.

[6] Sansonetti, P. J., Ryter, A., Clerc, P., Maurelli, A. T. & Mounier, J. (1986). Multiplication of *Shigella flexneri* within HeLa cells: lysis of the phagocytic vacuole and plasmid-mediated contact hemolysis. *Infect. Immun., 51,* 461-469.

[7]   Clerc, P., Ryter, A., Mounier, J. & Sansonetti, P. J. (1987). Plasmid-mediated early killing of eukaryotic cells by *Shigella flexneri* as studied by infection of J774 macrophages. *Infect. Immun., 55*, 521-527.

[8]   Bernardini, M. L., Mounier, J., d'Hauteville, H., Coquis Rondon, M. & Sansonetti, P. J. (1989). Identification of icsA, a plasmid locus *Shigella flexneri* that governs intra-and intercellular spread through interaction with F-actin. *Proc. Natl. Acad. Sci. USA, 86*, 3867-3871.

[9]   Mathan, M. M. & Mathan, V. I. (1991). Morphology of rectal mucosa of patients with shigellosis. *Rev. Infect. Dis., 13 (Suppl.4)*, 314-318.

[10]   Replogle, M. L., Fleming, D. W. & Cieslak, P. R. (2000). Emergence of antimicrobial-resistant shigellosis in Oregon. *Clin. Infect. Dis., 30*, 515-519.

[11]   Blake, MS; Gotschlich EC. Functional and immunological properties of pathogenic Neisserial surface proteins. In: Inouye M, editor. Bacterial outer membranes as model systems. New York: John Wiley & sons, Inc.; 1986; 377-400.

[12]   Benz, R. (1988). Structure and function of porins from gram-negative bacteria. *Annu. Rev. Microbiol., 42*, 359-393.

[13]   Nikaido, H. & Saier, Jr, M. H. (1992). Transport proteins in bacteria: common themes in their design. *Science, 258*, 936-942.

[14]   Galdiero, F., deLero, G. C., Benedetto, N., Galdiero, M. & Tufano, M. A. (1993). Release of cytokines induced by *Salmonella typhimurium* porins. *Infect. Immun., 61*, 155-161.

[15]   Galdiero, F., Tufano, M. A., Galdiero, M., Masiello, S. & DiRosa, M. (1990). Inflammatory effect of *Salmonella typhimurium* porins. *Infect. Immun., 58*, 3183-3188.

[16]   Weinberg, J. B., Ribi, E. & Wheat, R. W. (1983). Enhancement of macrophage-mediated tumore cell killing by bacterial outer membrane proteins (porins). *Infect. Immun., 42*, 219-223.

[17]   Lowell, G. H., Ballou, W. R., Smith, L. F., Wirtz, R. A., Zollinger, W. D., & Hockmeyer, W. T. (1988). Proteosome-lipopeptide vaccines: enhancement of immunogenicity for malaria CS peptides. *Science, 240*, 800-802.

[18]   Zollinger, WD. New and improved vaccines against meningococcal disease. In: Woodrow GC, Levine MM, editors. New generation vaccines. New York: Marcel Dekker, Inc.; 1990; 325-344.

[19]   Wetzler, L. M., Blake, M. S., Barry, K. & Gotschlich E. C. (1992). Neisserial porins induce B lymphocytes to express costimulatory B7-2 molecules and to proliferate. *J. Infect. Dis., 166*, 551-555.

[20]   Donnelly, J. J., Deck, R. R. & Liu, M. A. (1990). Immunogenicity of a *Haemophilus influenzae* polysaccharide-*Neisseria meningitidis* outer membrane protein complex conjugate vaccine. *J. Immunol., 145*, 3071-3079.

[21]   Livingston, P. O., Calves, M. J., Helling, F., Zollinger, W. D., Blake, M. S. & Lowell, G. H. (1993). GD3/proteosome vaccines induce consistent IgM antibodies against the ganglioside GD3. *Vaccine, 11*, 1199-1204.

[22]   Roy, S., Das, A. B., Ghosh, A. N. & Biswas, T. (1994). Purification, pore-forming ability, and antigenic relatedness of the major outer membrane protein of *Shigella dysenteriae* type 1. *Infect. Immun., 62*, 4333-4338.

[23] Ray, A., Chatterjee, N. S., Bhattacharya, S. K. & Biswas, T. (2003). Porin of *Shigella dysenteriae* enhances mRNA levels for Toll-like receptor 2 and MyD88, up-regulates CD80 of murine macrophage, and induces the release of interleukin-12. *FEMS Immunol. Med. Microbiol., 39*, 213-219.

[24] Isibasi, A., Ortiz-Navarrete, V., Paniagua, J., Pelayo, R., Gonzalez, C. R., Garcia, J. A. & Kumate, J. (1992). Active protection of mice against *Salmonella typhi* by immunization with strain-specific porins. *Vaccine, 10*, 811-813.

[25] Aderem, A., & Ulevitch, R. J. (2000). Toll-like receptors in the induction of the innate immune response. *Nature, 406*, 782-787.

[26] Kantor, A. B., Stall, A. M., Adams, S., Herzenberg, L. A. & Herzenberg, L. (1992). Differential development of progenitor activity for three B-cell lineages. *Proc. Natl. Acad. Sci. USA, 89*, 3320-3324.

[27] Kantor, A. B. & Herzenberg, L. A. (1993). Origin of murine B cell lineages. *Annu. Rev. Immunol., 11*, 501-538.

[28] Hayakawa, K., Hardy, R. R., Parks, D. R. & Herzenberg, L. A. (1983). The "Ly-1 B" cell subpopulation in normal immunodefective, and autoimmune mice. *J. Exp. Med., 157*, 202-218.

[29] Gardby, E., Lane, P. & Lycke, N. Y. (1998). Requirements for B7-CD28 costimulation in mucosal IgA responses: paradoxes observed in CTLA4-H gamma 1 transgenic mice. *J. Immunol., 161*, 49-59.

[30] Fagarasan, S. & Honjo, T. (2000). T-Independent immune response: new aspects of B cell biology. *Science, 290*, 89-92.

[31] Berland, R. & Wortis, H. H. (1998). The ontogeny and function of B-1 cells. *Mucosal Immunol. Update, 12*, 5-10.

[32] Pecquet, S. S., Ehrat, C. & Ernst, P. B. (1992). Enhancement of mucosal antibody responses to *Salmonella typhimurium* and the microbial hapten phosphorylcholine in mice with X-linked immunodeficiency by B-cell precursors from the peritoneal cavity. *Infect. Immun., 60*, 503-509.

[33] Lemaitre, B., Nicolas, E., Michaut, L., Reichhart, J. M. & Hoffmann, J. A. (1996). The dorsoventral regulatory gene cassette *spatzle/Toll/cactus* controls the potent antifungal response in *Drosophila* adults. *Cell, 86*, 973-983.

[34] Medzhitov, R., Preston-Hurlburt, P. & Janeway, Jr, C. A. (1997). A human homologue of the *Drosophila* Toll protein signals activation of adaptive immunity. *Nature, 388*, 394-397.

[35] Janeway, Jr, C. A. (1992). The immune system evolved to discriminate infectious nonself from noninfectious self. *Immunol. Today, 13*, 11-156.

[36] Medzhitov, R. & Janeway, Jr, C. A. (1997). Innate immunity: the virtues of a nonclonal system of recognition. *Cell, 91*, 295-298.

[37] Anderson, K. V. (2000). Toll signaling pathways in the innate immune response. *Curr. Opin. Immunol., 12*, 13-19.

[38] Lien, E., Means, T. K., Heine, H., Yoshimura, A., Kusumoto, S., Fukase, K., Fenton, M. J., Oikawa, M., Qureshi, N., Monks, B., Finberg, R. W., Ingalls, R. R. & Golenbock, D. T. (2000). Toll-like receptor 4 imparts ligand–specific recognition of bacterial lipopolysaccharide. *J. Clin. Invest., 105*, 497-504.

[39] Medzhitov, R., Preston-Hurlburt, P., Kopp, E., Stadler, A., Chen, C., Ghosh, S. & Janeway, Jr, C. A. (1998). MyD88 is an adaptor protein in the hToll/IL-1 receptor family signaling pathways. *Mol. Cell., 2*, 253-258.

[40] Muzio, M., Bosisio, D., Polentarutti, N., D'amico, G., Stoppacciaro, A., Mancinelli, R., van't Veer, C., Penton-Rol, G., Ruco, L. P., Allavena, P. & Mantovani, A. (2000). Differential expression and regulation of Toll-like receptors (TLR) in human leukocytes: selective expression of TLR3 in dendritic cells. *J. Immunol., 164*, 5998-6004.

[41] Akira, S., Takeda, K. & Kaisho, T. (2001). Toll-like receptors: critical proteins linking innate and acquired immunity. *Nat. Immunol., 2*, 675-680.

[42] Massari, P., Henneke, P., Ho, Y., Latz, E., Golenbock, D. T., & Wetzler, L. M. (2002). Immune stimulation by Neisserial porins is Toll-like receptor 2 and MyD88 dependent. *J. Immunol., 168*, 1533-1537.

[43] Takeda, K., Kaisho, T. & Akira, S. (2003). Toll-like receptors. *Ann. Rev. Immunol., 21*, 335-376.

[44] Krieg, A. M. & Davis, H. L. (2001). Enhancing vaccines with immune stimulatory CpG DNA. *Curr. Opin. Mol. Ther., 3*, 15-24.

[45] Ray, A., Karmakar, P. & Biswas, T. (2004). Up-regulation of CD80-CD86 and IgA on mouse peritoneal B-1 cells by porin of *Shigella dysenteriae* is Toll-like receptors 2 and 6 dependent. *Mol. Immunol., 41*, 1167-1175.

[46] Ray, A. & Biswas, T. (2005). Porin of *Shigella dysenteriae* enhances Toll-like receptors 2 and 6 of mouse peritoneal B-2 cells and induces the expression of immunoglobulin M, immunoglobulin G2a and immunoglobulin A. *Immunology, 114*, 94-100.

[47] Beg, A. A. & Baltimore, D. (1996). An essential role for NF-κB in preventing TNF-α-induced cell death. *Science, 274*, 782-784.

[48] Van Antwerp, D. J., Martin, S. J., Kafri, T., Green, D. R. & Verma, I. M. (1996). Suppression of TNF-α-induced apoptosis by NF-κB. *Science, 274*, 787-789.

[49] Wang, C. Y., Mayo, W. & Baldwin, A. S. (1996). TNF-α and cancer therapy-induced apoptosis potentiation by inhibition of NF-κB. *Science, 274*, 784-787.

[50] Aliprantis, A. O., Yang, R. B., Mark, M. R., Suggett, S. Devaux, B., Radolf, J. D., Klimpel, G. R., Godowski, P. & Zychlinsky, A. (1999). Cell activation and apoptosis by bacterial lipoproteins through Toll-like receptor-2. *Science, 285*, 736-739.

[51] Brightbill, H. D., Libraty, D. H., Krutzik, S. R., Yang, R. B., Belisle, J. T., Bleharski, J. R., Maitland, M., Norgard, M. V., Plevy, S. E., Smale, S. T., Brennan, P. J., Bloom, B. R., Godowski, P. J. & Modlin, R. L. (1999). Host defence mechanism triggered by microbial lipoproteins through Toll-like receptors. *Science, 285*, 732-736.

[52] Liu, Y., Wang, Y., Yamakuchi, M., Isowaki, S., Nagata, E., Kanmura, Y., Kitajima, I. & Maruyama, I. (2001). Upregulation of Toll-like receptor 2 gene expression in macrophage response to polyglycan and high concentration of lipopolysaccharide is involved in NF-κB activation. *Infect. Immun., 69*, 2788-2796.

[53] Constant, S. L. & Bottomly, K. (1997). Induction of Th1 and Th2 CD4[+] T cell responses, the alternative approaches. *Annu. Rev. Immunol., 15*, 297-322.

[54] Linsley, P. S. & Ledbetter, J. A. (1993). The role of the CD28 receptor during T cell responses to antigen. *Annu. Rev. Immunol., 11*, 191-212.

[55] Freeman, G. J., Borriello, F., Hodes, R. J., Reiser, V., Hathcock, K. S., Laszlo, G., McKnight, A. J., Kim, J., Du, L., Lombard, D. B., Gray, G. S., Nadler, L. M. & Sharpe, A. H., (1993). Uncovering of functional alternative CTLA-4 counter-receptor in B7-deficient mice. *Science, 262*, 907-909.

[56] Ozinsky, A., Underhill, D. M., Fontenot, J. D., Hajjar, A. M., Smith, K. D., Wilson, C. B., Schroeder, L. & Aderem, A., (2000). The repertoire for pattern recognition of pathogens by the innate immune system is defined by cooperation between Toll-like receptors. *Proc. Natl. Acad. Sci. USA, 97*, 13766-13771.

[57] Wetzler, L. M., Ho, Y., Reiser, H. & Wetzler, L. W. (1996). Neisserial porins induce B lymphocytes to express costimulatory B7-2 molecules and to proliferate. *J. Exp. Med., 183*, 1151-1159.

[58] Cong, Y., Weaver, C. T. & Elson, C. O. (1997). The mucosal adjuvanticity of cholera toxin involves enhancement of costimulatory activity by selective up-regulation of B7-2 expression. *J. Immunol., 159*, 5301-5308.

[59] Hathcock, K. S. & Hodes, R. J. (1995). Role of the CD28-B7 costimulatory pathways in T cell-dependent B cell responses. *Adv. Immunol., 62*, 131-166.

[60] Kuchroo, V. K., Das, M. P., Brown, J. A., Ranger, A. M., Zamvil, S. S., Sobel, H., Weiner, L., Nabavi, N. & Glimcher, L. H. (1995). B7-1 and B7-2 costimulatory molecules activate differentially the Th1/Th2 development pathways: application to autoimmune disease therapy. *Cell, 80*, 707-718.

[61] Ray, A., Chattopadhyay, K., Banerjee, K. K. & Biswas, T. (2003). Macrophage distinguishes *Vibrio cholerae* hemolysin from its protease insensitive oligomer by time dependent and selective expression of CD80-CD86. *Immunol. Lett., 89*, 143-147.

[62] Mackinnon, F. G., Ho, Y., Blake, M. S., Michon, F., Chandraker, A., Sayegh, M. H. & Wetzler, L. M. (1999). The role of B/T costimulatory signals in the immunopotentiating activity of Neisserial porin. *J. Infect. Dis., 180*, 755-761.

[63] Biswas, T. (2000). Role of porin of *Shigella dysenteriae* type 1 in modulation of lipopolysaccharide mediated nitric oxide and interleukin-1 release by murine peritoneal macrophages. *FEMS Immunol. Med. Microbiol., 29*, 129-136.

[64] Trinchieri, G. (1998). Interleukin-12: a cytokine at the interface of inflammation and immunity. *Adv. Immunol., 70*, 83-243.

[65] Dekruyff, R. H., Gieni, R. S. & Umetsu, D. T. (1997). Antigen-driven but not lipopolysaccharide-driven IL-12 production in macrophages requires triggering of CD40. *J. Immunol., 58*, 359-366.

[66] Hsieh, C. S., Macatonia, S. E., Tripp, C. S., Wolf, S. F., O'Garra, A. & Murphy, K. M. (1993). Development of Th1 CD4[+] T cells through IL-12 produced by *Listeria*-induced macrophages. *Science, 260*, 547-549.

[67] Schmitt, E., Hoehn, P., Huels, C., Goedert, S., Palm, N., Rude, E. & Germann, T. (1994). T helper type 1 development of naive CD4[+] T cells requires the coordinate action of interleukin–12 and interferon–gamma and is inhibited by transforming growth factor-beta. *Eur. J. Immunol., 24*, 793-798.

[68] Marinaro, M., Staats, H. F., Hiroi, T., Jackson, R. J., Coste, M., Boyaka. P. N., Okahashi, N., Yamamoto, M., Kiyono, H., Bluethmann, H., Fujihashi, K. & Mcghee, J. R. (1995). Mucosal adjuvant effect of cholera toxin in mice results from induction of T helper 2 (Th2) cells and IL-4. *J. Immunol., 155*, 4621-4629.

[69] Elson, CO; Beagley, KW. Cytokines and immune mediators. In: Johnson LR, editor. Physiology of the gastrointestinal tract. New York: Raven Press; 1994; 243-256.

[70] Kroese, F. G., Butcher, E. C., Stall, A. M. & Herzenberg, L. A. (1989). A major peritoneal reservoir of precursors for intestinal IgA plasma cells. *Immunol. Invest., 18*, 47-58.

[71] Kroese, F. G., Butcher, E. C., Stall, A. M., Lalor, P. A., Adams, S. & Herzenberg, L. A. (1989). Many of the IgA producing plasma cells in murine gut are derived from self-replenishing precursors in the peritoneal cavity. *Int. Immunol., 1*, 75-84.

[72] Macpherson, A. J., Gatto, D., Sainsbury, E., Harriman, G. R., Hengartner, H. & Zinkernagel, R. M. (2000). A primitive T cell-independent mechanism of intestinal mucosal IgA responses to commensal bacteria. *Science, 288*, 2222-2226.

[73] Akira, S., Taga, T. & Kishimoto, T. (1993). Interleukin-6 in biology and medicine. *Adv. Immunol., 54*, 1-78.

[74] Ramsay, A. J., Husband, A. J., Ramshaw, I. A., Bao, S., Matthaei, K. I., Koehler, G. & Kopf, M. (1994). The role of interleukin-6 in mucosal IgA antibody responses in vivo. *Science, 264*, 561-563.

[75] Beagley, K. W., Eldridge, J. H., Lee, F., Kiyono, H., Everson, M. P., Koopman, W. J., Hirano, T., Kishimoto, T. & McGhee, J. R. (1989). Interleukins and IgA synthesis. Human and murine interleukin 6 induce high rate IgA secretion in IgA-committed B cells. *J. Exp. Med., 169*, 2133-2148.

[76] Hoffmann, J. A., Kafatos, F. C., Janeway, C. A. & Ezekowitz, R. A. (1999). Phylogenetic perspectives in innate immunity. *Science, 284*, 1313-1318.

[77] Stevens, T. L., Bossie, A., Sanders, V. M., Fernandez-Botran, R., Coffman, R. L., Mosmann, T. R. & Vitetta, E. S. (1988). Regulation of antibody isotype secretion by subsets of antigen-specific helper T cells. *Nature 334*, 255-258.

Index

F

G

J

K

N

O

P

R

W

X

Y

Z